60

Advances in Biochemical Engineering/Biotechnology

Managing Editor: T. Scheper

Springer-Verlag Berlin Heidelberg GmbH

Relation Between Morphology and Process Performances

Volume Editor: K. Schügerl

With contributions by
K.-H. Bellgardt, S. R. Gerlach, R. King, P. Krabben,
J. Nielsen, G. C. Paul, M.-N. Pons, K. Schügerl,
D. Siedenberg, C. R. Thomas, H. Vivier

 Springer

This series represents critical reviews on the present and future trends in Biochemical Engineering/Biotechnology, including microbiology, genetics, biochemistry, chemistry, computer science and chemical engineering. It is adressed to all scientists at universities and in industry who wish to keep up-to-date in this extremly fast developing area of science.

In general, special volumes are dedicated to selected topics and are edited by well known guest editors. The managing editor and publisher will however always be pleased to receive suggestions and supplementary information. Manuscripts are accepted in English.

In references Advances in Biochemical Engineering/Biotechnology is abbreviated as Adv. Biochem. Engin./Biotechnol. as a journal.

ISBN 978-3-662-14777-1   ISBN 978-3-540-68082-6 (eBook)
DOI 10.1007/978-3-540-68082-6

Library of Congress Catalog Card Number 72-152360

© Springer-Verlag Berlin Heidelberg 1998
Originally published by Springer-Verlag Berlin Heidelberg New York in 1998
Softcover reprint of the hardcover 1st edition 1998

Typesetting: Fotosatz-Service Köhler OHG, Würzburg
Cover: Design & Production, Heidelberg
SPIN: 10554467   02/3020 – 5 4 3 2 1 0 – Printed on acid-free paper

Attention all "Enzyme Handbook" Users:

A file with the complete volume indexes Vols. 1 through 11 in delimited ASCII format is available for downloading at no charge from the Springer EARN mailbox. Delimited ASCII format can be imported into most databanks.

The file has been compressed using the popular shareware program "PKZIP" (Trademark of PKware INc., PKZIP is available from most BBS and shareware distributors).

This file distributed without any expressed or implied warranty.

To receive this file send an e-mail message to:
SVSERV@DHDSPRI6.BITNET.
The message must be: "GET/ENZHB/ENZ_HB.ZIP".

SPSERV is an automatic data distribution system. It responds to your message. The following commands are available:

HELP	returns a detailed instruction set for the use of SVSERV,
DIR (*name*)	returns a list of files available in the directory "name",
INDEX (*name*)	same as "DIR"
CD <*name*>	changes to directroy "name",
SEND <*filename*>	invokes a message with the file "filename"
GET <*filename*>	same as "SEND".

Preface

Microorganisms have been characterized by their morphology ever since the microscope became a standard instrument in microbiology. Bacteria, yeasts and algae usually have a simple geometrical shape and they can be characterized by few features. Fungi and actomycetes, however, grow in a filamentous or pellet form. For their characterization, many parameters are needed. Microbiologists may use this information for species identification.

About twenty years ago, N. W. F. Kossen and coworkers started to evaluate the morphology of *Penicillium chrysogenum* for biochemical engineering purposes. Due to the laborious image analysis required, only a few engineering researchers dealt with the morphology of this fungus. Now that suitable instruments are at our disposal for digital image analysis, the number of research reports have considerably increased. Colin Thomas was the first to develop suitable software for digital image analysis of fungi and later software for the automatic evaluation of their images. In the last ten years, other researchers have started to characterize filamentous microorganisms and used them to develop mathematical models for the morphological development of fungi and streptomycetes. A large number of excellent papers have been published in this field during the last few years. The aim of this special volume is to present the state of the art of this research in relation to the usual process parameters.

The first contribution by G. C. Paul and C. R. Thomas presents a general review on the materials and methods of digital image analysis. Several examples are given by them for the charaterization of the morphology of various mycelium-forming microorganisms. The review deals with fungal spore germination and hyphal differentiation as well.

The second contribution by M. N. Pons and H. Vivier gives an overview on the application of digital image analysis for non-mycelium-forming microorganisms. It discusses the extension of the morphological measurements by staining the hyphae to gain information on their physiology as well.

The contribution of R. King covers experimental results and mathematical modeling and simulation of hyphal elongation and branching, as well as of pellet formation from a single spore and population balances in submerged cultures.

P. Krabben and J. Nielsen present a generalized morphological model including spore germination, tip extension, and fragmentation. They also deal with the average property models and various population models as well.

In his contribution, K.-H. Bellgardt concentrates on the production of β-lactam antibiotics He gives a review on morphological models of *Penicillium chrysoge-*

num and discusses the mathematical description of the development of pellets from spores through hyphae and compares the simulated pellet structures with the measured ones. By simplification of the pellet model, he was able to develop a pellet population model that generates predictions in good agreement with measurements. He presents structured and segregated models as well as artificial neural network approach for the description of Cephalosporin C production by *Acremonium chrysogenum*.

In the sixth contribution, K. Schügerl, S.R. Gerlach and D. Siedenberg compare the experimentally evaluated morphology and enzyme formation of *Aspergillus awamori* cultivated in various reactors under different operation conditions. They relate the enzyme productivity to the morphology of the fungus and both of them to the reaction engineering parameters of the cultivation in the classical manner.

These six contributions give a good picture of the state of the art of biochemical engineering research on the morphology of microorganisms in relation to their cultivation.

Hannover, June 1997   K. Schügerl

Contents

Contents

Characterisation of Mycelial Morphology Using Image Analysis

Gopal C. Paul · Colin R. Thomas*

Centre for Bioprocess Engineering, School of Chemical Engineering
University of Birmingham, Edgbaston, Birmingham, B15 2TT, UK
E-mail: G.C.Paul@bham.ac.uk; C.R.Thomas@bham.ac.uk

Image analysis is now well established in quantifying and characterising microorganisms from fermentation samples. In filamentous fermentations it has become an invaluable tool for characterising complex mycelial morphologies, although it is not yet used extensively in industry. Recent method developments include characterisation of spore germination from the inoculum stage and of the subsequent dispersed and pellet forms. Further methods include characterising vacuolation and simple structural differentiation of mycelia, also from submerged cultures. Image analysis can provide better understanding of the development of mycelial morphology, of the physiological states of the microorganisms in the fermenter, and of their interactions with the fermentation conditions. This understanding should lead to improved design and operation of mycelial fermentations.

---

* To whom all correspondence should be addressed.

Advances in Biochemical Engineering/
Biotechnology, Vol. 60
Managing Editor: Th. Scheper
© Springer-Verlag Berlin Heidelberg 1998

1
Introduction

Many important industrial fermentation processes utilise filamentous micro-organisms such as fungi and *Actinomycetes* for production of commercially important products, including most antibiotics, and some enzymes and organic acids, in addition to microbial biomass itself which is used in some foodstuffs. The growth of filamentous microorganisms is more complex than that of uni-cellular bacteria and yeasts. The development of the filamentous form usually starts from the germination of a spore, the germ tubes of which are progressively transformed into mycelia, i.e. networks of hyphae, through hyphal extension and branch formation. A fungal or *Actinomycete* hypha may be up to several hundred microns long and is typically $3-10\,\mu m$ in diameter for fungi and $0.5-1.5\,\mu m$ for *Actinomycetes* [1].

The filamentous form of growth allows the organism to increase in size without altering the protoplasmic volume: surface area ratio, which makes these microorganisms well adapted to colonisation of solid substrates [1]. However, for economic production, most processes require submerged culture in large fermenters, often with intense agitation and aeration. Under these conditions, a wide variation in gross morphology (shape) is found, varying from discrete or loosely entangled filaments, the dispersed form, to pellets (spherical colonies of highly entangled hyphal mass). In many cases, productivity seems to depend on morphology [2, 3]. In addition, the dispersed form of growth can lead to hyphal entanglement in the fermentation broth. This type of morphology at high biomass concentration can lead to an adverse effect on the rheological properties of the broth, making it highly viscous and pseudoplastic, and thus difficult to mix in a large fermenter, so that inhomogeneities may arise. Further-more, oxygen transfer from sparged air bubbles to the microorganisms de-creases with increasing viscosity. In severe cases, this can lead to oxygen limita-tion. Heat transfer for fermenter cooling may also be reduced under such condi-tions. Such problems can lower productivity. They cannot always be overcome by the use of high agitation speeds because of the potential "shear" damage to mycelia, also with adverse effects on productivity. Rheological problems are less with the pelleted forms, but the centres of large pellets might be starved of oxygen and autolyse, again affecting productivity. Clearly the morphology of these filamentous microorganisms is a matter of profound concern for fermentation scientists.

Other types of structural variation can also occur in filamentous microorganisms. Some fungi and all *Actinomycetes* are septate (i.e. they have cross-walls which divide the hyphae into a series of compartments). As growth occurs only at the hyphal tips [1], these hyphal compartments vary greatly in age and in physiological state, and can show various forms of structural and biochemical differentiation [4–8]. For example, vacuoles can appear in the hyphae during later stages of growth (Fig. 1). In general, the further the compartment is from the apex (i.e. the older the compartment), the larger the vacuoles. Hyphal vacuolation has been used to quantify the physiological structure of fungal biomass [7, 8] and has been related to antibiotic production from *Streptomyces spectabilis* [9] and *Penicillium chrysogenum* [10]. The expression of genes associated with secondary metabolic activities such as antibiotic production might be related to the process of differentiation.

Morphology and differentiation are often studied using microscopy, which yields either qualitative information, or quantitative information through labour-intensive and sometimes subjective analysis by the observer. The power of image analysis lies in extracting quantitative information from such observations in a reproducible, (semi-) automated manner. A semi-automatic image analysis for characterising mycelial morphology was developed in 1988 [11]. Since then, the power and capability of image analysers have been increasing rapidly, and so have the applications in this area. Image analysis methods have been reported for automatic characterisation of mycelial morphology [12, 13]

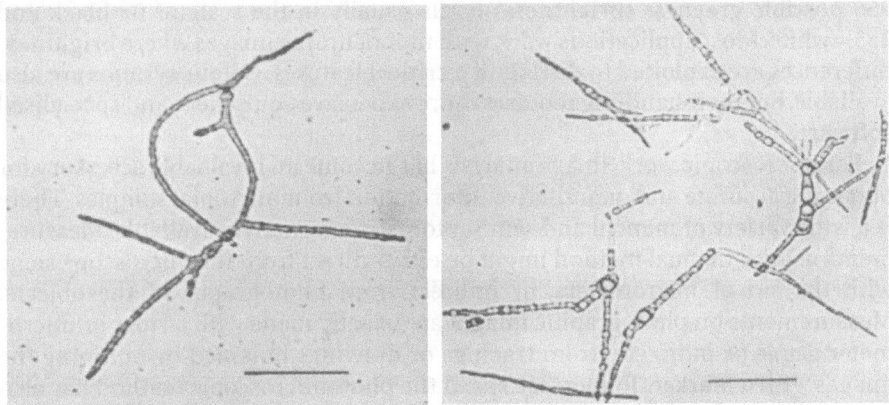

a   b

Fig. 1 a, b. Vacuolation of *P. chrysogenum* mycelia in fed-batch fermentation. Photograph of mycelial during: **a** the rapid growth phase (15 h); **b** the production phase (87 h). Bar length = 40 μm

and pellet structure [14]. Automatic image analysis has also been reported for characterising swelling and germination characteristics of fungal spores in inoculum fermentations [15]. Substantial evidence suggests that the quality of the initial spore preparation (e.g. spore age, concentration) and medium conditions can influence subsequent mycelial growth, morphology and productivity [3, 15, 16]. Recently an image analysis method has been developed to characterise hyphal vacuolation [7] and a method for the simple differentiation of *P. chrysogenum* is now available [8]. The resulting quantitative information about the biomass structure has permitted a powerful structured model for the penicillin fermentation to be built [17].

Many of these applications of image analysis were reviewed by Thomas [18]. A broader review of image analysis applications in biotechnology has been published by Vecht-Lifshitz and Ison [19] and more recently a review on applications of image analysis in cell technology has been published by Thomas and Paul [20]. The present work provides detailed descriptions of the main methods used to characterise mycelial morphology and differentiation. A complementary discussion concerning staining and more detailed physiological characterisation of *P. chrysogenum* can be found in Chap. 2 by Pons and Vivier.

2
Image Analysis Systems

A typical image analysis system consists of a personal computer (PC) or a work station, sometimes with specialised dedicated hardware, which can capture an image or picture and can be programmed to extract information from the image or to enhance features of interest within it, such as microorganisms. The source of the image is usually from a camera mounted on a microscope or from another video source or an electron microscope. The image is digitised in the computer both in space and tone to produce an array of picture elements or "pixels". A typical image consists of 512×512 pixels, each of which usually has one of 256 possible greyness (brightness) levels usually in the scale of $0 =$ black and $255 =$ white. Most applications work with monochrome images where brightness differences are exploited to distinguish critical features. Colour systems are also available, but their handling requires more expensive equipment and specialised software.

For microscopic work, image analysis has become an invaluable accessory for obtaining accurate and quantitative information from microbial samples. There is a wide variety of manual and semi-automatic alternatives available. Measurements using a manual method might be either direct from the microscope stage with the aid of micrometers, or indirect from photographs of the objects. Measurements on photographic images are usually made with a ruler or micrometer gauge or indirectly from tracings or drawings obtained by outlining the images with a marker. To increase speed the photomicroscopic method has also been combined with the use of an electronic digitiser attached to a computer [21]. With complicated objects more sophisticated measurements are necessary. Furthermore, studies on microorganisms often involve the measurement of hundreds of objects to obtain statistically sensible results. Such large numbers of measurements are too tedious to collect by manual means, and image analysis is often preferable, especially as it can easily be (semi)automated. Adams and Thomas [11] proposed the use of such semi-automatic image analysis for characterising the morphology of mycelial cultures. They compared their image analysis method with the electronic digitising method used by Metz et al. [21], using *Streptomyces clavuligerus* samples from submerged cultures. It was shown that the semi-automatic method is more precise and faster than the digitising method. An additional advantage of image analysis is its application of strictly consistent criteria to measurements, which should be independent of the operator. This can remove most of the subjectivity and inconsistency associated with manual methods. Finally, some measurements can also be considered with image analysis that would otherwise be unrealistic, or too costly.

An image analysis system consists of two major components: hardware and software, and these are discussed in the following sections.

2.1
Hardware

The basic hardware components of image processing systems are: input devices (microscope, macroviewer or similar device with a video camera attached; video recorder; scanner; etc.), and a high performance personal computer or workstation with optional image processing microprocessors. Additional items may include a mouse, light pen or joystick, a massive storage device, and normal and/or video printers. For automatic operation using a microscope, this should be fitted with a motorised stage, a controllable illumination system, and autofocus capability. Figure 2 shows the elements of image analysis systems which can be selected to give varying degrees of sophistication.

A motorised stage allows scanning of a microscope slide. An extra large automated stage can be loaded with several slides to allow extended operation. Field brightness can be controlled by a lamp controller using the video signal from the

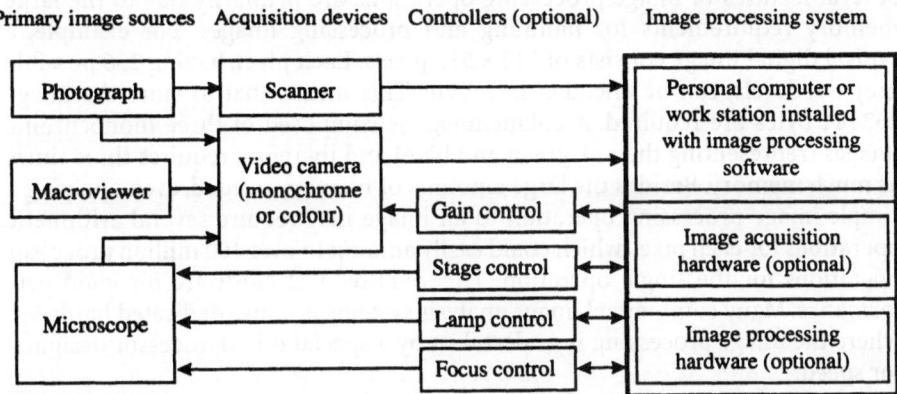

Fig. 2. Hardware components of typical image analysis systems

camera to control the lamp brightness. An automatic focus actuator is usually used in conjunction with the motorised stage where normal variations in surface height of the sample will cause the specimen to move out of focus on scanning.

Different types of devices such as video cameras and scanners are used in image acquisition. There has been a recent specialised review on different types of image acquisition devices and their general characteristics with respect to their use in advanced image analysis work [22]. Such devices translate video images or photographic images into electrical signals with amplitude proportional to the image brightness at any given point. A scanner perform a point-by-point scan from a picture to produce an array of spots representing the original picture. Scanners are usually attached to host computers, i.e. they do not include specialised hardware to acquire images rapidly. Advanced image processing systems use two main types of video camera: tube cameras and solid-state charged coupled device or CCD cameras. Tube cameras are older but they are still used for their high sensitivity especially for applications in fluorescence and luminescence (intensified cameras) where the light levels are very low. The CCD cameras are easier to use and are becoming more and more available at affordable prices. The analogue signal from either type of camera is converted to digital format by a high speed analog to digital converter (ADC) to give discrete grey levels. Usually the ADC converter is built into the image analysis hardware. Most video cameras produce square pixels (or picture elements) but some give rectangular or quasi-square pixels. In the final discretisation which takes place in the image analysis hardware such rectangular or quasi-square pixels are usually corrected to give square pixel based images. In an image the larger the size of the array, the higher the resolution, and the finer the spatial details that may be detected. It is of course necessary to match as closely as possible the array size of the camera and that of the image being handled by the image analyser. Usually 256×256 to 1024×1024 arrays of pixels are common. Other dimensions such as 768×576 pixels based images are also available; these reflect camera technology.

The digitised image can now be processed to extract the desired information. Several features of image processing operations are primarily due to the large memory requirements for handling and processing images. For example, a typical digital image consists of 512×512 pixels. Each pixel, having 256 possible greyness levels, can be encoded in a byte. This means that to store the image 262 144 bytes are required. A colour image is composed of three monochrome images (representing the red, green and blue) and therefore requires three times as much memory. Besides the large amounts of memory needed, even applying a simple image processing operation to an image may require several arithmetic operations for each pixel, which could easily amount to over 100 million processor operations for the single operation. This requires fast hardware for good performance. Many commercial image analysis systems now use dedicated hardware where the image processing is undertaken by a special microprocessor designed for speed.

A fast PC or a workstation installed with image analysis software is cheap but has limited abilities and speed. With additional dedicated image processing hardware the system can be more functional in terms of power and speed but becomes more costly. To base the complicated or time intensive operations on hardware, and the relatively simple operations on software relying on the host central processing unit (CPU) speed, is an acceptable compromise for some applications. This could be an attractive approach, particularly to fermentation technologists or microscopists who still rely on manual methods, and who also need a cheap route to image analysis. The rapid increase of CPU speed of PCs and workstations and the reduction of overall computer costs is expected to encourage many commercial firms to take advantage of these developments to provide more affordable solutions to image analysis problems. On the other hand, high performance systems consisting of on-board hardware and a library of functions for image processing also implemented in hardware allow complicated applications to proceed at high speed, saving the operator time and permitting more rapid research progress.

2.2
Software

Software is a vital element of an image processing system. It consists of programs that control the hardware to carry out a variety of enhancement, manipulation, filtering and smoothing operations, and measurements. Software provides the users with a set of tools which can be applied to develop a program appropriate to a particular application. The required sequence of image processing operations can be encoded in such a program to allow fully automatic repeatable analysis and measurements. Most image processing packages have been designed for application developers rather than the final users and the image analysis routines come as a library supported by a common programming language.

A number of distinct steps are involved in image analysis applications: image acquisition, grey image enhancement, image detection, binary image processing, image editing, measurements and calculations, and data analysis. Some of the functions are described briefly here, and will be referred to in describing specific applications in Sect. 3. More detailed explanations can be found in the thesis by Paul [23].

2.2.1
Image Acquisition

Image acquisition covers selection of fields of view, microscopic magnification, and adjustment of lamp brightness and focus to give sufficient contrast to enable the image analyser to identify the objects of interest. Finally the image is captured into digital form ready for processing.

2.2.2
Grey Scale Processing

There are a large number of image processing functions available to improve the quality of grey images. Each involves a grey scale transformation where the grey level of each pixel is transformed to another value using rules that take into account the original values for that pixel and its neighbours, or its spatial relationship to many other pixels. Sometimes images may be subtracted or added together to remove some spurious artefacts or to improve images that are poor in quality. There are certain filtering techniques that select only some kinds of image data, for example edges and boundaries.

Grey processing operations can be divided into two categories: point operations and neighbourhood operations. Point operations are performed on individual pixels. For instance, if the original greyness values covered only a small fraction of the total range available, then new values covering the whole range could be substituted pixel by pixel to increase the image contrast. Neighbourhood operations are carried out on groups of adjacent pixels. For example each pixel in a square array is surrounded by eight neighbours. Such a group of pixels is known as a "structuring element" (SE). SEs are often square but could be other shapes. By setting the central pixel greyness of a square SE, using its neighbours' greyness values and applying some rules relating the former to the latter, changes in intensity across small parts of the image can be achieved, e.g. for smoothing or to remove noise from an image. The terminology most often used to describe these operations is "morphological transforms". In "mathematical morphology" [24] the relationship between the value of each pixel and that of its neighbours is expressed in logical terms, using AND and OR for binary images and "greater than" and "less than" for grey images. The particular grey scale transformation functions used in a number of applications discussed in Sect. 3 are described briefly here.

"Erosion" and "dilation" are the most common morphological operations used to process both grey images and binary images. In grey erosion the pixels covered by the SE are examined for the one with the lowest numerical value. The central pixel in the SE is set to this minimum value in the output image. This produces two effects: first, the resulting image becomes darker than the original image; and second, any brighter details that are smaller than the SE are eliminated. Exactly opposite effects are observed in grey dilation, i.e. the maximum value covered by the SE is chosen and copied to the output image. By this operation the general brightness of the output image is raised and dark details are removed.

These two basic functions are used to build a large number of secondary functions designed to extract or separate details in a grey image. "Delineation" and "top hat" are two highly useful functions. Delineation selects a local minimum or maximum grey value of neighbouring pixels by a combination of erosion and dilation operations. This improves the contrast of a grey image particularly around object boundaries (see for example applications described in Sects. 3.1 and 3.2). Figure 3 shows the grey level variations across an object boundary before and after the delineation operation. By a top hat operation either the white details (using "white top hat") or dark details (using a "black top hat") can be extracted from an image. For example, this function has been used to extract halos from around hyphae in an image (Sect. 3.4). Figure 4 illustrates how halos around an object's boundaries can be extracted using the white top hat operation.

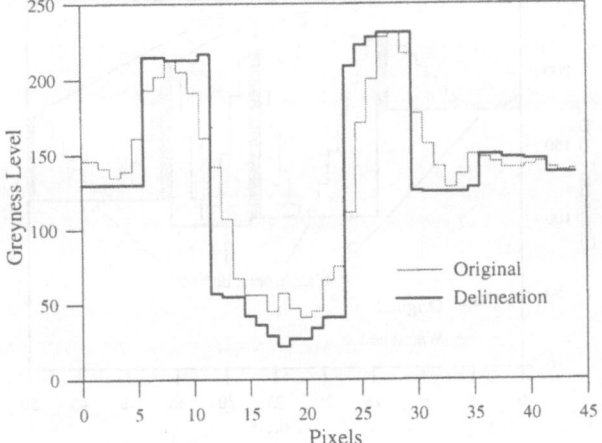

Fig. 3. Effect of delineation on a grey level profile

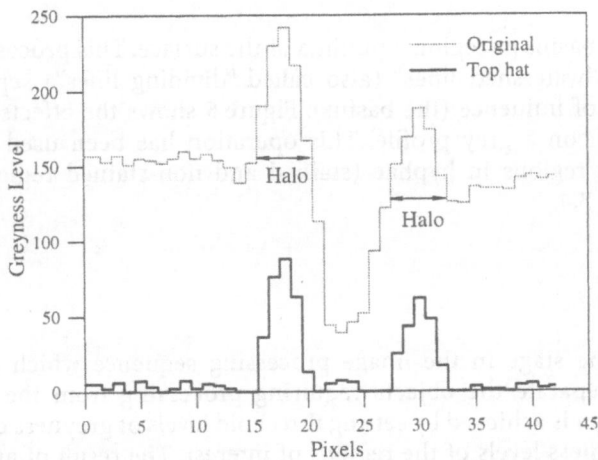

Fig. 4. Extraction of white details (e.g. halos) across object boundaries using the white top hat operation

A complex grey image sometimes contains many features touching each other at numerous points, and the zone of influence corresponding to each feature may not be easily distinguishable or separable. A "watershed" is a grey transformation which is often used to separate features from such images. The watershed segments a grey image into its "catchment basins". A grey image can be seen as a topographic surface. According to the law of gravitation, if a drop of water falls on such a surface, it will run down a hillside until it reaches the sea or a lake. All places from which such a drop will reach a given lake are in that lake's catchment basin. If a grey image is considered to be a topographic surface with height represented by the greyness level of any given pixel, a watershed will find

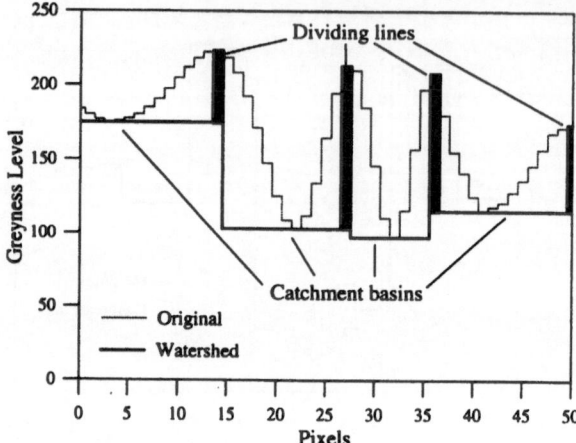

Fig. 5. Watershed of a grey profile. Local minima are expanded until the dividing lines are reached

the catchment basins or regional minima of the surface. This process highlights the ridges or "watershed lines" (also called "dividing lines") separating the various zones of influence (the basins). Figure 5 shows the effects of a watershed operation on a grey profile. This operation has been used to separate closely spaced regions in hyphae (stained and non-stained regions), as discussed in Sect. 3.4.

2.2.3
Object Detection

Detection is the stage in the image processing sequence which attempts to identify and separate the objects requiring processing from the rest of the image. Detection is achieved by setting threshold levels of greyness corresponding to the greyness levels of the regions of interest. The result of applying this process to a grey image is to create a simpler binary image, which has only two values, everything that is selected is "white" and the rest is "black". This is a data reduction step as these images contain less information than grey ones. However, they are smaller and therefore need less computer memory, and processing on binary images is much faster than on grey images. If the extra information in the grey image is not required, then detection is a useful operation.

2.2.4
Binary Image Processing

The binary image may not perfectly represent all the objects required. This may be due to noise or other imperfections in the original image, to inadequate processing methods, or sometimes to features inherent in the image itself such as touching objects. Normally, these kinds of situations are dealt with by manipulating the binary images to improve their quality or to achieve object separation, for example, using binary "erosion" or "dilation" operations to smooth object outlines, join broken or discontinuous objects, and to separate touching ones.

By a binary erosion operation a layer of pixels is removed from the boundary of the object in the binary image. Narrow protuberances on objects are removed, and objects connected by a narrow strand have that removed by this operation. Dilation is the reverse of erosion, causing features to grow in size by the addition of pixels at the boundary, which fills in small breaks, internal voids, or small indentations along the object edges.

Erosion and dilation may sometimes be used in tandem. The sequence of erosion followed by dilation is called an "opening". The initial erosion removes

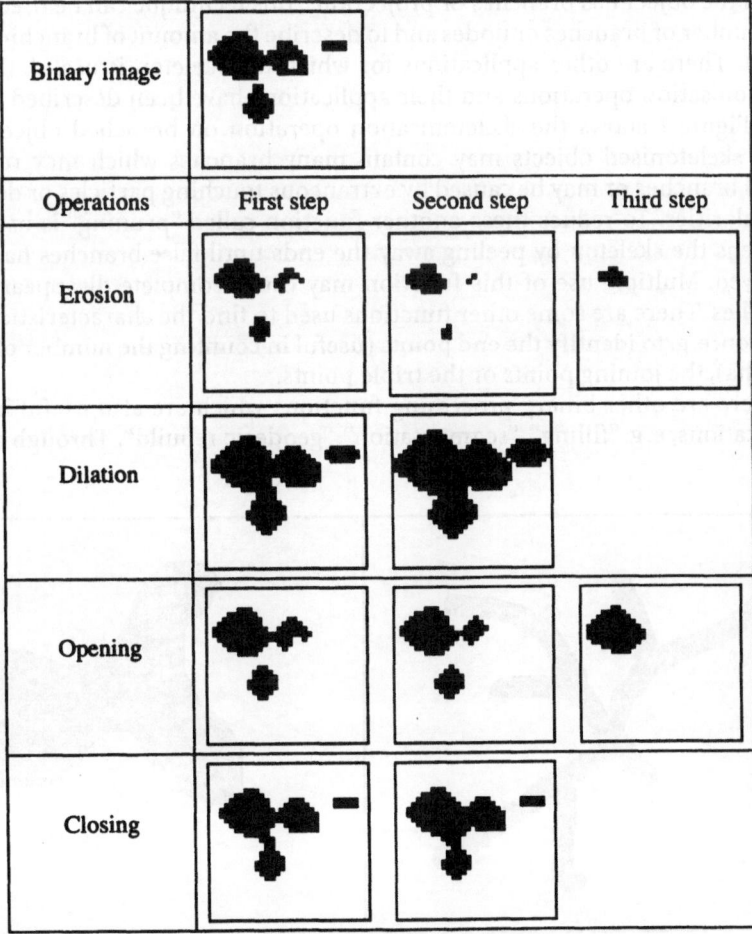

Fig. 6. Expanded view of binary erosion, dilation, opening and closing operations

small features which may represent noise, and also sharp protuberances. The subsequent dilation cannot restore these small features, which have permanently disappeared, but does fill in any small indentations in the outlines. The overall result is that the object size (number of pixels) is restored to nearly its original value, while the shape is modified to become more rounded, smooth and less noisy. When the sequence is reversed (i.e. dilation followed by erosion) it is referred to as "closing". In some respects it is the opposite to opening, because small features are not erased. Small voids in objects are still filled in, and breaks or gaps are joined. Multiple applications of opening and closing produce more extreme smoothing or shape modification, and more complete removal (in opening) or joining (in closing) of objects. Figure 6 illustrates by diagrammatic presentations how erosion, dilation and the derived functions opening and closing operate on different types of objects and the features attached to each.

One of the most interesting specialised uses of erosion is "skeletonisation". The skeleton consists of the lines of pixels that mark the midlines of an object. When the object has branches or projections, this technique can be used to find the number of branches or nodes and to describe the amount of branching in the object. There are other applications for which the skeleton is useful. Different skeletonisation operations and their applications have been described by Russ [25]. Figure 7 shows the skeletonisation operation on branched objects. Very often skeletonised objects may contain many branches which may represent actual branches or may be caused by extraneous touching particles or detection irregularities. To reduce these another function called "pruning" is used. This shortens the skeleton by peeling away the ends until false branches have been removed. Multiple use of this function may cause complete disappearance of branches. There are some other functions used to find the characteristics of the skeleton, e.g. to identify the end points (useful in counting the number of tips in a hypha), the joining points or the triple points.

There are other binary processing functions which are also useful in many applications, e.g. "filling", "segmentation", "geodesic rebuild". Through a filling

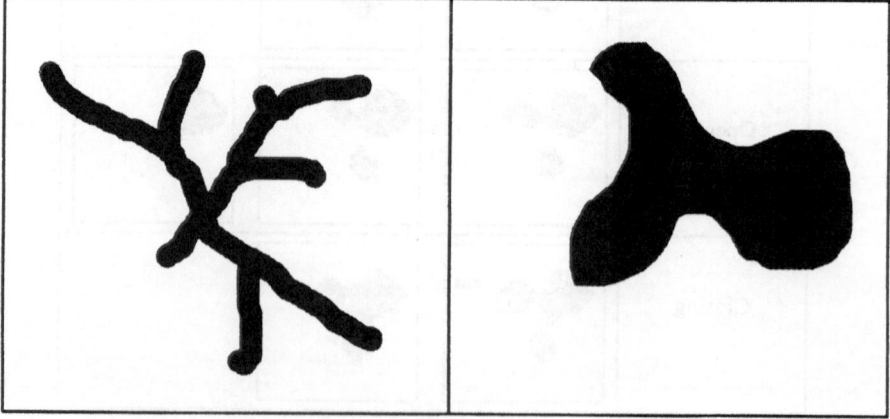

Fig. 7. Skeletonisation of binary image to obtain branches

operation the internal voids or holes of an object are filled (without affecting the outer boundary) giving a solid object. The segmentation operation is applied to separate touching objects which involves many iterations of binary erosion and dilation operations with reference to a map of greyness level variations of the original grey image. Many alternative algorithms have been proposed for obtaining the best separation [25, 26]. A geodesic rebuild is the reconstruction of an object by dilation of "seeds" or markers within the object until it corresponds with a mask of the original object. Objects without seeds cannot be rebuilt. This operation is often used after eroding away smaller objects in order to restore larger objects to their original shapes and sizes without reintroducing the smaller ones. If the smaller objects are desired, they can be found later by image subtraction.

2.2.5
Image Editing

The final image produced after complete (or a stage of) processing may contain false details and other unwanted objects which cannot be identified. It can then become desirable to modify the image manually before final measurements or before another stage of image processing is started. Using edit functions the operator can modify the image to:

- specify particular regions of interest;
- correct object details, e. g. draw in a partially or completely missing feature;
- reject unwanted regions of an object or a whole object;
- separate touching features or objects.

2.2.6
Measurements and Calculations

Four types of parameters can be obtained from image measurements: size, shape, position, and grey scale (brightness) information. Within each of these categories, there is a variety of individual parameters that can be measured or calculated from others. Table 1 summarises some of the parameters that are most often used in applications like characterising mycelial morphology, spore germination and differentiation.

2.2.7
Data Analysis

Once the measurements and calculations have been made, the data can be classified and analysed to give the final results, which might be in tabular form, graphical or might be stored in a file to be processed later using stand alone statistical or other software.

Table 1. Some important parameters obtained by image analysis

Parameters	Definition/Formula
Object size:	
Area or projected area (A)	Area of the projection of a 3-dimensional object into a two-dimensional image. This is often found by a pixel (picture element) count, but is expressed as an actual area by multiplying by a calibration constant squared
Perimeter (P)	Length of the boundary of an object. As pixels lie on a rectilinear grid, it is necessary to include diagonal inter-pixel distances where appropriate, otherwise a square and its inscribed circle would appear to have the same perimeter
Convex perimeter (P_C)	Length of the perimeter obtained by joining the outer points of an object, i. e. by filling in all the concavities in an object
Convex area (A_C)	Area inside the convex perimeter.
Length or fibre length (L)	The length of a rectangular object having the same area and perimeter as the measured object. This is derived from area and perimeter as $$L = \frac{P + \sqrt{P^2 - 16A}}{4}$$
Width or fibre width (W)	The width of a rectangular object having the same area and perimeter as the measured object. The width of an object is derived from area and perimeter as $$W = \frac{P - \sqrt{P^2 - 16A}}{4} \quad \text{or} \quad W = \frac{A}{L}$$
Equivalent circular diameter (D)	The diameter of a circle having the same area as the measured feature. Derived from area as $$D = \sqrt{\frac{4A}{\pi}}$$
Breadth and height	These parameters are obtained by feret measurements. A feret is equivalent to a diameter measured using a pair of callipers. Breadth gives the length of the shortest feret and height the longest feret
Maximum distance	The longest feret across the convex area of a mycelial particle gives the maximum distance. This parameter is used to characterise a mycelial clump
Object count:	
No. objects	Number of objects per field of view and cumulative counts for a sample

Table 1 (continued)

Parameters	Definition/Formula
Hyphal tips	Number of tips per hypha, mean number of tips per hypha for a sample
Object shape:	
Circularity (C)	A shape factor describing the deviation of an object in an image from a true circle. This is derived from area and perimeter $$c = \frac{P^2}{4\pi A}$$ This gives a minimum value of 1 for a circle, larger values for shapes having a higher ratio of perimeter to area
Roughness (R)	A measure of the irregularity of the perimeter of an object. It is obtained from the circularity measurement around an object boundary
Compactness or fullness (F)	This a measure of the voidage of a particle and is used to characterise mycelial clump and pellet structure. It is the ratio of the actual area of the particle to the convex area. For a pellet without hairy regions $F \cong 1$, and for a loose clump $F < 1$
Object position:	
Coordinates (X,Y)	x-coordinates, y-coordinates, often implemented with the origin at the pixel
Feature count point (X-FCP,Y-FCP)	The FCP is the last rightmost pixel of the lowest scan line contained in the object. FCP with a guard frame is used to avoid object truncation in automatic methods
Object brightness:	
Grey intensity (G)	A measure of the brightness of a part of the image. Two types of grey intensity are used: mean grey intensity (MG) giving the mean of grey levels of pixels overlaid by a binary feature and integrated grey intensity (IG) giving the sums of the grey levels of pixels overlaid by a binary feature

3
Image Analysis Methods

This section describes four main image analysis applications in fermentations using mycelial microorganisms. The image analysis methods were developed on a Quantimet 570 image analyser (Leica Cambridge Ltd, Cambridge, UK. – see Fig. 8). The Quantimet 570 is a general purpose image processing and analysis

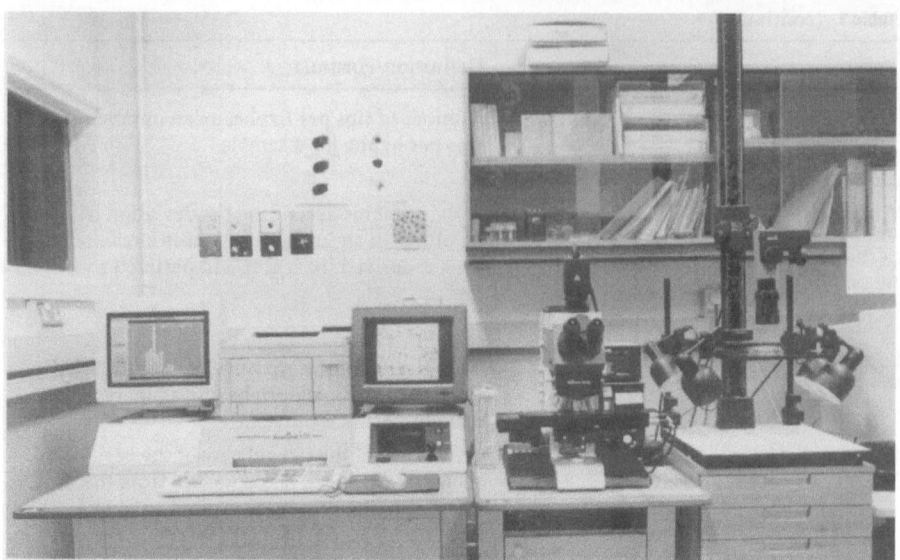

Fig. 8. Photograph of the Quantimet 570 image analysis system

system consisting of a Motorola 68000-based monitoring unit, supervising a morphological processor unit, graphic overlay and binary image memory, a measurement processor, the acquisition and display system, and peripheral control interfaces. An 80486-based PC host computer serves as the user's interface. The overall image processing speed of the Quantimet 570 has been improved by incorporating programmable logic cell array technology, based on concepts of mathematical morphology, into the morphological processing unit [24]. A Polyvar II optical microscope (Reichert Jung, Optische Werke AG, Wien, Austria) fitted with a colour 3CCD camera (Model XC-007, Sony, Japan) and macroviewer (supplied by Leica Cambridge Ltd) fitted with a CC-TV monochrome camera (Sanyo Electric Co. Ltd, Basel, Switzerland) were connected to the image analyser. For automatic operation the microscope was fitted with an eight-slide motorised stage (supplied by Leica Cambridge Ltd), a controllable illumination system, and autofocus capability. The stage is drivable in 3-dimensions with a 2.5 μm step size in the x-y directions for scanning on the microscope slide and a 0.05 μm step size in the vertical direction for automatic focus control.

The image analysis routines have been implemented on the Quantimet 570 by means of its interactive basic programming environment, QBASIC, or using a C compiler and its image processing library, ADSOFT 1.01 (Advanced Software Developer's Kit, Leica Cambridge Ltd, Cambridge, UK). ADSOFT consists of a C library of commands to control stage utilities (e.g. initialisation, scanning, and autofocus), lamp brightness, and the image processing hardware. The image processing library consists of an extensive range of grey, binary, and colour image processing routines (including those described already in Sect. 2.2) and facilities for image measurements. All the programs described here were initially developed using QBASIC [7, 13 – 15]. Except for the pellet program described in Sect. 3.2, all were then extended and enhanced [8, 27, 29] using C and C++ (Borland Inc., Scotts Valley, CA, USA) and ADSOFT. Although the routines were developed on the Quantimet 570, they ought to be implementable on other systems. However, the resulting performance will largely depend on the image processing hardware and the image acquisition optics used.

Image analysis begins with specimen preparation. In most cases, the specimen is carefully prepared for optical microscopy in order to achieve optimum image quality and ease of analysis. The fermentation samples are often diluted and shaken to disperse the mycelia, pellets, or spores, and stained to enhance the contrast of the microscopic images. Many methods use a special stain [8] or a number of stains to pick out hyphal regions of interest. This is described further in Sect. 3.4. Slide preparation needs to be performed carefully to avoid inclusion

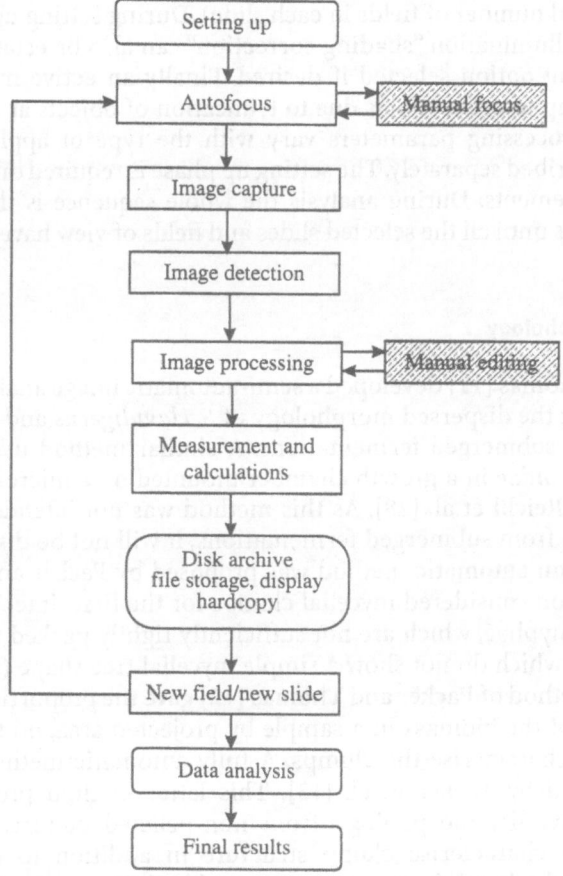

Fig. 9. General structure of the software used for automatic and semi-automatic image analysis. Semi-automatic image analysis uses manual focus adjustment and manual editing

of many possible artefacts, e. g. dirt, air bubbles. Bad slide preparation can make subsequent image analysis more difficult and increase the load of operator time through requiring unnecessary manual editing of images.

All the image analysis routines presented here follow the general structure shown in Fig. 9. A routine consists of seven phases (some of which were described in Sect. 2.2): setting up, image acquisition, grey image processing, image detection, binary image processing, image measurement and data analysis. Grey and binary image processing stages consist of a combination of image processing operations, chosen to achieve a particular result; therefore these vary with the application. Detailed descriptions of these application-specific phases are given in Sects. 3.1 – 3.4 below.

Each program requires two types of parameters to be set by the user: (a) hardware parameters, and (b) image processing parameters. The setting up for the hardware parameters include the microscope lamp brightness, autofocus parameters, the calibration factor for the objective to be used, and control parameters for automatic microscope stage movement (number and position of slides and total number of fields in each slide). During setting up, correction for uneven slide illumination "shading correction" can also be established, and the manual editing option selected if desired. Finally an active measuring frame might be set up to prevent bias due to truncation of objects at the image edge. The image processing parameters vary with the type of application and are therefore described separately. The setting up phase is required only once for each set of measurements. During analysis the whole sequence is then executed in repeated cycles until all the selected slides and fields of view have been analysed.

3.1
Dispersed Morphology

Adams and Thomas [11] developed a semi-automatic image analysis method for characterising the dispersed morphology of S. clavuligerus and P. chrysogenum samples from submerged fermentations. A similar method used to study the mycelia of S. tendae in a growth chamber mounted on a microscope stage was developed by Reichl et al. [28]. As this method was not intended for studying samples taken from submerged fermentations, it will not be discussed further. Subsequently an automatic method was proposed by Packer and Thomas [12] and this method considered mycelial clumps for the first time. These are loose aggregates of hyphae, which are not sufficiently tightly packed to be described as pellets, but which do not show a simple mycelial tree shape (see Fig. 10). Although the method of Packer and Thomas [12] gave the proportion of clumps as a percentage of the biomass in a sample by projected area, no further attempt was made to characterise the clumps. A fully automatic method has recently been published by Tucker et al. [13]. This latter method provided detailed analysis of mycelial morphology from non-pelleted cultures and included parameters to characterise clump structure in addition to the proportion measurement obtainable by the earlier method [12]. The method of Tucker et al. [13] has been further improved [29] and has been applied to a wide range of morphological forms from submerged Actinomycete and other mycelial

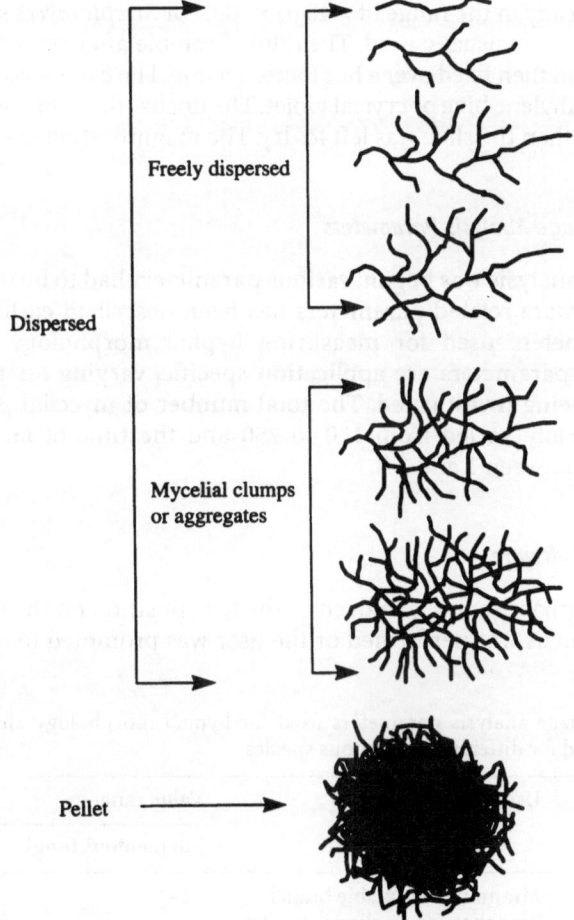

Fig. 10. Classification of mycelial morphology by image analysis

fermentations. It is described in more detail below. It does not characterise pellets and a separate method for this purpose is presented in Sect. 3.2.

3.1.1
Sample and Slide Preparation

Samples from fungal fermentations were fixed with an equal volume of fixative containing 13 ml 40% formaldehyde, 5 ml glacial acetic acid and 200 ml 50 vol.% ethanol. Before slide preparation the fixed sample was diluted with 20% (w/v) sucrose solution to prevent quick evaporation during analysis and stained with a few drops of lactophenol cotton blue (Fluka Chemika-BioChemika, Dorset, UK). The biomass concentration after dilution was usually kept below 1.0 g/l. The diluted sample was then placed on a slide and covered with a cover slip. The magnification was typically in the range of ×40 to ×100. For *Streptomyces* spp. the Gram-staining method was usually used. The diluted sample after spreading on a slide was air-dried and then fixed over a hot Bunsen flame. The fixed mycelia were stained either by methylene blue or crystal violet. The unabsorbed stain was washed off with water and then the slide was left to dry. The magnification used was ×100.

3.1.2
Setting Up of Image Analysis Parameters

Before images analysis was begun, various parameters had to be set. The setting up of the hardware related parameters has been described earlier. The image analysis parameters used for measuring hyphal morphology are listed in Table 2. These parameters are application specific, varying for the particular fermentation being investigated. The total number of mycelial particles to be measured typically varied from 150 to 250 and the time of analysis for one sample varied between 1 and 2 h.

3.1.3
Image Analysis Software

In the analysis phase the stage moved to the first position on the slide and then either an autofocus was performed or the user was prompted to adjust the fine

Table 2. List of image analysis parameters used for hyphal morphology analysis and their typical values used for different filamentous species

Parameters	Use in analysis stage	Value ranges	
		Filamentous fungi	Actinomycetes
Minimum branch length, µm	Minimum acceptable branch branchessize, to eliminate false touching debris	4–5	2–4
Mean hyphal width, µm	Mean width is used to derive a minimum area, to reject debris based on size	4–5	2–3
Minimum length, µm	Minimum length is used to derive a minimum area, to reject debris based on size	20–30	10–20
Number of loops	To re-classify mycelia into freely dispersed and clumps based on the number of loops. Loops are caused by simple overlapping of branches	3	3
Maximum fullness ratio	The value of the fullness ratio above which particles are classi- fied as debris or large media particles	0.6–0.7	0.6–0.7
Circularity factor	Used to remove rounder media particles and debris	2–3	2–3

focus. The image was captured and stored in digital format. This grey image was enhanced by delineation. It was then detected to obtain a binary image. The binary image was then subject to a single closing operation to consolidate the detection. The subsequent major image processing stages are described diagrammatically below.

1. Diagrammatic representation of a binary image obtained after detection of a grey image followed by closing operation. The image might contain many small background artefacts and medium particles which were removed by a single opening operation.

2. Most of the smaller non-hyphal particles not eliminated by the opening operation of step 1 were removed using a size (area) filter derived from the specified mean width and the minimum length parameters.

3. A binary filling operation was applied to fill the internal holes or voids of each object.

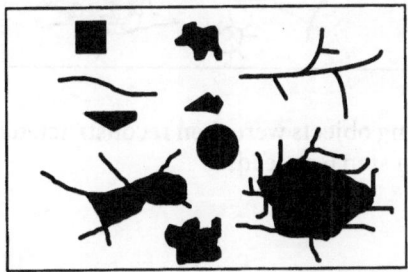

4. A circularity (shape) factor filter was applied to remove media particles and debris which appear more round than mycelial particles. Remaining objects were restored to their unfilled forms.

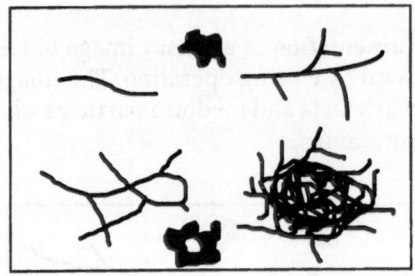

5. The image was skeletonised.

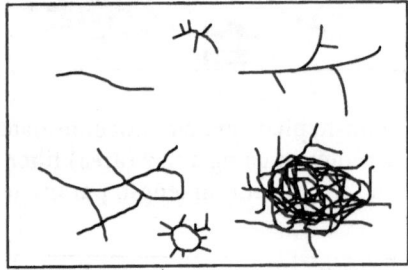

6. An exhaustive pruning was applied to the skeletonised image which removed all skeletons not containing a loop, i.e. unbranched and branched mycelia, and debris appearing as unholed.

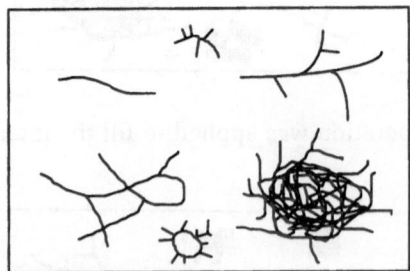

7. The loop-containing objects were then reconstructed from the image in step 4 using loops from step 6 as seeds.

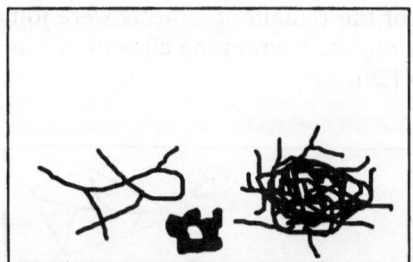

8. By subtraction of image 7 from image 4, an image containing unbranched and branched mycelia including some debris was obtained.

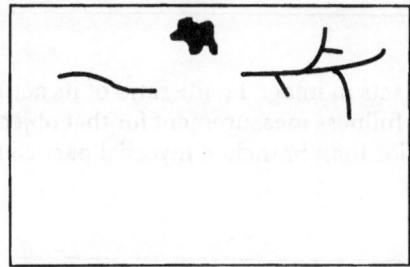

9. The image was then re-skeletonised.

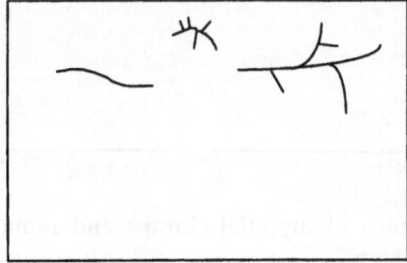

10. The unbranched hyphae were separated based on the number of branching points. Before this operation the skeletons were pruned to remove any false branches or small touching particles of lengths less than the minimum specified.

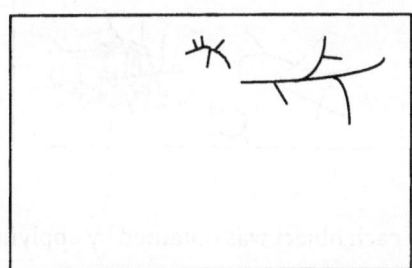

11. The outer points of the remaining objects were joined to give the convex
    perimeter of each object. A wrapping algorithm was used to find and join
    these outer points [29].

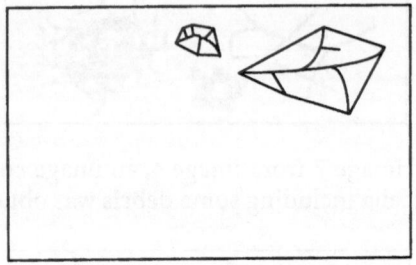

12. For each of the objects in image 11, the ratio of its actual area (step 8) to the
    convex area gave a fullness measurement for that object. Debris usually gave
    higher fullness ratios than branched mycelial particles, and could be elimi-
    nated on this basis.

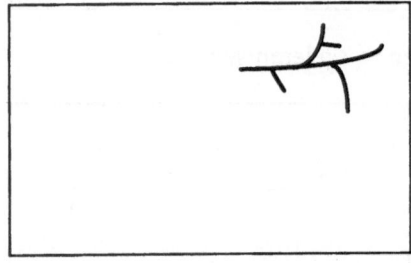

13. A skeletonised image of mycelial clumps and remaining debris, derived
    from the image of step 7.

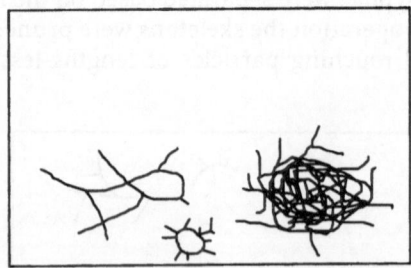

14. The convex area of each object was obtained by applying the wrapping algo-
    rithm.

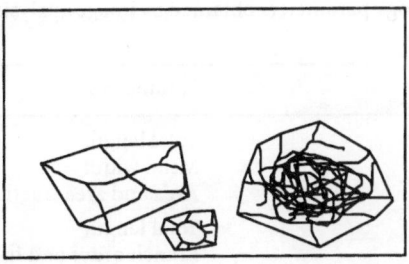

15. The non-hyphal particles were then eliminated using the fullness criterion. A further classification of clumps was based on the number of holes. A lesser number of holes usually indicates artefactual overlapping rather than a real clump. This latter class is termed as simple clumps or "entanglements" (MN Paris; personal communication) and are considered to be part of the class of freely dispersed mycelia.

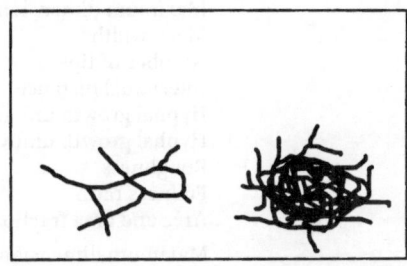

Following the elimination of non-hyphal particles and classification of all the mycelial elements, measurements were performed. The image containing branched hyphae could be processed further to identify individual branches by branch order. The detailed description of image processing operations for this step can be found in a paper by Tucker et al. [13]. This optional measurement of individual branches is not usually done, in order to increase the speed of analysis. The hyphal measurement parameters are listed in Table 3. Information on each particle was stored in a file for detailed statistical analysis, e.g. mean and standard deviation calculations, and for generating histograms.

Figure 11 illustrates the different types of morphology of *S. clavuligerus* in a submerged batch fermentation obtained by using different inoculum spore concentrations [30]. In general, fungal species also show a similar range of morphological forms in submerged fermentations. Image analysis [3, 31, 32, 34] has recently been used to replace manual method [33] in studying the effect of the initial spore concentration, medium pH and medium composition on the growth and morphology of fungal fermentations. The developed morphology in the fermenter is also influenced by the agitation, which can lead to even more heterogeneous forms and sizes. Figure 12 shows the morphological parameters of the freely dispersed hyphae, i.e. unbranched and branched hyphae and entanglements (combined together), throughout a fed-batch penicillin fermentation in

Table 3. Hyphal morphology parameters obtained by image analysis (see Table 1 for definition of these parameters)

Mycelial classes	Parameters
Unbranched	Total length Mean width Area and area fraction
Branched	Total length Branch order and individual branch length Longest length Maximum dimension Number of tips Mean width Internodal distance Hyphal growth unit length Hyphal growth unit volume Area and area fraction
Simple clumps (containing 1 – 3 holes) or "entanglements"	Total length Maximum dimension Mean width Number of tips Internodal distance Hyphal growth unit length Hyphal growth unit volume Roughness Fullness ratio Area and area fraction
Clumps	Maximum dimension Roughness Fullness ratio Area and area fraction

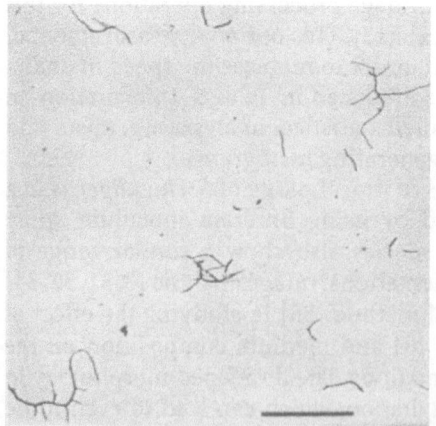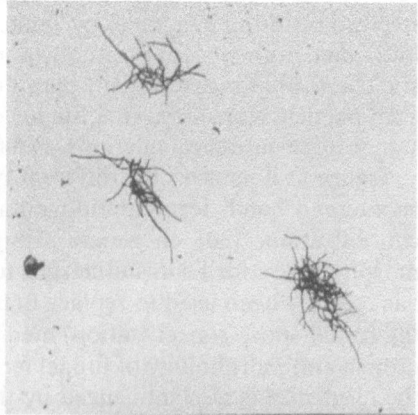

Fig. 11 a, b. Different classes of morphology from a submerged *S. clavuligerus* fermentation: **a** freely dispersed mycelia; **b** small clumps; **c** bigger clumps; **d** a pellet (bar length = 60 μm)

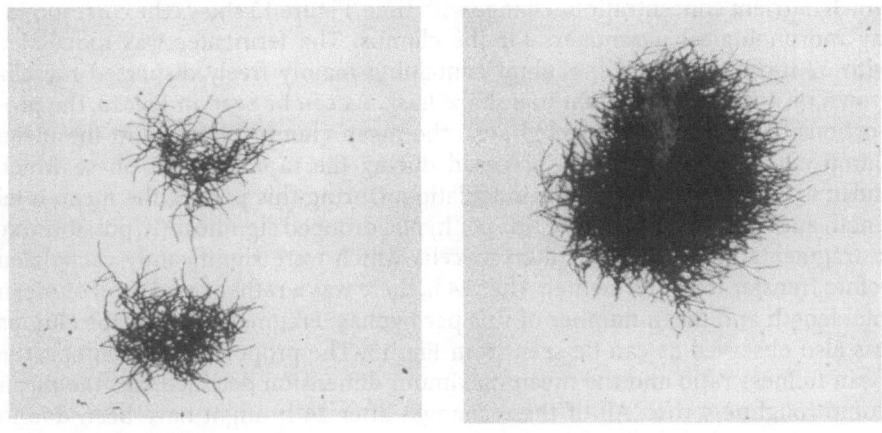

c

d

Fig. 11 c, d

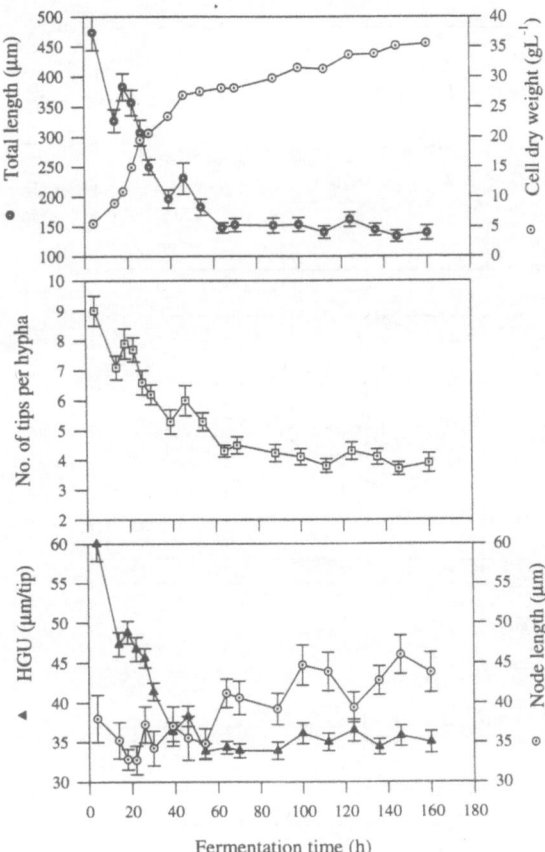

Fermentation time (h)

Fig. 12. Time profiles of cell dry weight and the mean values of total length, tips per mycelium, hyphal growth unit and node length for freely dispersed hyphae from a typical fed-batch *P. chrysogenum* fermentation with constant glucose feed rate

which nutrient concentrations change with time. Figure 13 shows the corresponding morphological parameters for the clumps. The fermenter was inoculated with 32 h-old vegetative inoculum containing mainly freely dispersed mycelia grown on a complex medium in a shake flask. As can be seen in Fig. 13, the proportions of clumps by projected area, the mean clump fullness and the mean clump maximum dimension increased during the rapid growth phase which ended approximately 24 h after inoculation. During this period, the mean total length and the mean number of tips per hypha dropped significantly, possibly due to fragmentation of the inoculum mycelia which were significantly vacuolated before transfer to the fermenter. After 24 h, there was a rather rapid drop of mean total length and mean number of tips per hyphae. Fragmentation of the clumps was also observed as can be seen from Fig. 13. The proportions of clumps, the mean fullness ratio and the mean maximum dimension declined and the mean clump roughness rose. All of these changes after 24 h might have been due to increased vacuolation of the hyphae (see Sect. 3.4) as the carbon source became

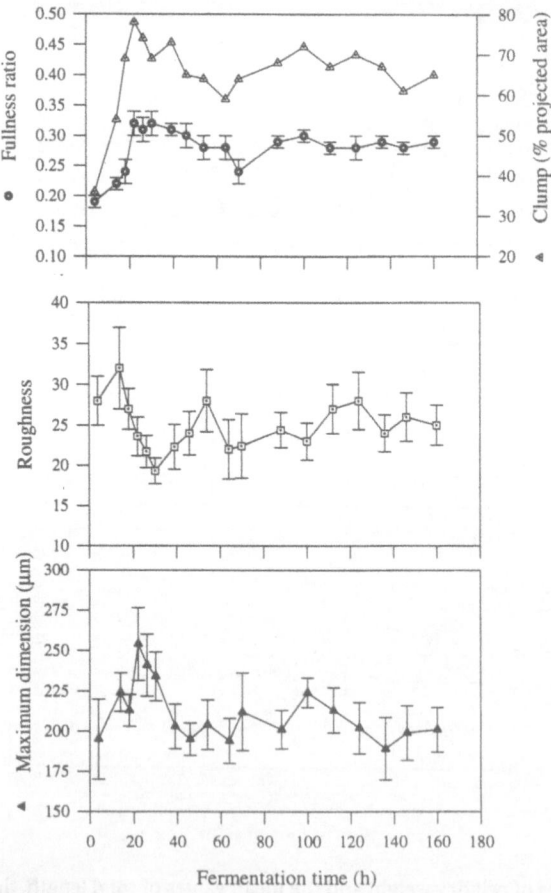

Fig. 13. Time profiles of the proportion of clumps and mean values of fullness ratio, roughness and maximum dimension for the fed-batch *P. chrysogenum* fermentation shown in Fig. 12

depleted at the end of the rapid growth phase [35]. It must be noted however that the apparent reduction of the hyphal lengths for the freely dispersed mycelia might be due, at least in part, to loss of fragments from clumps, fragments which reappear in the freely dispersed category affecting its size distributions. This needs further study.

For most industrial mycelial fermentations, dispersed growth is preferred and for such fermentations image analysis can have an important role in characterising morphology. It should be noted here that a high degree of clump formation in submerged fermentations implies that classical morphological measurements on the freely dispersed class alone provide only partial information which is based on only a small fraction of the total biomass. There is little point in making measurements on only the freely dispersed form, especially as evidence exists that the hyphal elements within clumps are of different morphology to those not aggregated [36]. Fully automatic characterisation of morphology including clumps is now possible, which opens up new research opportunities in engineering and physiological studies of mycelial fermentations.

Modelling mycelial morphological development is very complex. Nevertheless there are several models available in the literature concerning hyphal growth and branch formation [37-41]. Chapter 4 gives a general review of morphological models of *Penicillium* species and Chap. 3 of *Streptomyces* species. Modelling in submerged fermentations still involves many unproven assumptions because of the complex interactions of fermentation environments and mycelial morphology and because of lack of knowledge, particularly of the kinetics of the formation and breakage of clumps in real fermenters. For this purpose fully automatic image analysis can be used to gather the large amounts of data needed for model development, extension and verification.

3.2
Pellet Morphology

There are a number of image analysis methods available to characterise pellets from submerged fermentations. Reichl et al. [42] used image analysis to study *S. tendae* pellets, measuring for each sample the frequency distribution of size, the mean size, the percentage of total mycelia (by projected area) that existed as pellets, and pellet shape. Pellets were distinguished from hyphal fragments and clumps by greyness level differences; they were relatively dark. This approach provides a possible definition of a pellet. The pellet-shape analysis relied on image-processing operations which affected the irregular outline of fluffy pellets more than that of smooth ones. The shape factor thus obtained was the first proposed for quantitative discrimination between pellet types. Durant et al. [43] discriminated between these zones by rinsing out a stain (i.e. by a diffusional criterion) which made image analysis much easier. Later, the method was improved using colour-image processing [44].

Pichon et al. [45] and Cox and Thomas [14] proposed image analysis methods to characterise pellets based on the presence of a central core. Each pellet was analysed using grey image processing giving darker regions in the pellet-centre and relatively brighter outer mycelial regions. Cox and Thomas [14] further classified the pellets into smooth and hairy types using automatic image analysis. The central core was identified by removing the lateral hyphae by image processing. The core size (projected area) and shape (circularity) were determined. The hairy annular region of lateral hyphae was also characterised in terms of size, fullness and roughness. The method was developed on the Quantimet 570 using simple image processing operations common to most commercial image analysis software packages and can therefore be implemented easily on other systems. This method is described here in a greater detail.

3.2.1
Sample and Slide Preparation

Samples from *A. niger* fermentations were fixed in an equal volume of fixative containing 9% (w/v) formaldehyde and 50% ethanol. The pellets and mycelia were stained with lactophenol cotton blue to enhance visualisation under bright field illumination. The three dimensional structure of the pellets was preserved by observing pellets in a cavity slide of depth 1 mm. The typical analysis time for a pellet is approximately 5 s [14], so that a few hundred pellets can usually be characterised in 15 min.

3.2.2
Image Analysis Parameters

Table 4 gives the list of image analysis parameters and their typical values used for *A. niger* pellet analysis.

3.2.3
Image Analysis Software

Pellets of sizes less than 600 μm equivalent diameter were viewed under a microscope (as described in Sect 3.1). For larger pellets a macroviewer was used. This consisted of a CCD camera fitted with a macroscopic zoom lens mounted on an adjustable camera stand allowing transmitted light examination of samples. Movement of the sample in the field of view was manual.

Table 4. List of image analysis parameters and their typical values

Parameter	Use	Typical range (cycles)
Erosion	To remove smaller non-hyphal debris and media particles	3–6
Opening	To remove lateral hyphae to find the pellet core	4–6

An image of pellets was acquired and delineated to increase the outline definition. The image was then detected to produce a binary mask of the pellets overlaying the grey image. The subsequent image analysis was on this binary image. Small debris and media particles including small mycelial fragments were first removed from the image by repeated erosion until they vanished, leaving the pellets and mycelial clumps, which are larger, not completely eroded away. A rebuilding operation (based on the original binary image) was then used to restore pellets and mycelial clumps to their original size. The analysis of any individual pellet involved the recognition of the solid core within it. The existence of a core was ascertained by an ultimate skeletonisation of the pellet image, i.e. the exhaustive removal of the outer pixel layers of the skeletonised image to identify a central single pixel marker. However, the presence of holes in the annular regions created by entangled lateral hyphae prevented this marker being located by reducing down to loops rather than a single point. If this occurred, lateral entangled branches were removed by an opening operation. Repeated opening cycles were done until an ultimate skeletonisation did give a single marker, or until the preset maximum number of openings was reached (Fig. 14). Once a marker was identified, the pellet was rebuilt from it. The pellet was then subjected to an opening operation to remove the annular hyphal region and separate out the core. The subtraction of the core from the whole pellet gave the annular regions. An object without a core identifiable with the preset maximum number of openings was

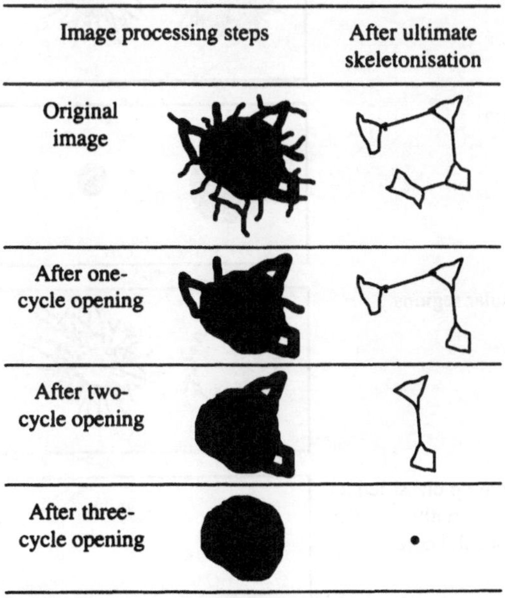

Fig. 14. Image processing operations to identify the core of a pellet. In the illustrated case, after three cycles of opening skeletonisation gives a single pixel marker. The core has then been identified

classified as a clump. Figure 15a–c shows different forms of morphology and their classification on the basis of structure.

Figure 16 shows a smooth and a hairy pellet of *A. niger* from a submerged citric acid fermentation. Figure 17a shows the biomass growth, mean pellet equivalent diameter and mean core equivalent diameter across the time course of an *A. niger* batch fermentation. Both the whole pellet diameter and the core diameter increased with time and followed the growth of biomass in the fermenter. After 50 h, the annular regions grew faster than the core. Figure 17b shows the mean projected area of the cores and the mean core circularity. More regular core shapes were found later in the fermentation. Figure 17c gives the corresponding mean annular area and the mean whole pellet fullness ratio. The coincidence of the increase of annular area and the decrease of fullness suggests that growth of the annular regions took place by the extension of hyphae in these regions, therefore increasing the hairyness of the pellet. This can be seen more clearly from Fig. 17d from the mean ratio of convex area of a core to that of the whole pellet. This suggests that there existed a transition of morphologies from smooth to hairy after 50 h in this fermentation.

Pellets of some strains are sensitive to agitation conditions. At high agitation levels pellets can be fragmented. Figure 18 shows the time profiles of mean pellet

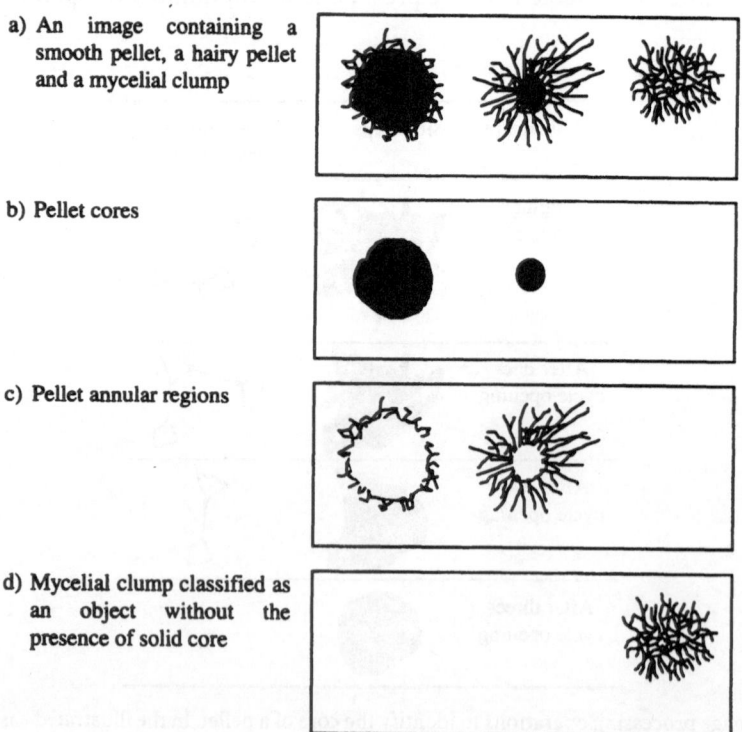

a) An image containing a smooth pellet, a hairy pellet and a mycelial clump

b) Pellet cores

c) Pellet annular regions

d) Mycelial clump classified as an object without the presence of solid core

Fig. 15. Diagrammatic representation of the classification and characterisation of pellets

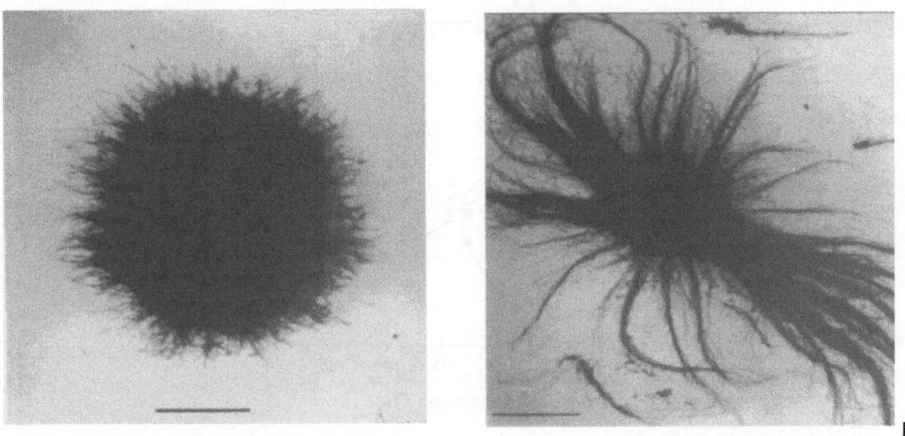

Fig. 16 a, b. *A. niger* pellets from a submerged citric acid fermentation: **a** smooth pellet; **b** hairy pellet (bar length = 0.25 mm)

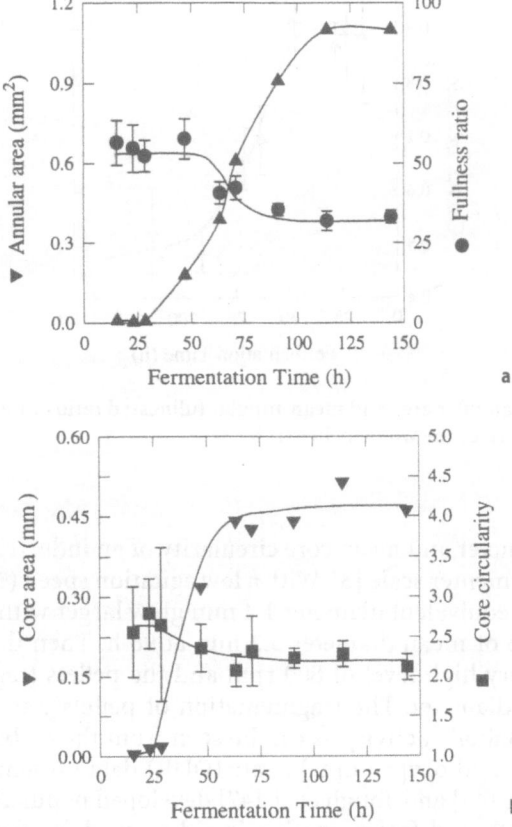

Fig. 17 a, b. Pellet growth characteristics in *A. niger* batch culture: **a** cell dry weight and the mean equivalent diameter of the whole pellet and core; **b** mean core area and mean core circularity

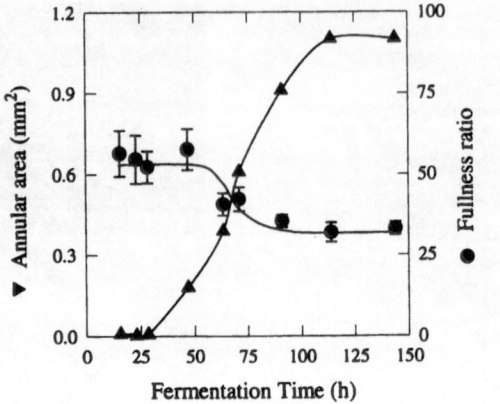

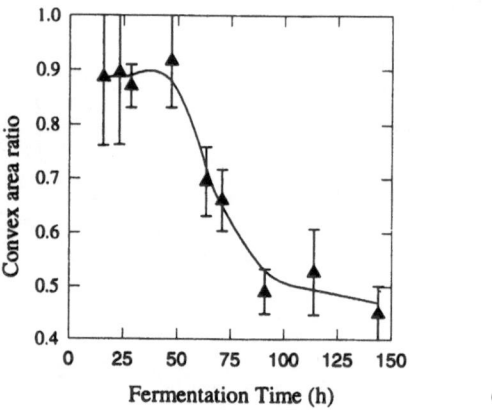

Fig. 17 c, d. c mean annular area and mean annular fullness; d ratio of mean convex area of core to mean convex area of the whole pellet [14]

equivalent diameter and mean core circularity of an industrial strain of *A. niger* grown at 5-l fermenter scale [3]. With a low agitation speed (300 rpm) the pellets of initial mean equivalent diameter 1.4 mm grew larger with fermentation time until they were of mean diameter 3.2 mm at 96 h. Then the agitation was increased to a very high level of 800 rpm and the pellets fragmented rapidly to 1.6 mm mean diameter. The fragmentation of pellets was associated with an increase of metabolic activity as can be seen from the carbon dioxide production rate (CPR) and oxygen uptake rate (OUR) data presented in Fig. 19. Edelstein and Hader [46] and Tough et al. [47] developed population balance models for pellet growth and fragmentation in submerged fermentations. Chapter 3 describes a review of the modelling of pellet growth and fragmentation of *Streptomyces* spp.

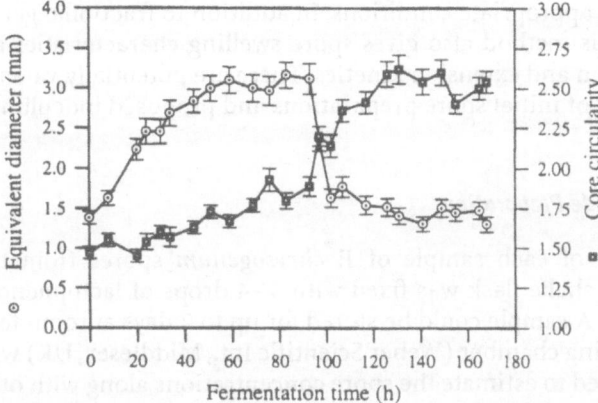

Fig. 18. Effect of agitation speed on the fragmentation of *A. niger* pellets. Agitation speed: 300 rpm from 0 to 12 h, 500 rpm 12 to 96 h, 800 rpm 96 to 168 h

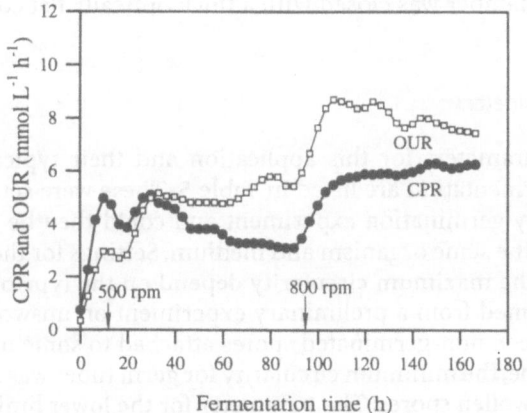

Fig. 19. The metabolic activity in the fermentation of Fig. 18 is influenced by pellet breakup following the increase of agitation speed

3.3
Fungal Spore Germination

An automatic image analysis method was developed by Paul et al. [15] to study the viability and germination characteristics of fungal spores for samples from submerged inoculum cultures. This was intended to replace the traditional plate count technique in which a dilute suspension of spore is spread on a solid agar medium and the fractional spore viability is assessed by counting the isolated colonies. Because the germination of spores depends on the composition of the medium as well as on the growth conditions the information obtained from a plate count cannot be applied directly to submerged culture. The image analysis method works on samples taken from the fermenter or shake flask, so the spores can be under appropriate conditions. In addition to fractional germination, the image analysis method also gives spore swelling characteristics and the germ tube formation and extension kinetics. These are potentially valuable in studies of the quality of initial spore preparations and proposed inoculum media.

3.3.1
Sample and Slide Preparation

Exactly 1 ml of each sample of *P. chrysogenum* spores from the inoculum fermenter or shake flask was fixed with 3–4 drops of lactophenol cotton blue stain (Fluka). A sample could be stored for up to 7 days at room temperature. A Helber counting chamber (Weber Scientific Int., Middlesex, UK) with a depth of 20 μm was used to estimate the spore concentrations along with other germination parameters. The chambers were supplied without grid lines to ease the image processing. When germination was significant, the culture was diluted with an equal volume of distilled water to separate objects on the slide before analysis. Approximately 0.04 ml of the (diluted) culture was placed on the counting surface. The chamber was closed with a thick optically flat coverslip.

3.3.2
Image Analysis Parameters

Image analysis parameters for this application and their typical values for a *P. chrysogenum* fermentation are listed in Table 5. These were set using samples from a preliminary germination experiment and could then be used for other experiments with the same organism and medium. Settings for the lower limit of diameter and for the maximum circularity depend on the type of spores in use and could be obtained from a preliminary experiment on unswollen spores. To discriminate between non-germinated spores attached to some other spore and a genuine germ tube, the minimum circularity for germ tubes was set equal to the maximum for unswollen spores. The parameter for the lower limit of germ tube length was set to eliminate artefacts caused by small debris touching the spores.

The four parameter values described above could be kept constant throughout a germination test on *P. chrysogenum* spores. The other two parameters were

Table 5. Image analysis parameters and their typical values used for the image analysis of *P. chrysogenum* spore germination

Parameters	Typical values	Values preferable at early stages	Values preferable at late stages
Minimum spore diameter, μm	2–3	2–3	2–3
Maximum spore diameter, μm	5–16	5–8	5–16
Minimum germ tube length, μm	2–3	2–3	2–3
Maximum germ tube length, μm	10–50	10–20	10–50
Circularity limit for spores	≤ 1.2	≤ 1.2	≤ 1.2
Circularity limit for germ tubes	> 1.2	> 1.2	> 1.2

settings for the maximum spore diameter and the maximum length of a germ tube. These parameters are used to distinguish between spores, germ tubes and debris. It was found that the accuracy of the measurements could be improved if different values of these parameters were used in the early and late stages of germination (Table 5). This was particularly important when large amounts of debris and media particles were present.

The magnification used for *P. chrysogenum* spores was × 200. The typical number of spores (non-germinated + germinated) per sample was 400. The total number of fields for analysis of this number of spores was in the range of 25 to 32, 17 to 13 spores per field respectively. The time of analysis for this number of spores varied between 10 and 22 min depending on the complexity of the samples and number of spores per field of view.

3.3.3
Image Analysis Software

An image of spores is captured from a microscope field of view and then stored. The image is then detected and all the image processing operations are done on the detected binary image.

1. A typical captured image of *P. chrysogenum* spores 16 h after inoculation in a defined medium. The image contains both non-germinated and germinated spores. It also contains many undesired objects and debris (particularly dead mycelia) and small particles.

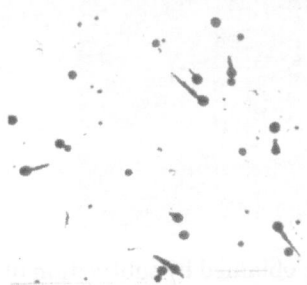

2. The image after detection of the original grey image. The subsequent image processing operations were on this binary image.

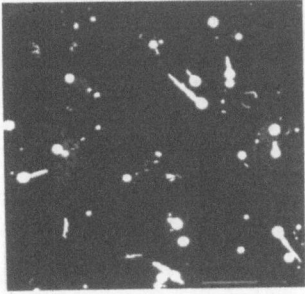

3. The binary image was subjected to a two-step erosion operation which removed most of the small particles. The image was then rebuilt to its original form, excluding those small extraneous objects. The larger unwanted objects could be identified by using a larger erosion (10–16 steps) which removed spores and similar sized debris, leaving only the cores of the large debris. After rebuilding the latter to its original size, it could be subtracted from the earlier image, leaving a new image which contains spores and germ tubes and some spore-sized debris.

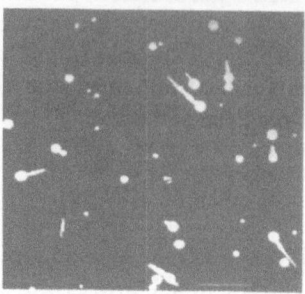

4. Objects in image shown above were then classified into two groups using a preset circularity parameter. The objects in this image are definitely non-germinated spores obtained by setting circularity ≤ 1.2.

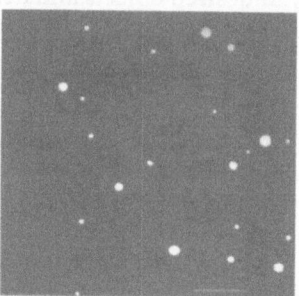

5. Objects in this image were obtained by subtraction of the image in step 4 from that in step 3. This is equivalent to applying circularity > 1.2 to objects in image 3. This image contains germinating spores, non-germinated spores artefactually attached to germ tubes of other spores, touching non-germinated spores, and non-germinated spores touching one of those germinating, as well as debris.

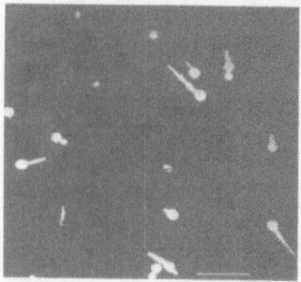

6. The germ tubes and the touching objects were then separated by a multistage opening operation. The opening removes the narrow touching objects including the germ tubes, leaving behind the wider germ tube spores. By this process objects (mostly debris) without a spore were eliminated. Subtraction of the germ tube spores from those shown in step 5 gave germ tubes and touching non-germinated spores. Germ tubes were identified using preset circularity and size criteria and are shown in image 6.

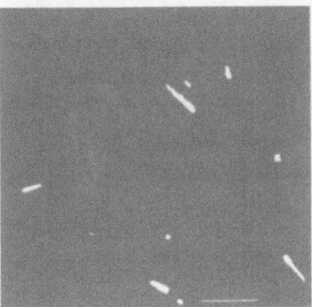

7. The germ tubes of the germinated spores identified by the procedure described above and then superimposed with the image in step 5 to identify the germinated spores.

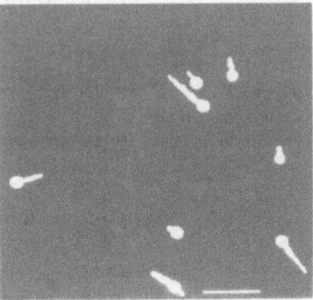

8. The germ tube spores of the germinated spores were identified by using the germ tubes as markers. The spores connected with the germ tubes are presented in this image.

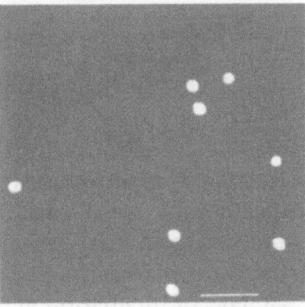

9. The touching or connected objects classified as nongerm tubes in step 6 were reclassified to non-germinated spores using a preset circularity parameter. These extra non-germinated spores were finally added to those obtained previously shown in step 4. This image contains all the non-germinated spores.

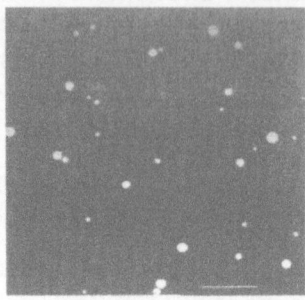

Table 6. Lists of parameters obtained from the measurement and calculation of spore samples from inoculum cultures

Classification	Parameters
Non-germinated spores	Total number
	Number fraction
	Concentration
	Equivalent diameter
	Circularity
	Area
	Volume
Germ tube spores	Total number
	Number fraction (fractional germination)
	Concentration
	Equivalent diameter
	Circularity
	Area
	Volume
Germ tubes	Total number
	Length
	Mean width
	Area
	Volume

3.3.4
Measurement

Measurements were performed on images shown in steps 9, 8 and 6 representing the non-germinated spores, germ tube spores and germ tubes respectively. Table 6 lists the parameters obtained from image measurement.

The method was tested using three media and two spore stocks of *P. chryso-genum*, one collected after 6–8 days of incubation at 25 °C while the other was obtained from a 40-day old culture stored at 4 °C after 6–8 days in the incubator at 25 °C. Table 7 lists the inoculum experiments [15] using different media and two spore ages. Figure 20 shows spores for samples taken at 20 h from shake flasks containing defined medium M2. The spores in Fig. 20a were grown from stock S1, whereas those of Fig. 20b were grown from stock S2. It is clear that the latter did not germinate as well as the former, indicating that the germinability of the spores deteriorated with the duration of storage. Figure 21 shows the percentage of germinated spores against incubation time in the three different media and with two ages of spores. These very different spore preparations combined with the range of media used gave an excellent test of the image analysis method.

The proportions of the spores which formed germ tubes and the time of germ tube formation were affected by medium and spore age. At 24 h, 89% of the spores had germinated in the complex medium M4, compared to only 16% in

Table 7. List of inoculum experiments using *P. chrysogenum*

Experiment	Medium	Spore type
S1M1	Defined medium [48].	Fresh spores (S1)
S1M2	Medium M1 with double the Fe^{++} salt and supplemented with 0.6 gl^{-1} Na-EDTA.	Fresh spores (S1)
S1M3	Complex medium [49].	Fresh spores (S1)
S1M4	Solids of M3 removed and supplemented with 1.0 gl^{-1} Na-EDTA.	Fresh spores (S1)
S2M2	Medium M2.	Old spores (S2)
S2M4	Medium M4.	Old spores (S2)

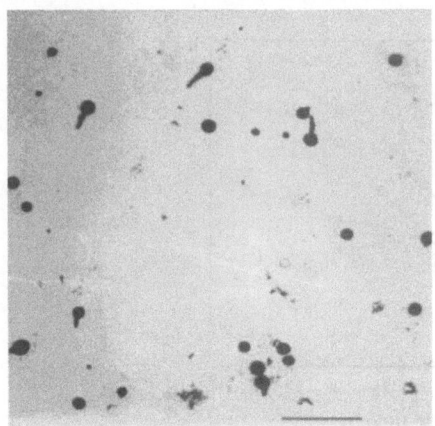 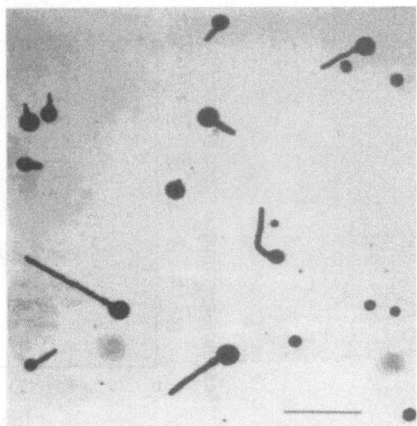

a   b

Fig. 20 a, b. *P. chrysogenum* spores taken at 20 h in medium M2 with spore stock: **a** S1; **b** S2. Photographs were taken from the image analyser display. Bar length = 50 µm [15]

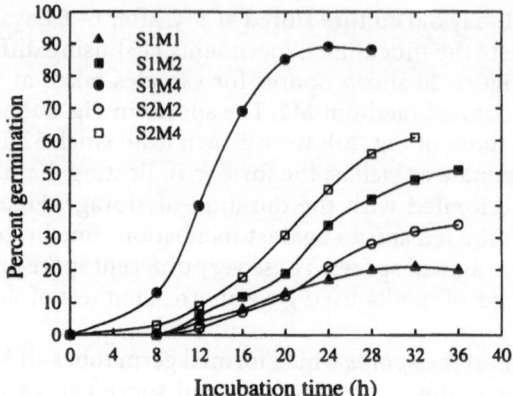

Fig. 21. Time course of germination of *P. chrysogenum* spores in different inoculum media and two ages of spores [15]

the defined medium M1. When the concentration of Fe^{++} in M1 was increased twofold (to give medium M2), the germination level increased from 16 to 33%. The defined media M1 and M2 required longer incubation times for germination than the complex medium. It can also be seen that germinability of spores deteriorated significantly with storage. The 40-day old spore (stock S2) had only germinated 21 and 45% in media M2 and M4, respectively after 24 h, compared to 33 and 89% germination with fresh spores (stock S1). Nielson and Krabben [39] used this image analysis information to fit their 5 parameter β-distribution model (see Chap. 4).

The mean volumes of non-germinated and germ tube spores during the course of the incubation are presented in Figs. 22 and 23, respectively. Spherical growth of the spores progressed with continuing incubation. The media had a marked influence on spore swelling and their eventual size, with higher swelling

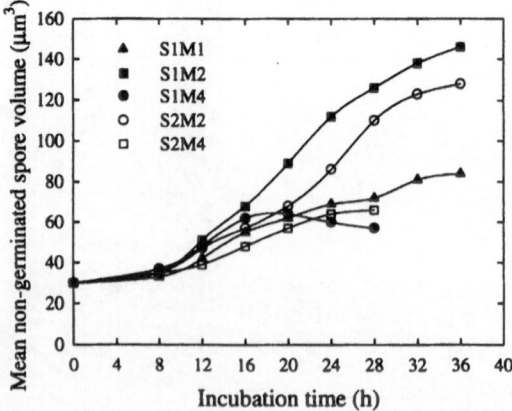

Fig. 22. Time course of mean volumes of non-germinated spores of *P. chrysogenum* in different inoculum media and using two ages of spores [15]

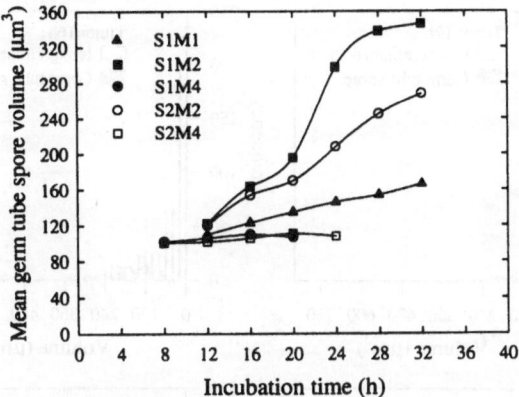

Fig. 23. Time course of mean volumes of germ tube spores of *P. chrysogenum* in different germination media and using two ages of spores [15]

in the defined media. It can be seen in this plot how the mean size of the non-germinated spores in the complex medium M4 was restricted by rapid germination which removed spores into the germinated class early in the swelling process. The mean volume of spores alone might not be adequate to represent the swelling process, as each sample contains a wide range of spore sizes, whether germinated or not. Image analysis can also provide the distributions of spore sizes as shown in Fig. 24 for defined medium M2. Together with percent germination and the spore size distribution a population balance model was proposed by Paul and Thomas [50] to describe the inoculum stage of fungal fermentations. However, such modelling is still not entirely satisfactory because the information on the physiology of spore germination that would pull together the relationship between spore swelling and germ tube emergence is not available.

Besides spore characteristics, germ tube lengths of the germinating spores could also be found by image analysis (Table 6). Figure 25 shows the mean length of the germ tubes during the course of incubation in different media with both spore stocks (S1 and S2). Mean germ tube length increased with incubation time. Differences of germ tube growth with different medium and spore ages are evident in this figure.

The method described here offers a number of advantages over photomicroscopy or colony counting (on solid media) to determine spore viability: it is rapid, is more accurate and consistent, can discriminate between non-germinated and just germinated spores, and in particular can be used on spores germinating in the actual submerged fermentation medium under appropriate agitation and aeration conditions. It provides measurements on spore preparations that are very appropriate for assessment of the quality of an inoculum. This method might be used to study the effects of medium formulation on germination and for the development of better inoculation methods as part of the optimisation of fungal fermentations. For complete studies of early hyphal extension and branching phenomena this method might be used in conjunction with that of Tucker et al. [13].

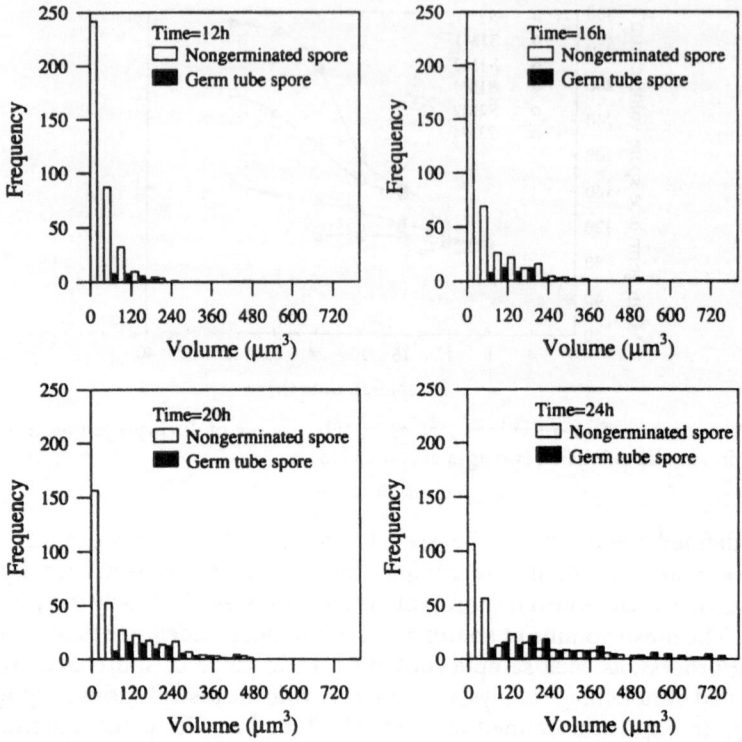

Fig. 24. Comparison of distributions of non-germinated spore and germ tube spore volumes of *P. chrysogenum* grown in medium M2 using spore S1 [15]

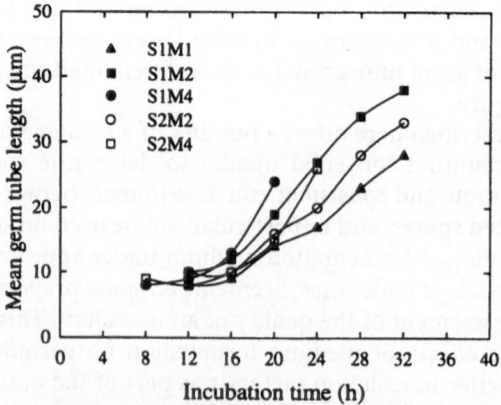

Fig. 25. Time course of mean volumes of germ tube lengths of *P. chrysogenum* in different germination media and using two ages of spores [15]

3.4
Hyphal Differentiation

For the filamentous form of fungi, differentiation of the hyphae may be characterised by image analysis. Packer et al. [51] identified two different regions of *P. chrysogenum*: (1) the cytoplasmic regions, and (2) degenerated regions including large vacuoles. The volume occupied by each of these regions in a fixed volume of sample could be estimated and hyphal-density values for each region could be used to estimate biomass concentrations from 0.03 to 38 g/l, even in the presence of up to 30 g/l of the solid ingredients often found in commercial media. This method gives some information needed to build a structured model such as that proposed by Nestaas and Wang [52]. It was, however, rather slow (several hours per sample), although no attempts were made to speed up the process.

Paul et al. [7] have developed a new image analysis method for a detailed characterisation of fungal vacuoles in terms of the percentage by volume of vacuoles and empty regions, and vacuole size and shape. Figure 26 shows by images how vacuoles can be identified by this method, which involves very complicated image processing operations reported in detail in Paul et al. [27]. Recently the method has been extended to identify and quantify the proportion of growing regions [8]. This section describes this image analysis method for complete characterisation of simple differentiation, i.e. growing regions (mainly apical), non-growing regions, vacuoles, and empty regions.

3.4.1
Sample and Slide Preparation

Samples from fermentations were diluted with 20% (w/v) sucrose solution in order to prevent evaporation from the slide during image analysis. Final bio-

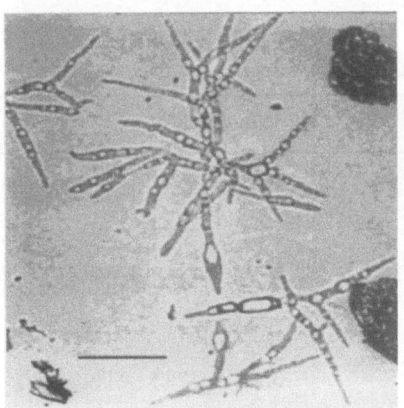
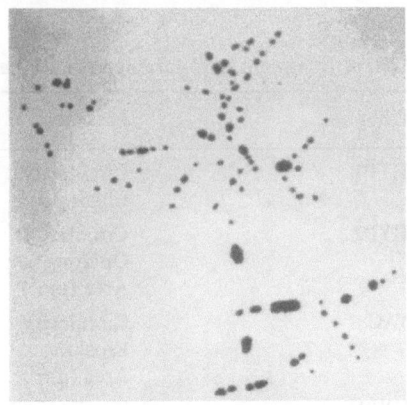

a
b

Fig. 26 a, b. Image analysis was used to characterise fungal vacuoles. a an original grey image; b a binary image of vacuoles and empty regions obtained after image processing operations [27]. (Bar length = 50 μm)

mass concentrations after dilution were 1 to 1.5 g/l. The pH of the diluted sample was adjusted to < 5.0 by the addition of dilute sulphuric acid. Approximately 0.1 ml of neutral red solution (BDH Ltd, Poole, UK) was added to 1 ml of the diluted sample. After 3–5 min a drop was pipetted onto a slide and covered by a cover slip. Analysis was done within 30 mins.

3.4.2
Setting Up of Image Analysis Parameters

The image analysis parameters used in the various filters described later and the typical range of their values are listed in Table 8. Usually a same set of values were used throughout analysis. With solid-containing complex medium a different set of values might be needed, particularly in the early stage of a fermentation when large quantities of solids might interfere with sample analysis.

3.4.3
Image Analysis Software

For each field of view to be analysed, a grey image was captured. Figure 27a shows a typical grey image of *P. chrysogenum* mycelia at ×200 magnification, stored by the image analyser in the digitised form. The growing regions (stained apices) are darker than the remaining cytoplasm and vacuoles, whereas the greyness of vacuoles is lighter than cytoplasm and growing regions. The image was delineated to remove the intermediate greyness levels from the boundaries of the objects, resulting in sharpening of greyness gradients in those regions (see Sect. 2.2 for the delineation operation). Objects of interest in the delineated image were detected by greyness level. Two detection levels (chosen during setting up of hardware parameters) were used: a detection level for hyphae, i.e.

Table 8. Image analysis parameters and their typical range of values

Filter	Parameter	Value/limit
HYP1	Circularity	$C > 2.0$
	Closing	1
HYP2	Circularity	$1.0 \leq C < 3.0$
	Opening	8–12
	Area (μm^2)	$A \leq 25$
VAC	Circularity	$C > 2.0$
	Erosion	6–8
	Area (μm^2)	$3.1 \leq A < 78.5$
HYPV	Erosion	4–6
TIP	Area (μm^2)	$A > 20.0$
	Greyness	$G < 90$

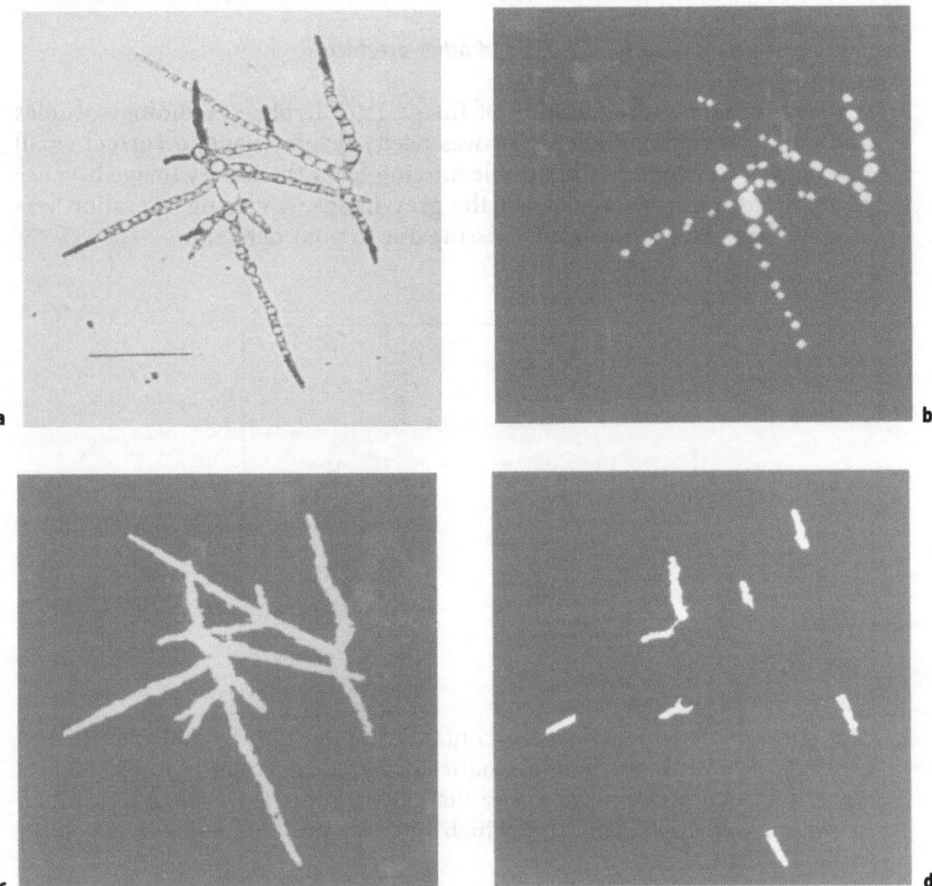

Fig. 27 a–d. *P. chrysogenum* mycelia at different image processing steps (bar length = 50 μm): **a** captured grey image; **b** binary image of vacuoles and empty regions; **c** binary image of hyphae including vacuoles and empty regions; **d** binary image of growing regions in the hyphae. The mycelia were from a typical fed-batch fermentation sampled during the production phase [8]

cytoplasmic regions and one for vacuoles and empty regions. These two binary images are referred to here as Image 1 (cytoplasmic regions) and Image 2 (vacuoles and empty regions) respectively. The microscopic images are not uniform in greyness which poses difficulties in separating features. The overall image processing steps were divided into three phases: Phase I consisted of elimination of debris, media particles and image artefacts; Phase II identification and separation of vacuoles and empty regions; and Phase III identification and separation of growing regions. Phases I and II are described using constructed images to illustrate the various stages of image processings involved and phase III using an actual image of *P. chrysogenum* hyphae.

Phase I.
Elimination of debris, media particles and other artefacts

1. A diagrammatic representation of Image 1 (of hyphae excluding vacuoles and empty regions). Filter HYP1 was used on this image to correct small image imperfections, mainly pixels missing from the binary image because of the non-uniform intensity of the grey image. A closing operation was applied first to fill in the pixels missing due to poor detection.

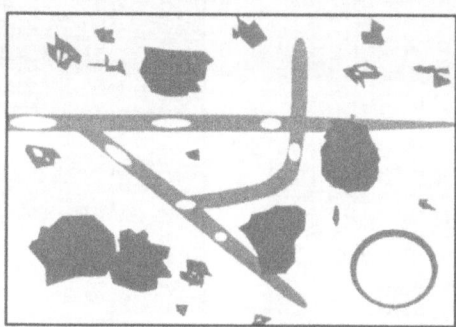

2. The consolidated binary image contained many non-hyphal objects and artefacts as illustrated diagrammatically here. These undesirable objects were identified and eliminated by the use of filter HYP2. An image filling operation was applied first by which internal voids or holes in an object could be filled.

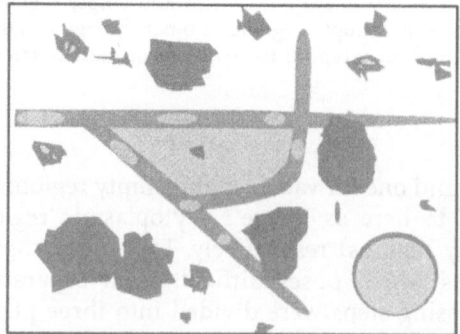

3. Most of the unwanted objects were identified as more circular than hyphae. This did not work on non-hyphal objects attached to hyphae because connected objects are processed as a whole.

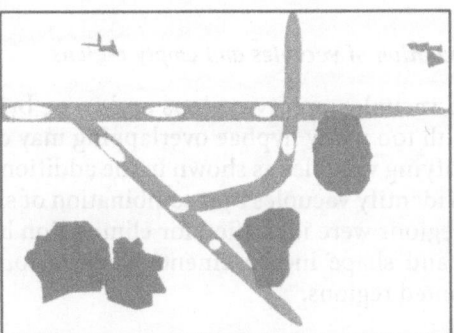

4. The attached non-hyphal objects were identified for elimination by a number (preset) of opening operations. This removed the filamentous hyphae whilst leaving the wider non-hyphal objects. Subtraction of the non-hyphal objects from the original gave the hyphae and remaining small artefacts.

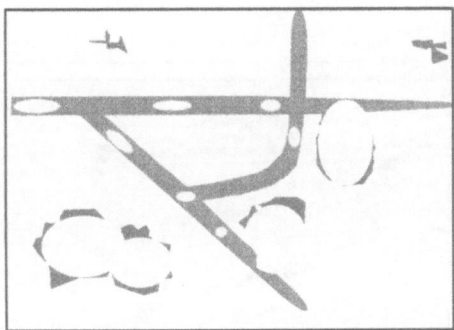

5. The smaller artefacts outside the hyphae created by the opening operation in step 4 were removed by the use of a preset size (area) filter.

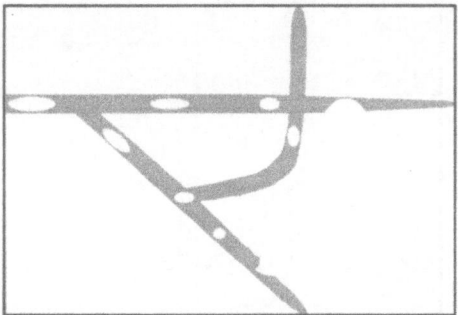

Phase II.
Identification and separation of vacuoles and empty regions

6. Image 2 masked vacuoles, empty regions and some bright background regions. Images with too many hyphae overlapping may cause additional difficulties in identifying vacuoles as shown in the additional illustration. Filter VAC was used to identify vacuoles by a combination of size and shape filters. The very large regions were identified for elimination by an erosion operation. Then size and shape measurements were performed to remove the remaining unwanted regions.

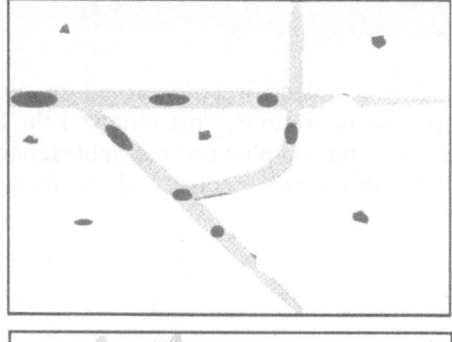

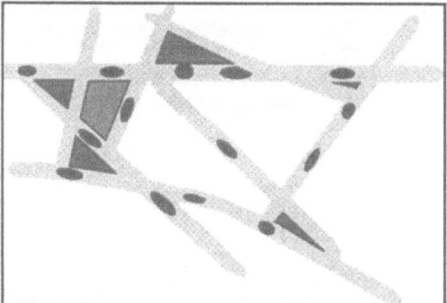

7. Many small background artefacts with similar shape and size to vacuoles as shown in this diagram could not be removed by filter VAC. This diagram shows these small background regions outside the hyphae.

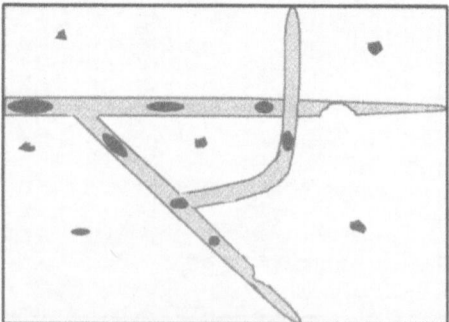

8. Filter HYPV was applied to remove the small non-vacuoles regions from the image. HYPV used the image that contains hyphae alone as a mask to enable artefacts outside the hyphae to be removed from the image. The image shown in step 2 was combined with that of hyphae alone. The smaller background regions were then eliminated by an erosion operation. An image was then reconstructed which contained hyphae and vacuoles.

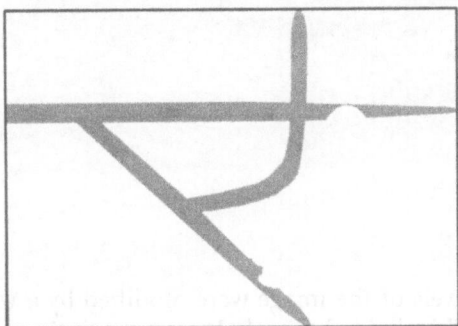

9. The difference between the combined image and the image containing the hyphae alone gave the vacuoles and empty regions.

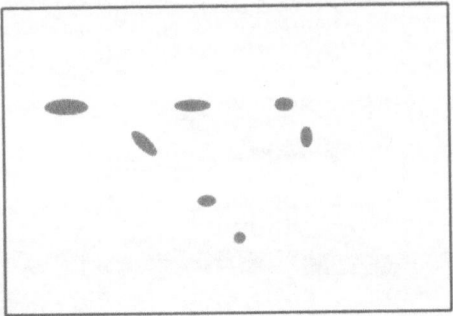

Phase III.
Identification and separation of growing regions

10. The growing regions in the hyphae show darker than the rest of the cytoplasmic regions. Identification and separation of these regions were achieved by a routine or filter called TIP. Some grey processing operations were applied to improve the original image. Firstly, the halos around the boundaries were identified for removal by a top hat operation (see Sect. 2.2). The result of this step is not shown here.

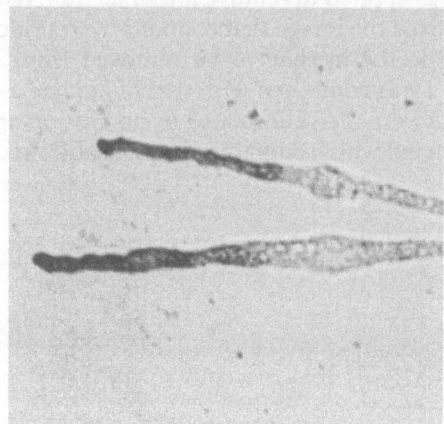

11. The greyness levels of the image were modified by a watershed operation (see Sect 2.2). This divided the whole image into tiny zones each having a uniform grey intensity. The zones are surrounded by single pixel width dividing lines. Watershedding resulted in a much clearer distinction between the growing regions and the rest of the hyphae.

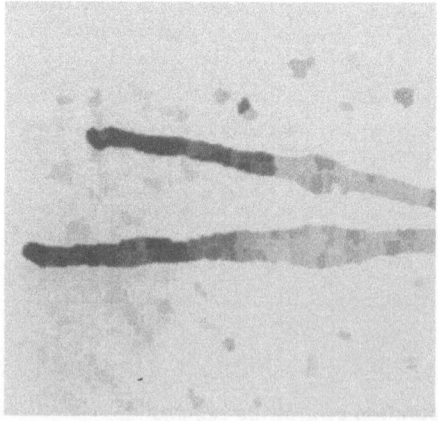

12. The growing regions were then picked out using a detection level which identified all the zones darker than that level, and gave a binary image of the selected zones. Because of the watershed these were separated by dividing lines one pixel wide. Neighbouring zones were then joined together into groups by a closing operation, each masking a growing region of the hyphae. Smaller isolated zones, probably artefacts, were removed by a size filter.

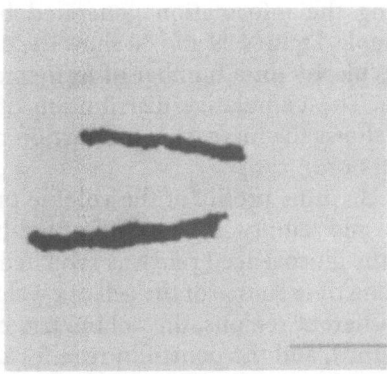

Figure 27b–d shows the binary images of vacuoles and empty regions, hyphae including vacuoles and empty regions (whole hyphae), and the growing regions corresponding to the grey image shown in Fig. 27a. The time of analysis of a sample varied between 25 and 35 min depending on the type of samples and the extent of manual editing required. Following the measurement of all the fields of view, the volumetric proportions of the growing and non-growing cytoplasmic regions (by subtraction of growing regions, and vacuoles and empty regions from total hyphae), and vacuoles and empty regions were obtained. Distributions of different parameters, e.g. vacuole volume, vacuole circularity, were obtained and mean widths and distributions of all the different regions were also found. The degenerated regions were estimated from the distribution of vacuole volume. It was found that vacuoles approximately larger than 30 μm^3 in volume constitute the degenerated regions of the hyphae [8]. The volume proportions of the degenerated regions were estimated. From the proportions by volume of these four regions, the amount of biomass in each was found using hyphal density values from the literature [52]. The classification of the four regions is shown in Fig. 28.

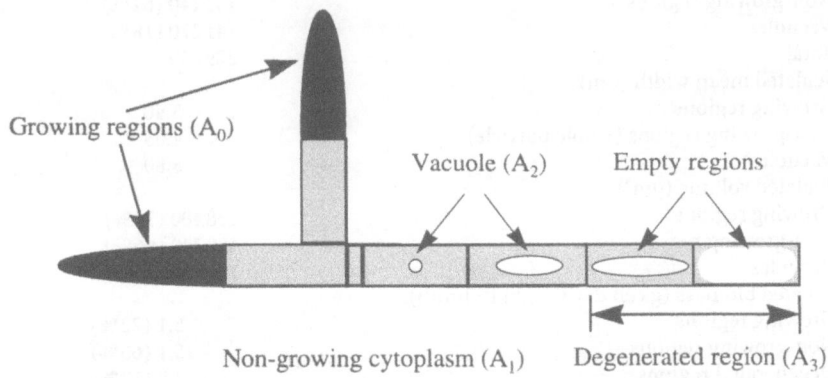

Fig. 28. The regions of hyphal differentiation measured by image analysis [8]

Table 9 gives the results for a sample taken from a fed-batch *P. chrysogenum* fermentation illustrating the information generated by the image analysis method on a typical sample. Figures 29 and 30 show the distributions of vacuole volume and vacuole circularity for a number of fermentation samples across a fed-batch fermentation. The cumulative distribution of vacuole volumes for three samples (Fig. 31) shows the increasing proportion of larger vacuoles with the duration of fermentation.

Figure 32 represents the time profile of the volume proportions of growing regions, and vacuoles and empty regions of a fed-batch *P. chrysogenum* fermentation in which the glucose feed rate was switched between a high and a low value (see Fig. 33). The time course of the cell dry weight, estimated biomass concentrations of the different regions, the volumetric concentration of small vacuoles (volume ≤ 30 μm^3), and the penicillin titre for the same fermentation are shown in Fig. 33. It can be seen from Figs. 32 and 33 that when the feed rate was reduced there was an increase in both penicillin production and non-growing regions and a decrease in growing regions. This suggests that by switching to a low feed rate some of the growing regions converted to non-growing regions. The results from this fermentation demonstrate the potential of this technique for obtaining better understanding of the penicillin fermentation, particularly the relationship between productivity and differentiation. Clearly, physiological changes caused by manipulating nutrient conditions could be monitored by image analysis.

Table 9. Summary of image analysis results for a *P. chrysogenum* sample taken from a fed-batch fermentation at 72 h [8]

Number of fields analysed	64
Total analysis time (min)	27
Number of features measured:	
Hyphae (total number of mycelial particles)	57
Vacuoles	1740
Growing regions	181
Measured area (μm^2):	
Growing regions	45850 (20%)
Non-growing regions	142140 (62%)
Vacuoles	41270 (18%)
Total	229260
Calculated mean width (μm):	
Growing regions	3.40
Non-growing regions (whole particle)	4.05
Vacuoles	3.80
Calculated volume (μm^3):	
Growing regions	138400 (19%)
Non-growing regions	480740 (66%)
Vacuoles	109260 (15%)
Estimated biomass (g cell dry weight l^{-1} broth):	
Growing regions	5.1 (22%)
Non-growing regions	15.1 (66%)
Degenerated regions	2.8 (12%)
Total	23.0

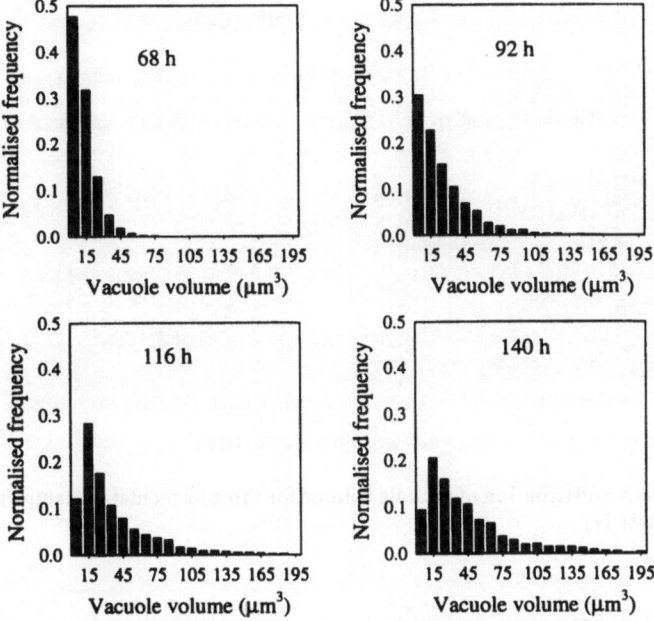

Fig. 29. Distribution of vacuole size during the production phase of fermentation using a defined medium and inoculated with spore inoculum

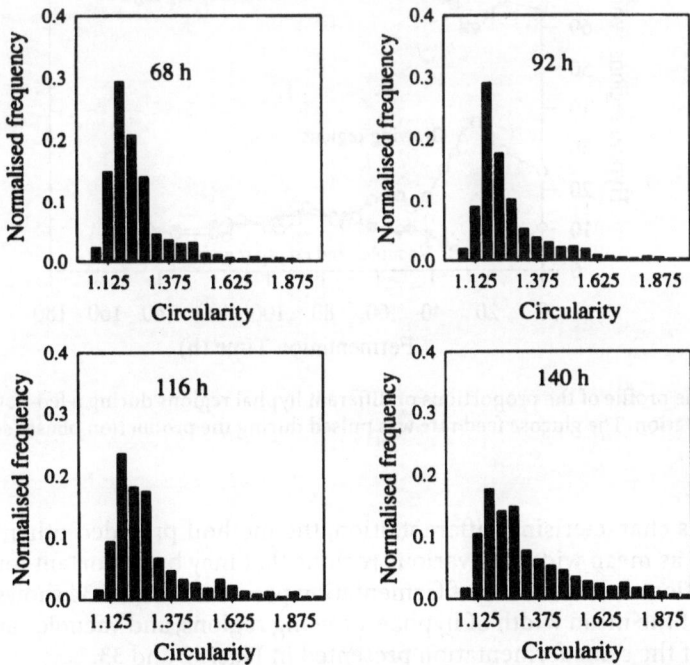

Fig. 30. Distribution of vacuole shape (circularity parameter) for four samples across a fed-batch penicillin fermentation

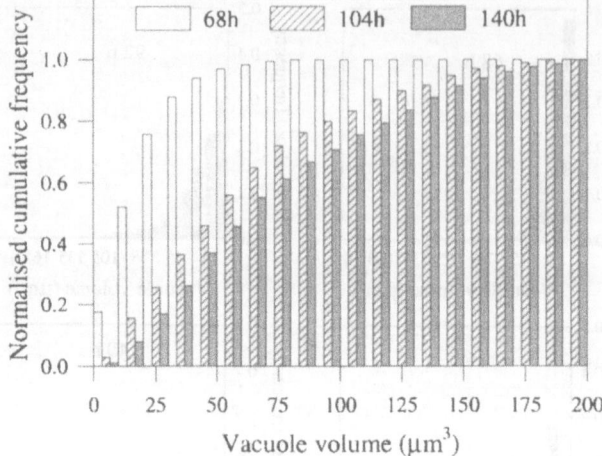

Fig. 31. Cumulative distribution of vacuole volume for three fermentation samples during the production phase [7]

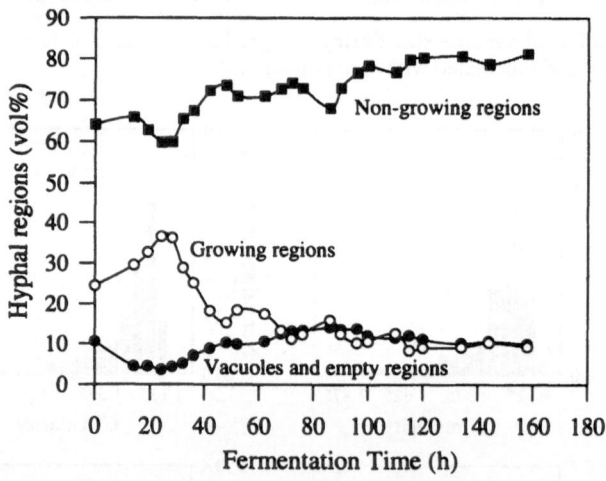

Fig. 32. Time profile of the proportions of different hyphal regions during a fed-batch penicillin fermentation. The glucose feed rate was pulsed during the production phase (see Fig. 33)

Besides characterising differentiation, the method provided other information such as mean widths of various regions that may be important for detailed investigation and modelling of fermentation processes. Figure 34 shows the time profiles of the mean width of hyphae, growing regions, and vacuoles and empty regions of the same fermentation presented in Figs. 32 and 33.

Based on the information of simple differentiation obtained by image analysis a structured model was developed [17]. With a number of fed-batch fermenta-

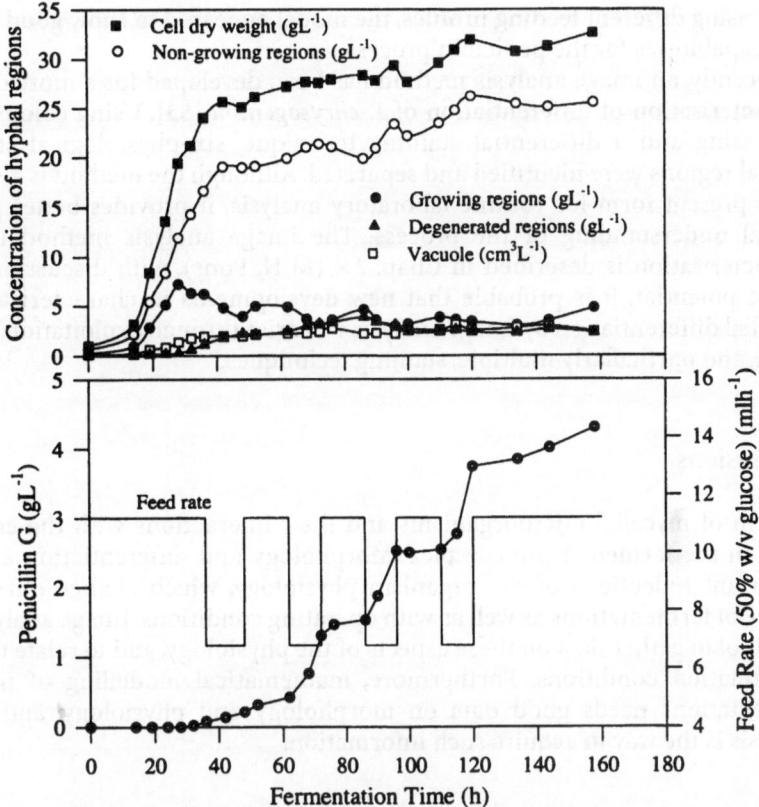

Fig. 33. Time profiles of the concentrations of different hyphal regions corresponding to the fed-batch penicillin fermentation shown in Fig. 32

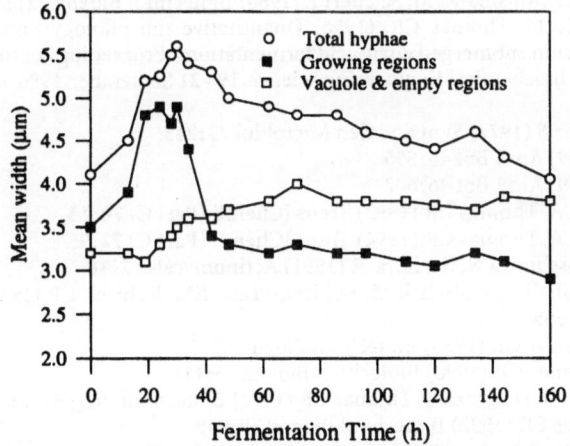

Fig. 34. Time profiles of the mean width of total hyphae, growing regions, and vacuoles and empty regions of the fed-batch fermentation of Fig. 32

tions using different feeding profiles, the model appeared to show good prediction capabilities for the penicillin process.

Recently, an image analysis method has been developed for a more detailed characterisation of differentiation of *P. chrysogenum* [53]. Using colour image processing and a differential staining technique, six physiological states of hyphal regions were identified and separated. Although the method is very slow in its present form for routine laboratory analysis, it provides better physiological understanding of the process. The image analysis method for this characterisation is described in Chap. 2× (M.N. Pons), with discussion of its future potential. It is probable that new developments in characterisation of mycelial differentiation by image analysis will arise through exploitation of such novel, and particularly multiple, staining techniques.

4
Conclusions

Growth of mycelial microorganisms and their interactions with the environment in the fermenter are complex. Morphology and differentiation are very important reflections of the organism physiology, which change during the course of fermentations as well as with operating conditions. Image analysis is a vital tool to gather data on these aspects of the physiology, and to relate them to fermentation conditions. Furthermore, mathematical modelling of mycelial fermentations needs good data on morphology and physiology, and image analysis is the way to acquire such information.

5
References

1. Oliver SG, Trinci APJ (1985) Modes of growth of bacteria and fungi. In: Bull AT, Dalton H (eds) Comprehensive Biotechnology, vol. 1. Pergamon, p 159
2. Ujcová E, Fencl Z, Musílková M, Seichert L (1980) Biotechnol Bioeng 22:237
3. Paul GC, Priede M, Thomas CR. (1996) Quantitative morphology and physiology of *Aspergillus niger* in submerged citric acid fermentations. Proceedings of the 1st European Symposium on Biochemical Engineering Science, 19–21 September, 1996. Dublin, Ireland, p 168
4. Bartnicki-Garcia S (1973) Sym Soc Gen Microbiol 23:245
5. Zalokar M (1959) Am J Bot 46:555
6. Zalokar M (1959) Am J Bot 46:602
7. Paul GC, Kent CA, Thomas CR (1992) Trans IChemE (Part C) 70:13
8. Paul GC, Kent CA, Thomas CR (1994) Trans IChemE (Part C) 72:95
9. Spassova D, Vesselinova N, Gesiieva, R (1991) Actinomycetes 2:18
10. Lendenfeld T, Ghali D, Wolsehek M, Kubicek-Pranz EM, Kubicek CP (1993) J Biological Chemistry 268:665
11. Adams HL, Thomas CR (1988) Biotechnol Bioeng 32:707
12. Packer HL, Thomas CR (1990) Biotechnol Bioeng 35:111
13. Tucker KG, Kelly T, Delgrazia P, Thomas CR (1992) Biotechnol Prog 8:353
14. Cox PW, Thomas CR (1992) Biotechnol Bioeng 39:945
15. Paul GC, Kent CA, Thomas CR (1993) Biotechnol Bioeng 42:11
16. Smith GM, Calam CT (1980) Biotechnol Lett 2:261
17. Paul GC, Thomas CR (1996) Biotechnol Bioeng 51:558

18. Thomas CR (1992) Trends in Biotechnol 10:343
19. Vecht-Lifshitz SE, Ison AP (1992) J Biotechnol 23:1
20. Thomas CR, Paul GC (1996) Current Opinion in Biotechnology 7:35
21. Metz B, de Bruijn EW, van Suijdam JC (1981) Biotechnol Bioeng 23:149
22. Ramm P (1994) J Neuroscience Methods 54:131
23. Paul GC (1993) Image analysis for characterising *Penicillium chrysogenum* differentiation, PhD thesis, University of Birmingham, UK
24. Klein JC, Collange F, Bilodeau M (1989) Programmable logic cell arrays: a new technology for image analysis. Cambridge Instrument Ltd, Cambridge, UK
25. Russ JC (1990) Computer assisted microscopy. The measurement and analysis. Plenum, New York
26. Serra J (1982) Image analysis and mathematical morphology. Academic Press, London
27. Paul GC, Kent CA, Thomas CR (1993) Binary 5:92
28. Reichl U, Buschulte TK, Gilles ED (1990) J Micros 158:55
29. Paul GC, Thomas CR (1995) An image processing algorithm for characterising mycelial aggregates grown in submerged fermentation. Proceedings of the 4th International Quantimet and Stereoscan User Conference, 2–5 Oct. 1995, Madingley Hall, Cambridge, UK
30. Paul KR, Paul GC, Thomas CR (1995) Effect of spore inoculum concentration on morphology of *Streptomyces clavuligerus*. Proceedings of the 1995 IChemE Research Event – First European Conference for Young Researchers in Chemical Engineering, 5–6 January, 1995, Edinburgh, UK, p 980
31. Tucker KG, Thomas CR (1992) Biotechnol Lett 14:1071
32. Tucker KG, Thomas CR (1994) Biotechnol Techniques 8:153
33. Galbraith JC, Smith JE (1969) Trans Brit Mycol Soc 52:237
34. Paul GC, Thomas CR (1995) Effect of fermentation medium on the morphology of *Penicillium chrysogenum* in submerged fermentation. Proceedings of the 1995 IChemE Research Event – First European Conference for Young Researchers in Chemical Engineering, 5–6 January, 1995, Edinburgh, UK, p 974
35. Paul GC, Kent CR, Thomas CR (1994) Biotechnol Bioeng 44:655
36. Nielsen J, Johansen CL, Jacobsen M, Krabben P, Villadsen J (1995) Biotechnol Prog 11:93
37. Prosser JI, Trinci APJ (1979) J Gen Microbiol 111:153
38. Prosser JI, Tough AJ (1991) Crit Rev Biotechnol 10:253
39. Nielsen J, Krabben P (1995) Biotechnol Bioeng 46:588
40. Yang H, Reichl U, King R, Gilles ED (1992) Biotechnol Bioeng 39:44
41. Yang H, King R, Reichl U, Gilles ED (1992) Biotechnol Bioeng 39:49
42. Reichl U, King R, Gilles ED (1992) Biotechnol Bioeng 39:164
43. Durant G, Crawley G, Formisyn P (1994) Biotechnol Techniques 8:395
44. Durant G, Cox PW, Formisyn P, Thomas CR (1994) Biotechnol Techniques 8:759
45. Pichon D, Vivier H, Pons MN (1993) Growth monitoring of filamentous microorganisms by image analysis. In: Karim MN, Stephanopoulos, G (eds), Modeling and control of biotechnological processes. Pergamon, p 307
46. Edelstein L, Hadar Y (1983) J Theor Biol 105:427
47. Tough AJ, Pulham J, Prosser JI (1995) Biotechnol Bioeng 46:561
48. Deo YM, Gaucher GM (1984) Biotechnol Bioeng 26:285
49. Mou D-G, Cooney CL (1983) Biotechnol Bioeng 25:225
50. Paul GC, Thomas CR (1996) A mathematical model of swelling and germination of fungal spores in inoculum cultures. Proceedings of the 5th World Congress of Chemical Engineering, 14–18 July, 1996, San Diego, CA, p 717
51. Packer HL, Kesharvarz-Moore E, Lilly MD, Thomas CR (1992) Biotechnol Bioeng 39:384
52. Nestaas E, Wang DIC (1983) Biotechnol Bioeng 25:781
53. Vanhoutte B, Pons MN, Thomas CR, Louvel L, Vivier H (1995) Biotechnol Bioeng 48:1

Received December 1996

Beyond Filamentous Species ...

Marie-Noëlle Pons · Hervé Vivier

Laboratoire des Sciences du Génie Chimique, CNRS-ENSIC-INPL,
1, rue Grandville BP 451, F-54001 NANCY cedex, France

1
Introduction

When Antonie van Leeuwenhock invented the first microscope in the seventeenth century, he made a wonderful gift to biologists by allowing them to discover bacteria, yeasts, spermatozoa, blood cells, etc. However, at that time, and even in Pasteur's time in the late nineteenth century, biologists had to rely on manual drawings to transcribe the information to their peers. Today the automatic analysis of numerical images, as captured by an electronic substitute of the human eye, i.e., a camera, enables to extract quickly quantitative information. We have seen in the previous chapters how filamentous microorganisms have been so far a particularly well suited support for automated image analysis, first because of their economic value, second because of their complexity. However, it would not be fair to restrict the potential use of automated image analysis to this single category of cells. Although not yet as well

Advances in Biochemical Engineering/
Biotechnology, Vol. 60
Managing Editor: Th. Scheper
© Springer-Verlag Berlin Heidelberg 1998

Table 1. Some microbial size characteristics

Microorganism	Type	Typical length diameter (μm)
E. coli	Bacterium	2
S. cerevisiae	Yeast	5-20
A. pullulans	Yeast	5-25
P. chrysogenum spores	Fungus	5
M. citrifolia	Plant cell	100
Birch torpedo embryo	Plant cell	400
C. vulgaris	Alga	5
CHO	Animal cell	10
Rat hepatocyte	Animal cell	20
Sf9	Insect cell	15

developed in the biotechnological context, these techniques can be applied to bacteria, yeasts, animal or plant cells. We will not mention here the many applications of automated image analysis in medicine.

There are four main domains of applications of image analysis: counting, sizing, morphology characterization, and physiology characterization using staining techniques. Clearly there must be a relationship between the size of the cell to be examined and its characteristics. Table 1 tabulates some typical cell sizes for various microorganisms. It is difficult to give absolute guidelines but the following rule of thumb could be used to obtain valid information for a convex and rather compact object (not filamentous!):

- counting: $> \approx 10$ pixels
- size: $> \approx 100$ pixels
- morphology, physiology: $> \approx 500$ pixels

In terms of effective size, it depends on the microscope magnification, the characteristics of the camera and the resolution of the image grabbing board. On our PC-based image analysis system equipped with a Leitz light microscope and a Vidicon-tube Bosch camera which provides 512×512 rectangular pixel images, a pixel represents a surface of approximately 0.04 μm^2 at the 1000x magnification. A *Saccharomyces cerevisiae* yeast cell, of approximately 10 μm in diameter will have an apparent surface of 2000 pixels. Even when working with a smaller image of 256×256 an apparent surface of 500 pixels will be obtained, which is enough for morphology characterization. But even with a 512×512 image an *E. coli* cell will have an apparent area of about 50 pixels!

2
Bacteria

Bacteria are among the smallest microorganisms observable by optical microscopy and therefore the measurement of their size is delicate. Fluorescent staining, by Acridine Orange [1] or DAPI [2] is often used for single bacterium detection. Special and expensive cameras with image intensifiers are generally necessary due to the low level of emitted light in fluorescence and luminescence. New low cost CCD cameras with steerable frame integration time are nowadays available for routine epifluorescence. In any case epifluorescence images give rise to a number of problems because of the halo around the objects. For detection of small plankton bacteria, at the limits of resolution of epifluorescence (< 0.3 μm) Schröder et al. [1] use an intensified camera but the concentration of the stain, Acridine Orange, should be adjusted. At low concentration, small bacteria cannot be detected out of the background noise. If the dye concentration is too large, cells are too bright, the background is discolored and debris also pick up the coloration. The optimal concentration depends on camera sensitivity and microscope characteristics. At the high magnification used for this type of application, good focusing is important (and difficult to obtain) especially when a population of cells of different sizes is observed. Image quality depends on exposition duration. Image segmentation is not immediate because the average gray level within each cell varies from one to another (depending on the size and the physiological state of the cell): an ordinary thresholding procedure is ineffective. Also the halo renders the object edges difficult to detect. Schröder et al. [1] suggest using gradient-based segmentation by convolution with a Mexican-hat type of matrix.

Counting bacteria is problematic, and even more so is their size and morphology determination which is of great help for their recognition and identification. Figure 1 presents some morphologies exhibited by lactic bacteria

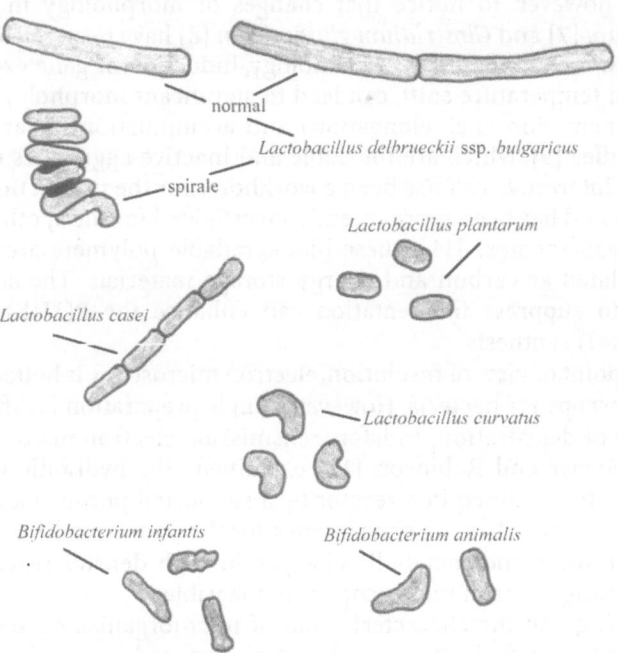

Fig. 1. Some morphologies of lactic bacteria

[3]. Conventional characterization of bacterial species by selective culture is a troublesome method and image analysis methods have been looked at to identify various strains on the basis of their morphology. Meijer et al. [4] have examined different bacterial strains (*Streptococcus puogenes, Escherichia coli, Streptococcus* Group D, *Klebsiella pneumoniae, Proteus mirabilis, Pseudomonas aeroginosa,* and *Staphylococcus aureus*). Primary parameters are the area, the perimeter, the area of the convex hull and the inertia moment. Secondary parameters deduced from these measurements are the size, the contour irregularity, the concavity and the global circularity of the shape. Multivariate analysis of variance is then applied to the set of secondary parameters for species recognition. Bacteria viability has been assessed by automated measurement of their elongation after incubation in a nalidixic acid solution, which causes the viable cells to elongate [5].

Discrimination between bacterial species has also been attempted by Dubuisson et al. [6]: the procedure separates and identifies *Methanospirillum hungatei* (filament) and *Methanospirillum mazei* (round shape) even in case of overlapping cells or contact between microorganisms. Two shape factors are used: Φ_1 is the ratio of the minimal and maximal eigenvalues of the covariance matrix ($\lambda_{min}/\lambda_{max}$), based on the second-order moments of the particle, and Φ_2 is the circularity. There is a good correlation between these two descriptors and the type of microbial species: if $\Phi_1/\Phi_2 > 30$, the microorganism is *Methanospirillum hungatei*.

These techniques were developed mainly to detect pathogenic species and marine plankton. Examples of applications to industrial bacteria are scarce. It is worthwhile, however, to notice that changes of morphology in *Clostridium acetobutylicum* [7] and *Clostridium glutamicum* [8] have been related to culture conditions. In recombinant DNA technology, induction of gene expression, for example by a temperature shift, can lead to significant morphological changes such as filamentation (i.e., elongation) and accumulation of large inclusion refractile bodies [9], which are insoluble and inactive aggregates containing a substance of interest. *E. coli* has been a workhorse for the production of various proteins [10] and has been more recently investigated for the synthesis of PHAs (polyhydroxyalkanoates) [11]: these biodegradable polymers are synthesized and accumulated as carbon and energy storage materials. The selection of a strain able to suppress filamentation can enhance the P(3HB) (poly(3-hydroxybutyrate)) synthesis.

From the point of view of resolution, electron microscopy is better suited than optical microscopy for bacteria. However, sample preparation is difficult, due to the necessity of dehydration, and, for transmission electron microscopy, due to sectioning. Fowler and Robinson [12] examined the hydraulic resistance of *E. coli* aggregates, retained in a reactor by a supported porous membrane. Although this type of work is of prime interest for the understanding of the fundamentals of transfer and metabolic changes in high-density reactors, on-line monitoring using electron microscopy is not feasible.

The counting and the characterization of microorganisms fixed on a solid surface present great difficulties. Fixed bacteria are used for metal recovery from mineral ores or wastes. *Thiobacillus ferroxidans* digest many metals such as iron, copper, zinc, and cadmium. The lixiviation process of iron pyrite is made difficult by the great number of mineral components present, the very special culture conditions, and the low cell density. Yeh et al. [13] have proposed to use epifluorescence microscopy and Acridine Orange staining to study the activity of bacteria fixed on pyrite grains. Bacteria dig galleries in pyrite and Bärtels et al. [14] monitor the bacterial attack of an ultrafine layer of iron sulfide. The pyrite layer, partially transparent under visible light, enables one to monitor the adhesion kinetics, the reproduction of bacterial surface and the density of active bacteria as well as the degradation of the mineral. Actually the mechanisms of pyrite and arsenopyrite bacterial oxidation are not yet well understood. A better knowledge of the behavior of adhering and free bacteria as well as the corrosion patterns should help model and improve the performances of bioleaching processes [15].

3
Yeasts

Applications on yeast have been focused on *Saccharomyces cerevisiae*. Berner and Gervais [16] simply assume that the cells are spherical in their study on the response of the cells to osmotic shift. But they are more generally considered to be prolate ellipsoids. In addition to number and size, budding rate is important in yeast. Hirano [17] developed a procedure to determine this parameter automatically by assuming that the projection of the non-budding cell on the image can be considered as an ellipse. A shape factor F_H is computed by

$$F_H = \frac{\pi(L+B)}{2.P}$$

where L is the "length" (given by the maximal Feret diameter), B is the "breadth" (given by the minimal Feret diameter) and P is the perimeter (Fig. 2). F is close to 1 for a circle or an ellipse (non-budding cell). When the shape is irregular (budding cell), F decreases ($0.7 \leq F < 1$). The budding rate is the ratio of budding cell number to total cell number.

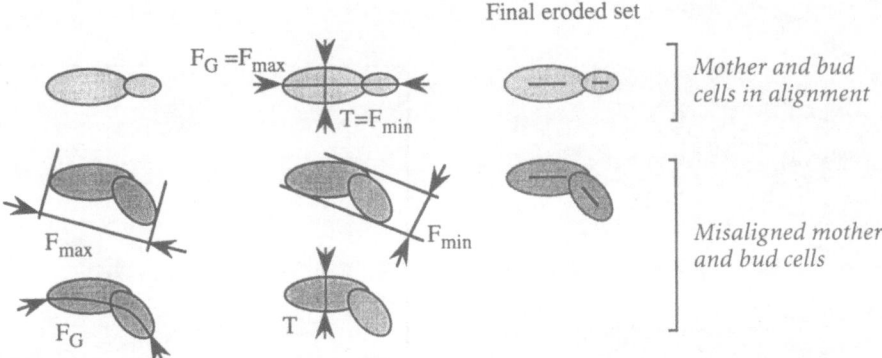

Fig. 2. Size measurements on *Saccharomyces cerevisiae* cells

Even this assumption of prolate ellipsoid is not exactly true when the cell is budding. Therefore, using the contour of the cell and after determination of the major axis of the ellipsoid, Huls et al. [18] determined the volume of *Saccharomyces cerevisiae* observed by epifluorescence. Their sophisticated volume integration technique should however be conducted only on isolated cells (budding or non-budding). Using a light pen the operator has to select the valid objects within an image. To avoid this problem Pons et al. [19] determine the number of elements in aggregates by computing their final eroded set. Aggregates with a final eroded set of two elements are considered as budding cells. The maximal and minimal Feret diameters of the cells (F_{max} and F_{min} respectively), as well as their geodesic diameter (F_G) and their thickness (T), are calculated (Fig. 2). The budding rate is evaluated by the elongation factors

$$E = \frac{F_{max}}{F_{min}}$$

and

$$E' = \frac{F_G}{T}$$

E and E' being different in case of misalignment of the mother and daughter cells. The circularity R is also computed by

$$R = \frac{P^2}{4\pi A}$$

where P is the "perimeter" and A the "projected area." Although the volume calculation is less precise than in Huls et al.'s method [18], it is more convenient for the operator who has to deal with a certain number of images per sample to obtain statistically valid results. Pons et al. [19] use 50 images per sample to obtain about 200 objects per sample. A typical grey-level image is shown in Fig. 3a, with the corresponding binary image in Fig. 3b.

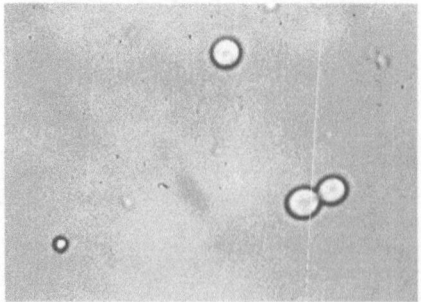 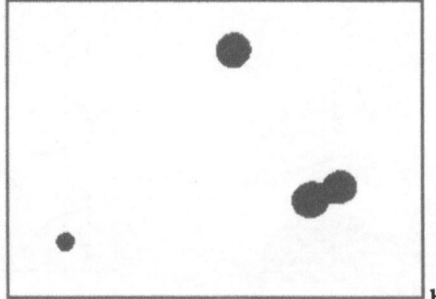

Fig. 3. *Saccharomyces cerevisiae*: **a** grey-level image; **b** corresponding binary image

To test the ability of the technique to discriminate between the various morphologies, test particles were submitted to the same shape characterization – a principal component analysis was conducted on three basic shape descriptors: circularity, elongation (E) and size of final eroded set. The two first principal components, f_1 and f_2, which are linear combinations of the basis descriptors, describe more than 85% of the total variance and the location of the test particles in the f_1-f_2 space are shown in Fig. 4. Generally the maximum values of R and E are of the order of 1.8 and 2.9 for a mother cell having a large bud not yet detached and a smaller bud already growing. However, under stress, there may exist elongated yeast cells as the ones pictured in Fig. 5. Depending upon the strain, aggregates of more than two elements may not be uncommon, Zalewski et al. [20] investigated the correlation between the maximum frequency of tetrades (four-cell aggregates) and maximum growth rate in aerobic batch growth of *Saccharomyces cerevisiae*.

As an example, we will describe the monitoring of the morphology of two *Saccharomyces cerevisiae* strains, noted B and K, during anaerobic fermentation

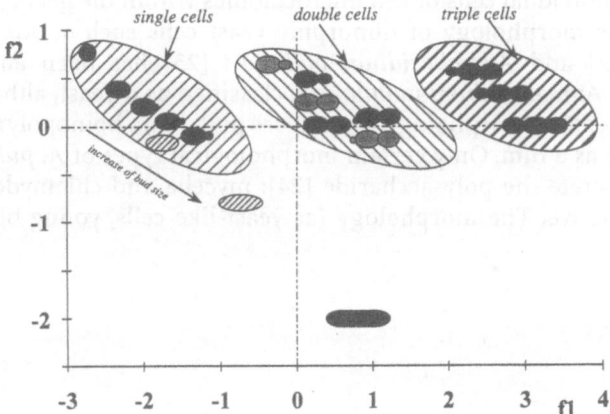

Fig. 4. Yeast-like test particles in the Principal Components Sub-Space f_1-f_2

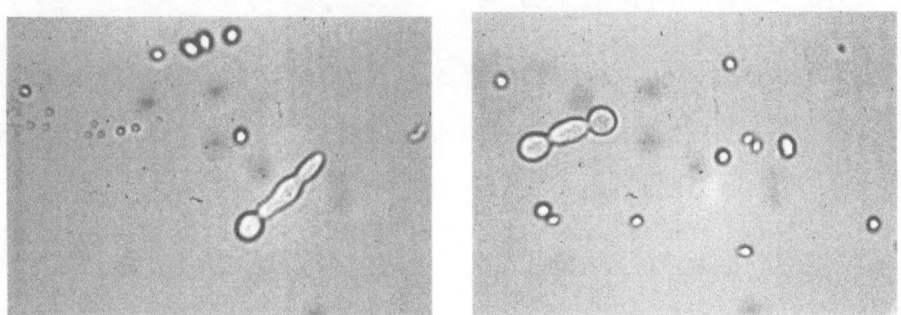

Fig. 5 a, b. Abnormal elongated *Saccharomyces cerevisiae* cells

on grape juice. For strain B, 100% of the theoretical ethanol yield was obtained
after 30 h and for strain K, only 60% of this yield was obtained after 48 h, although the inoculation rates were identical. This indicates that K cells were less
active than B cells. During both runs broth samples were regularly taken and
observed under phase contrast light microscopy (x1000 magnification), without
any staining. The 256 × 170 square pixels images were analyzed on a workstation, on which VISILOG 4.1.4 (Noesis, Orsay, France) was implemented. A
minimal number of 70 images were analyzed per sample. The locations in sub-
space f_1–f_2 of the different cell populations, at three different fermentation
times, are shown in Fig. 6. A higher proportion of double cells was observed for
strain B and a higher proportion of abnormal cells was observed for strain K (up
to 5% when the run was stopped), in agreement with the higher activity observed for strain B.

Although *Saccharomyces cerevisiae* cells are largely cultivated in suspension,
there are processes in which they are immobilized in a matrix such as alginate
or carrageenan gel particles. During process optimization numerous questions
arise, due to the problem of diffusion of the substrates (carbohydrates, oxygen,
etc.) and metabolites in and out of the carriers, as well as the localization, size
and shape of individual cells or cell microcolonies within the gel beads [21].

Recently the morphology of dimorphic yeast cells such as *Kluyveromyces
marxianus* [22] and *Aureobasidium pullulans* [23] has been automatically
characterized. *Aureobasidium pullulans* is classified as a yeast, although it can
exhibit a filamentous morphology. It produces pullulan, a homopolymer used as
an adhesive or as a film. Only certain morphological types of *A. pullulans* syn-
thesize and excrete the polysaccharide [24]: mycelia and chlamydospores are
scarcely productive. The morphology (as yeast-like cells, young blastospores,

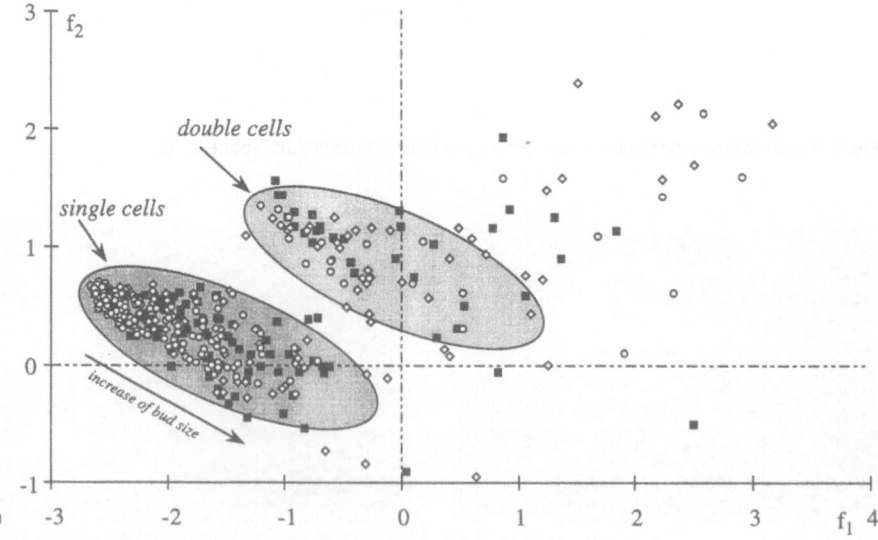

Fig. 6 a. Locations of: a strain B

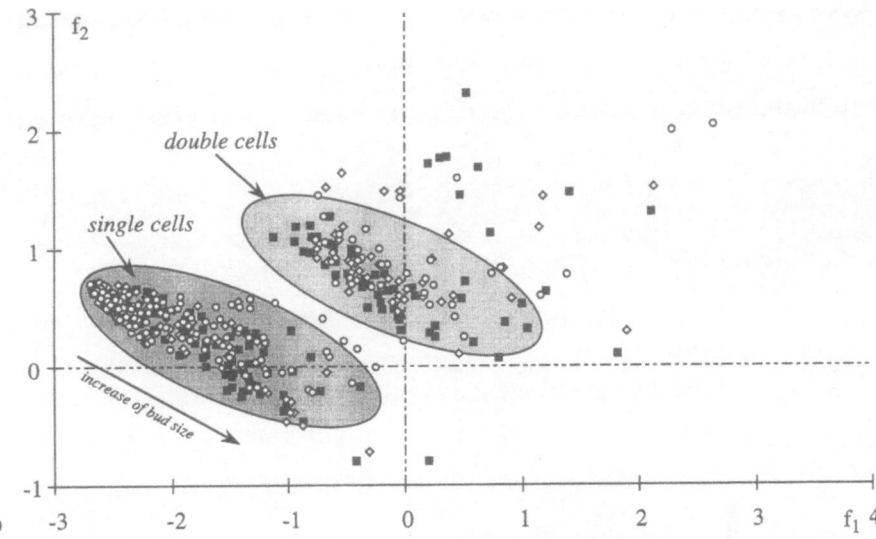

Fig. 6 b. strain K cells in Principal Components Sub-Space f_1–f_2; (■) 2 h after inoculation, (◇) 8 h after inoculation, (○) end of fermentation (24 h for strain B and 48 h for strain K)

swollen blastospores, chlamydospores and mycelia) (Fig. 7), is known to depend upon culture conditions [25, 26]. A Self-Tuning Vision System has been developed [23, 27] to monitor the morphological changes during a culture. After a pre-processing stage to remove noise, enhance cell edges and improve image contrast, a segmentation step isolates the cells from the background and discards debris. Different features are then calculated to characterize the cells: size features (length, area, Feret diameters), shape features (ratio of minimal to maximal Feret diameters), density features (histogram of gray levels within cells). The features are fed to a classifier consisting of a two-layer feed-forward neural network. In the future such an automated system could be used to optimize and control the process by keeping the proportion of some morphological types below or above certain limits.

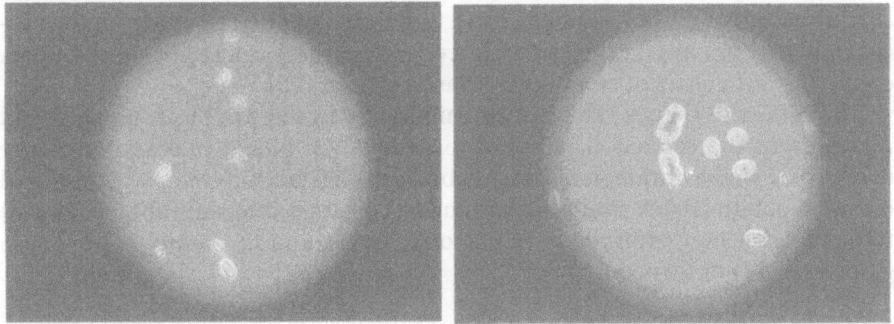

Fig. 7 a, b.

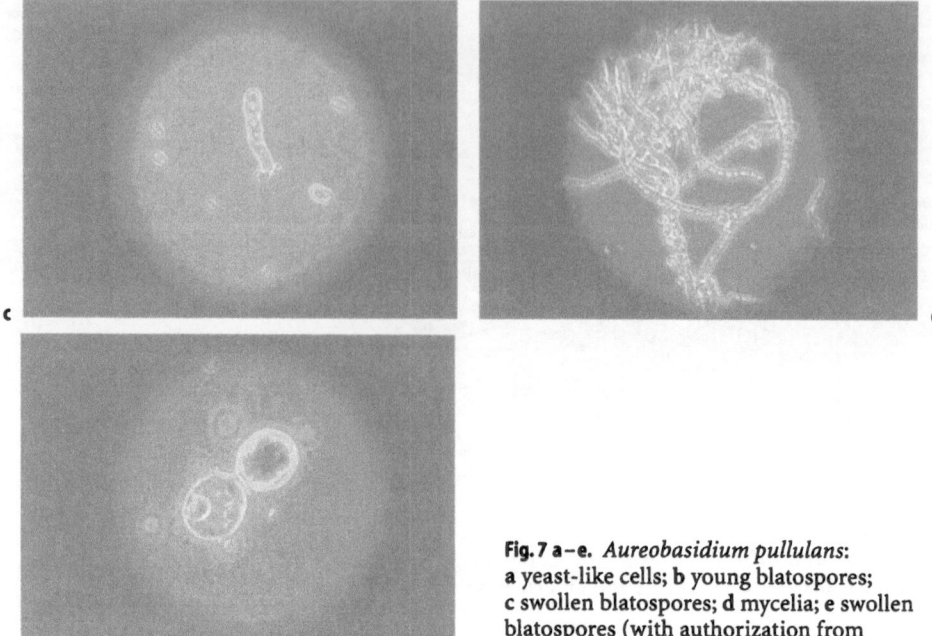

c

d

e

Fig. 7 a–e. *Aureobasidium pullulans*:
a yeast-like cells; b young blatospores;
c swollen blatospores; d mycelia; e swollen
blatospores (with authorization from
Shabtai et al.)

4
Plant Cells

Somatic embryogenesis has aroused great interest for the large-scale micro-
propagation of plants. About 200 different plant species have already been tested
among which are carrot, coffee, alfalfa, and sweet potato. For the bioreactor control a rapid and operator-independent quantitative method of assessing the
morphology of the embryo population is of utmost interest. In most cases the
population is largely heterogeneous with embryos at various stages of development, abnormal embryos, and callus. Sorting of normal and mature embryos by
an operator is a tedious and subjective task, with a low degree of statistical confidence. Higher overall process efficiency could be achieved by automated
sorting based on a vision system. Such a tool would not be useful only at the production level but also at the research level where there is a serious lack of data
due to the difficulties of comparable classifications [28].

The variety of shapes exhibited by embryos is depicted in Fig. 8. Points A and
A' represent the cotyledon joints in torpedoes. An image analysis procedure
using basic shape parameters (length, breadth) has been developed and tested
on sweet potato [29]. A similar technique has been used by Cazzulino et al. [30]
to distinguish non-embryos from carrot embryos and to classify the latter as
globular, heart or torpedo:

$$E = \frac{L}{B}$$

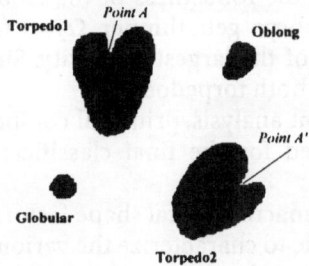

Fig. 8. Some plant-cells embryo shapes

$$R = \frac{P^2}{4\pi A} \quad \text{or} \quad R' = \frac{2P}{\pi(L+B)}$$

$$F = \frac{P(L+B)}{8A}$$

where L is the "length", B the "breadth", P the "perimeter" and A the "projected area". As in the case of yeast cells these descriptors are invariant with respect to the position of the embryo in the image and are easily available in any image analyzer. Table 2 gives the 2D-descriptors of the embryos shown in Fig. 8. Ω_1 and Ω_2 are two descriptors helping to compare the embryo or any kind of particle with respect to its convex hull [31], as depicted in Fig. 9. Ω_1 is a descriptor in the

Table 2. Shape characteristics of plant cell embryos

Embryo	R	E	L/D_{eq}	Ω_1	Ω_2
Torpedo 1	1.48	1.41	1.27	0.71	0.13
Oblong	1.16	1.41	1.22	0.77	0.09
Globular	1.07	1.20	1.12	0.82	0.12
Torpedo2	1.35	1.33	1.22	0.71	0.13

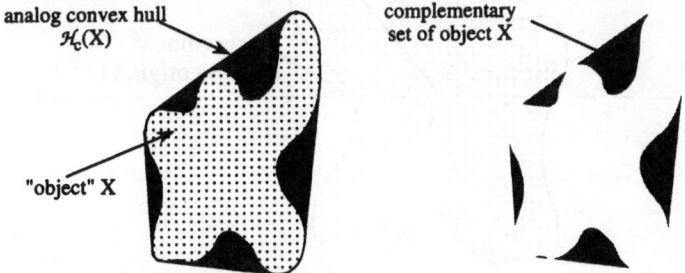

Fig. 9. Particle convex hull and complementary set

range 0–1, which describes the robustness of the embryo: $\Omega_1 = 1$ for a square and decreases when the embryo gets thinner. $\Omega_2 = 0$ for a true convex object and increases with the size of the largest concavity. Similar values for the five descriptors are obtained for both torpedoes.

Classificatory discriminant analysis, principal component analysis or neural networks [32] are then used for the final classification based on such 2D descriptors.

Instead of using classical macroscopical shape factors, Hämäläinen et al. [33] developed a special technique to characterize the various possible developmental stages for birch somatic embryos. Ten features were extracted from the signature s (s) of the projected contour (Fig. 10):

$$\sigma(s) = R = GM$$

where G is the barycenter, M a point on the contour and s = OM is the curvilinear abscissa of origin O along the contour. A classification decision tree based on tests on Mahalanobis distance (for sets of features assumed to follow the multivariate normal distribution) or on simple lower/upper bounds tests has been built.

Cazzulino et al. [30] developed a method based on Fourier analysis of the contour, considered as a periodical signal. This method is known to work well for globular shaped objects (i. e., exhibiting no sharp angles). However it is sensitive to size and position of the embryo in the image fields. This is illustrated in Fig. 11 which presents the signatures of the embryos of Fig. 8: it can be seen in Fig. 11a that, due to the size difference, the signature levels are different for the torpedo and the globular and oblong embryos. Although a visual similarity can be found in the signatures of both torpedoes (Fig. 11b), points A and A' are not located at the same place. In the case of Cazzulino et al. [30], an equal angle increment was used to obtained the set of points on which a Fast Fourier Transform algorithm was applied: this causes a problem when the embryo has a concave shape

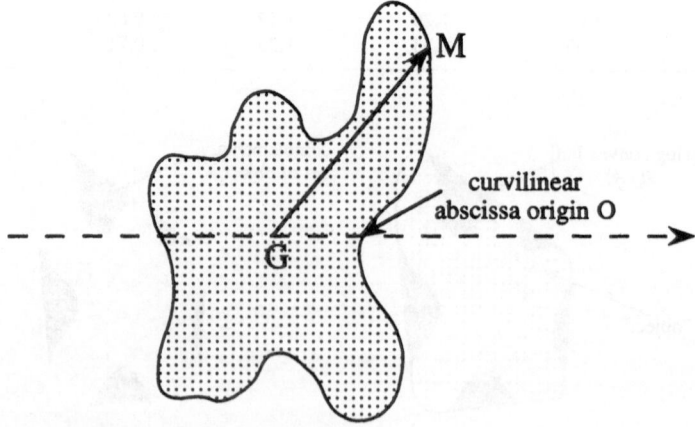

Fig. 10. Particle signature

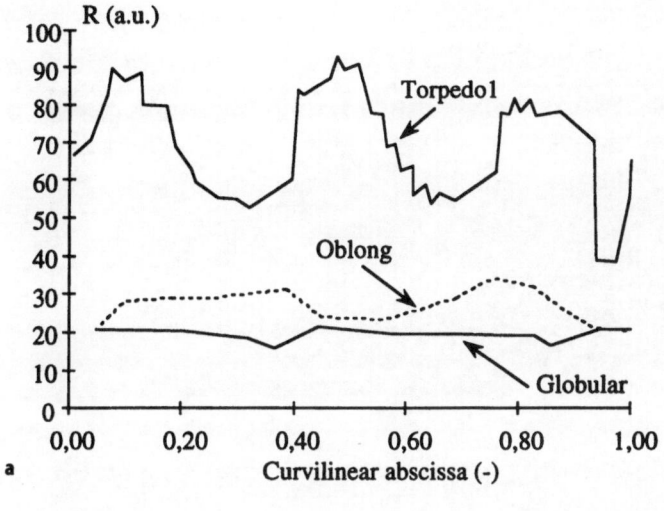

a

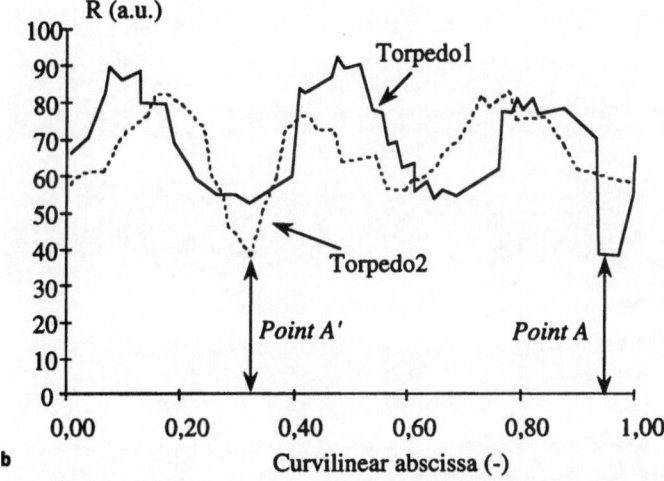

b

Fig. 11 a, b. Signatures of: a torpedo, oblong and globular embryos; b two torpedos

(possible folding of the signature). To avoid that, Chi et al. [34–36] used an equal arc-length sampling scheme: a heuristic decision tree and a k.NN algorithm (k-Nearest Neighbor) were then applied to the 32 Fourier features.

Most applications of image analysis in plant cell cultures have been devoted to somatic embryo characterization: the problem is made complex because of the variety of shapes but once made available the procedures should allow one to characterize the effects of operation conditions in order to improve the bioprocesses. There are also some attempts to monitor cell morphology in suspension cultures: Kieran et al. [37] examined the shear susceptibility of *Morinda citrifolia*, whose cell chain length and cell size are affected by the mixing rate

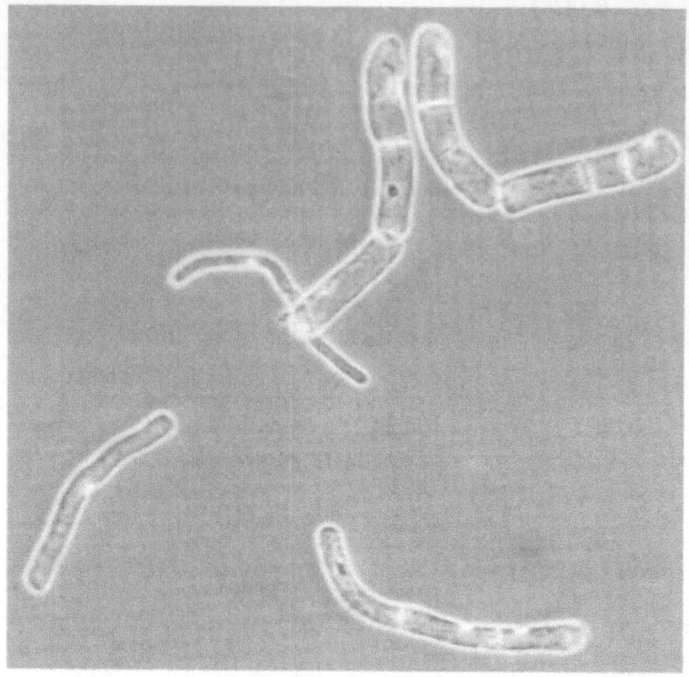

a

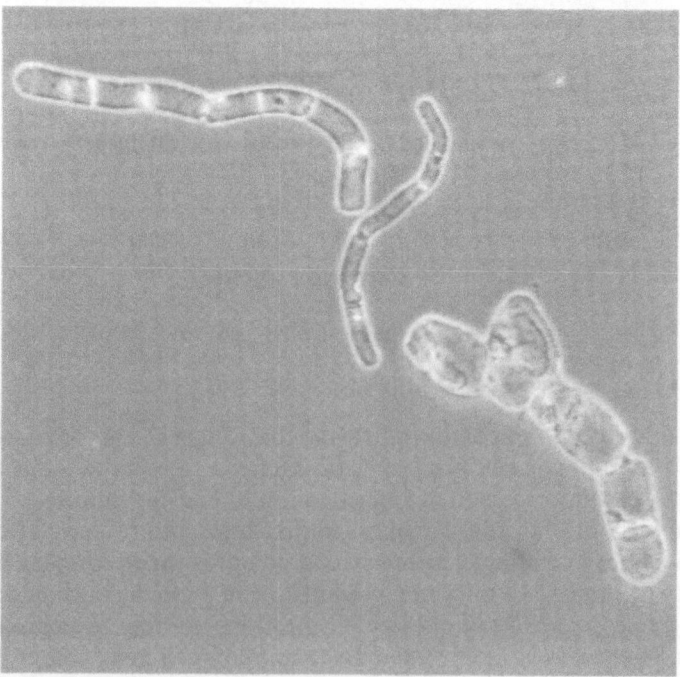

b

Fig. 12 a, b. Examples of *Morinda citrifolia* chains (with authorization from P. Kieran)

(Fig. 12). Natural anthocyanin (red) pigments can be produced by the suspension culture of purple-leafed *Ajuga pyramidalis* cells. Here image analysis can be used at two stages [38]. Large and friable callus colonies, growing on a solid medium, can be selected automatically. Once transferred into the liquid suspension *A. pyramidalis* grow as isolated cells or cell aggregates, each with different degrees of pigmentation. The discrimination between cells require an analysis of color images: the classical Red-Blue-Green approach proved to be unsatisfactory and a segmentation procedure based on the Hue-Saturation-Intensity color system gave the best results.

5
Mammalian Cells

Morphological assessment is also of interest in mammalian cell cultures. In the case of hepatocytes, their morphology changes dramatically depending on the substrate and there is evidence of a strong relationship between cell morphology and function [39]. A current priority for the pharmaceutical industry is the elimination of serum or other animal-derived substances which are expensive, can be contaminated by pathogenic impurities, and their consistency from batch to batch is low. Different ways to achieve that goal are tested. Renner et al. [40] described the effect of mitogenic stimulation on the morphology of Chinese Hamster Ovary cells. Gutsche et al. [41] demonstrated that different cell morphologies (in their case the cell was rat hepatocyte) may be obtained on different cultivation substrate support. Serum-free media used for anchorage-dependent cell lines should be supplemented with attachment factors, usually found in serum. Wierzba et al. [42] have developed an automated image analysis procedure to help them in their search for an efficient attachment factor. Cells cultivated in vitro can be characterized and their proliferation monitored [43]. Such data would also be useful to validate individual cell growth models [44].

Microcarrier cultures are mainly used to produce, on a large scale, important medicines such as vaccines, hormones, etc. by means of anchorage-dependent mammalian cell lines. In such cultures the cells grow on the surface of small spheres, which are usually suspended in culture medium by gentle stirring. After reactor inoculation, a few cells (less than ten depending on the cell line) are attached to the microcarriers. During the first hours of culture, the cells progressively cover the available surface. The coverage can be partial or complete, while some of the microcarriers do not show any sign of colonization. Because of the cost of the growth medium and of the microcarriers, optimization of the culture requires homogeneous use of the spheres, which can be solid or have a macroporous structure.

In the case where the solid microcarrier size distribution is gaussian, the number of cells per microcarrier varies according to the microcarrier diameter (120–200 mm), assuming that the number of cells per surface unit is constant [45]. The cell proliferation depends strongly on the initial conditions, as there is little chance for a living cell to travel from one microcarrier to another due to stirring. Quantification of cell growth is usually achieved by direct counting (manually or automatically) of the cells after trypsinization or by counting their nuclei after membrane disruption and nuclei staining. However, no information is obtained on the quality of microcarrier colonization.

Coulter counter techniques have been tested [46]: the successive layers attached to the microcarriers increase the total diameter, so its measurement provides an estimation of cell growth. The method is really sensitive only when a large number of microcarriers are colonized, due to the somewhat broad initial size distribution. Furthermore the stress induced by the circulation in the apparatus could cause a detachment of the cells from their support.

To overcome these problems Pichon et al. [47, 48] have proposed an image analysis-based method to monitor the colonization by counting the cells attached to transparent microcarriers and by measuring the surface they cover. The microcarriers have a lower gray level than the background. Because they are transparent, cells attached to both faces of the sphere are visible. As seen on the gray-level image of Fig. 13a, the cells attached to the lower half-sphere exhibit gray levels higher than those on the upper half-sphere and they are easily discarded by thresholding. The image treatment produces two images: A with the silhouette of the microcarriers and B with the cells, or more precisely, with the cell clusters as it may be difficult to distinguish individual cells when they are close to each other (Fig. 13b). The coverage ratio is calculated taking into

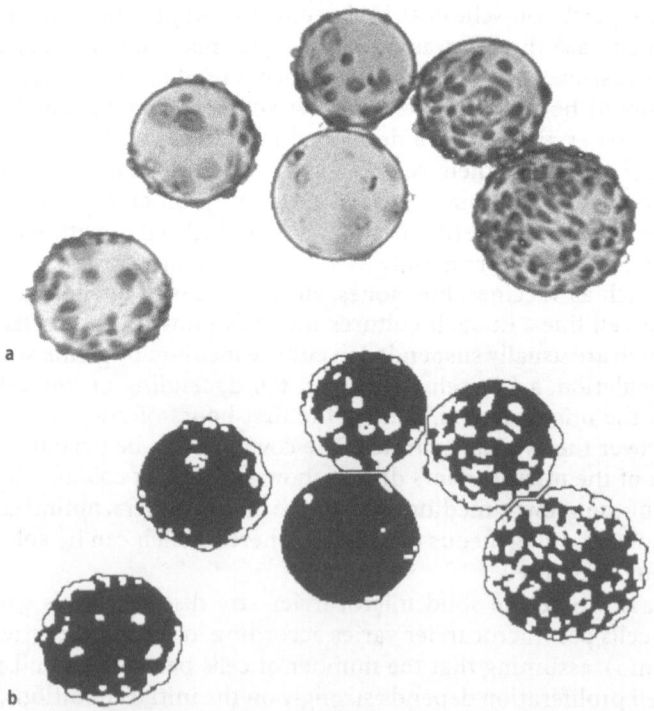

Fig. 13 a, b. CHO cells growing on transparent microcarriers: a gray-level image; b corresponding binary image with localization of cell cluster

account the spherical shape of microcarriers. It may also happen that two or more microcarriers are in contact one with another. An automated segmentation procedure has been implemented to disconnect the beads.

Figure 14 presents the kinetics of a batch culture of Chinese Hamster Ovary cells (CHO) in terms of the number of cell clusters per microcarrier (Fig. 14a) and in terms of surface coverage ratio (Fig. 14b). In Fig. 15 the average number of cell clusters per microcarrier and the average coverage ratio are plotted vs time with a 95% confidence interval. Approximately 80 microcarriers are examined per sample, which is sufficient for an estimation of the coverage ratio (Fig. 16). Linear correlations with results provided by the classical cell nuclei counting technique are obtained with a correlation coefficient of 0.9 (Fig. 17) [49].

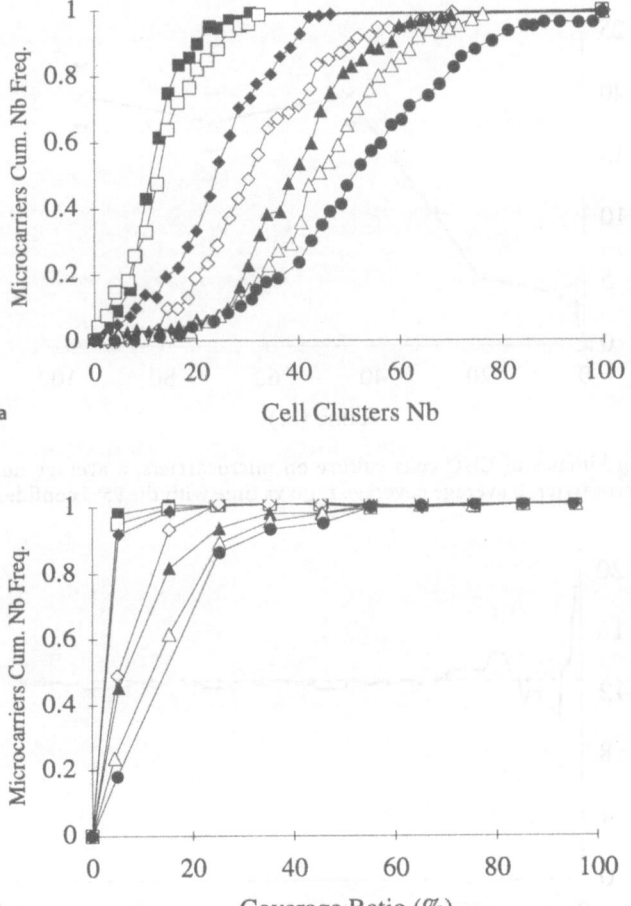

Fig. 14 a, b. Batch kinetics of CHO cells culture on microcarrier: a number of cell clusters per microcarrier; b coverage ratio; (■) 2 h, (□) 6 h, (♦) 20 h, (◊) 29 h, (▲) 44 h, (△) 70 h, (●) 94 h

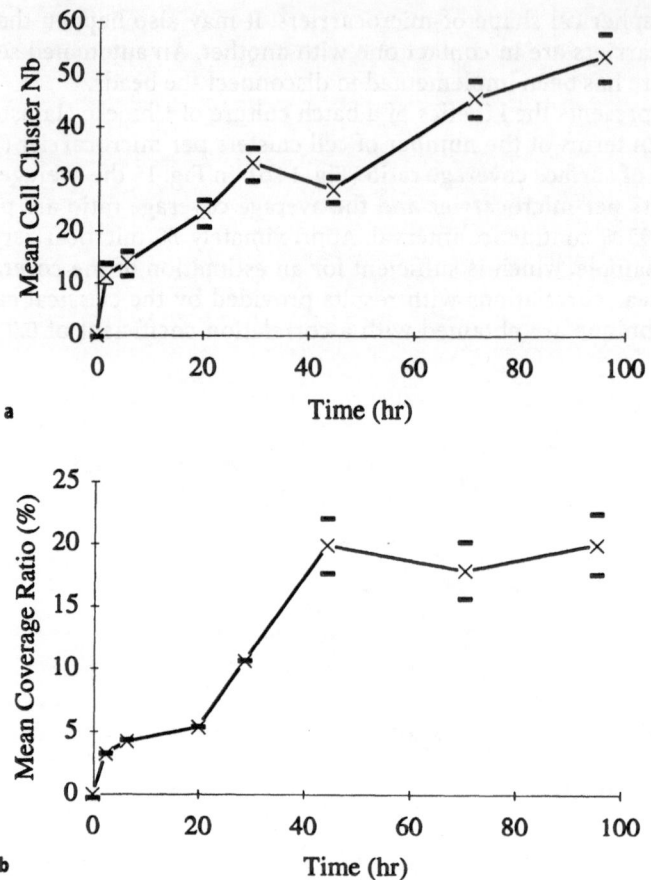

a

b

Fig. 15 a, b. Batch kinetics of CHO cells culture on microcarriers: **a** average number of cell clusters per microcarrier; **b** average coverage ratio vs time with the 95% confidence interval

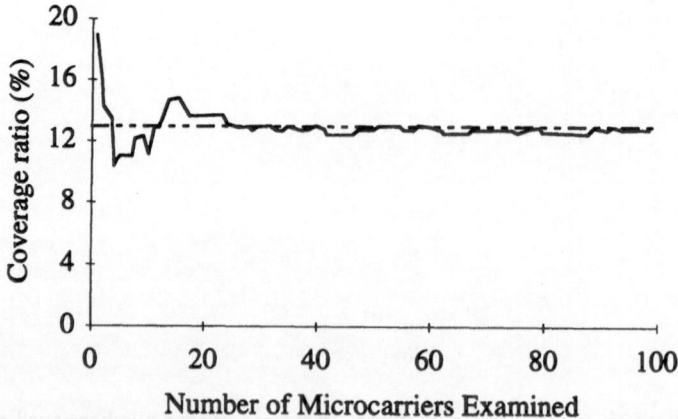

Fig. 16. Evaluation of the sample size

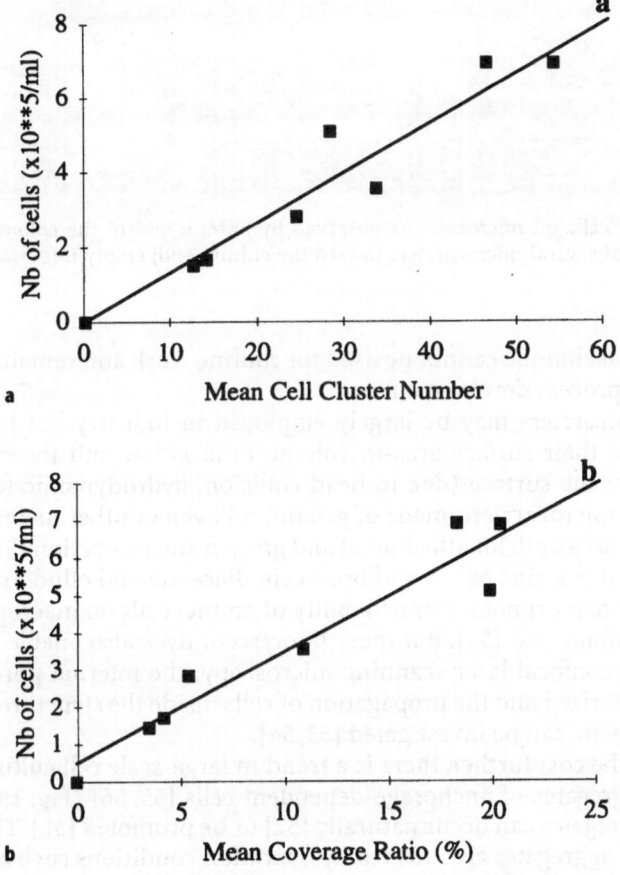

Fig. 17 a, b. Correlation between the average number of cells clusters: **a** the coverage ratio; **b** the cell count

The technique is easy and rapid to use but is restricted to the first stage of the culture, up to the moment when the cell monolayer is complete. At later stages the evolution of the diameter distribution of the colonized microcarriers can be compared to the diameter distribution of empty microcarriers, helping to monitor the successive build-up of cell layers.

Animal cells on microcarriers can also be observed by means of scanning electron microscopy (Fig. 18). Absolute information on cell size cannot be obtained because dehydration causes shrinkage (about 30%). However size proportions are preserved and it could be established that the average cell size changes slightly during the microcarrier colonization process [50]. The cells tend to spread on the available surface when the microcarrier is not yet fully covered. The cell size decreases slightly when no space is left and remains constant for the rest of the culture, even when aggregates of several microcarriers are formed. However, due to the use of SEM and the difficulty of sample pre-

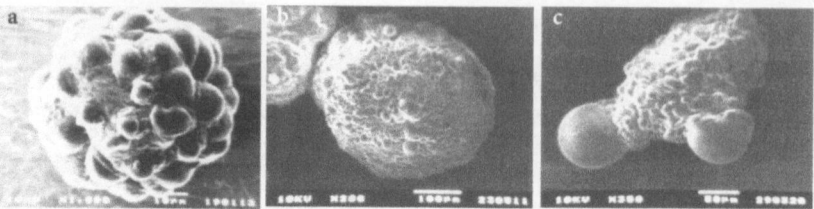

Fig. 18a–c. CHO cells on microcarriers observed by SEM: **a** end of the colonization stage; **b, c** aggregates of several microcarriers toward the culture end; empty microcarriers can be noticed on **c**

paration, this technique cannot be used for routine work and remains a tool for research and process development.

Solid microcarriers may be largely employed in industry but have several disadvantages: their surface area-to-volume ratio is low and the cells can be detached from the surface (due to bead collision, hydrodynamic forces, etc.). Macroporous microcarriers, made of gelatin, collagen or other materials offer a larger surface area both for attachment and growth and protection [51]. By using fluorescent viable stains based on fluorescein diacetate and ethidium bromide, the distribution, morphology and viability of animal cells on macroporous carriers can be monitored [52]. But these fluorescent dyes also enable the optical sectioning by confocal laser scanning microscopy; the internal pore structure can be characterized and the propagation of cells inside the structure, by migration or by growth, can be investigated [53, 54].

To reduce the cost further, there is a trend in large-scale cell cultures toward the use of aggregates of anchorage-dependent cells [55, 56] (Fig. 19). The formation of aggregates can occur naturally [57] or be promoted [58]. The size and the density of aggregates are affected by operation conditions such as agitation [59]. The problem is much simpler than with microcarriers as the cell aggregates can simply be observed by phase contrast light microscopy and their size deduced from their projected area. In some cases they may also be grown in suspension [56] and the assessment of their viability by trypan blue exclusion, combined with size and shape measurement, can be automated [60].

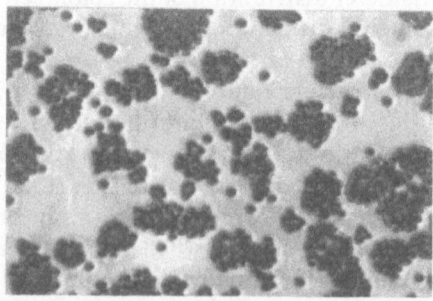

Fig. 19. CHO cells growing as suspended aggregates (By comtesy of Dr A. Marc)

In the field of mammalian cells, as in that of plant cells, image analysis should help provide more valuable data for research [61–63] and be the basis for the routine assessment of the culture quality, when morphology is a key parameter.

6
Algae

Seaweeds and macroalgae are considered to be potentially rich and diverse sources of new medicinal substances. However the culture of algal cells or tissues in bioreactors is not yet well developed.

As plant cells and bacteria, algal species can differ in size and shape, and these characteristics are used to identify and enumerate them [64]. Vynalek [65] studied the morphology of zoospores (daughter cells) of green algae (Chlamydomonas geitheri) growing in synchronous suspension culture with a 16 h light/8 h dark sequence by a semi-automated image analyzer. Zoospores appear at the end of the dark period, during which C. geitleri cells develop regularly and at the beginning of the light period. The automatic identification of the cells was not possible because the contrast with respect to the background was not good enough and because the daughter cells are attached to the mother algae. An optical pen enabled the author to detect four points on the cell borderline. Based on these points an ellipse was fitted to the borderline and the cell volume was computed from the corresponding ellipsoid. The average gray level within the domain defined by the points gave an estimation of the starch concentration in the cell after correction of the background level. The cell volume increased considerably during the light phase. According to Vynalek [65], if image analysis was in this case a very valuable tool, real problems remained for the automated detection of the cells and the calibration between the gray level and the starch concentration.

Recently Acrosiphonia coalita, a macroalgae known to express pharmacologically active substances, was successfully cultivated in a bioreactor [66]: the culture consisted of linear or lightly branched filaments in a liquid suspension. The tools developed for the quantification of the morphology of filamentous microorganisms could also be used in this context.

7
Back to Filamentous Species

The previous chapters have focused on the interests of a morphological approach to understand and to model the growth of filamentous species, mostly in suspension cultures. Applications to solid-state fermentations could also be of interest (Fig. 18) However on some occasions the relation between morphology and differentiation is unclear: new tools need to be developed to describe the structural and biochemical differences observed in different parts of the hyphae. In the case of Penicillium chrysogenum at higher magnification than the one used for morphological analysis, some internal structural elements of the filaments can be clearly seen by optical microscopy. Packer et al. [67] estimated the volume occupied by the cytoplasm and the degenerated regions containing vacuoles. Still higher magnification allowed Paul et al. [68] to characterize fully the size and shape of vacuoles, as explained in Chap. 1. Using neutral red, apical segments could also be detected, and they have been associated with the growing tips (Fig. 20). A complex differential staining procedure based on methylene blue and Ziehl fuschin made it possible to investigate further the differentiation process of *P. chrysogenum* [69]. Six physiological states could be detected, based on color and granulation: the growing tips, three active types of segments, highly vacuolized zones and dead compartments (Fig. 21).

The color images contain a large amount of information, which is counterbalanced by the image system memory requirement and the present lack of well-developed color image analysis tools. The initial color image, captured as separate Red-Green-Blue images, is transformed into the V, R/V, B/V, normalized system where V = max (R,G,B). The problem to be solved may be stated as:

- detection of the hyphae is followed by systematic elimination of possible sources of bias, such as halo and debris;
- hyphal segmentation into different zones, each characterized by a set of parameters: a partition of the filament into homogeneous zones (tesselation) is

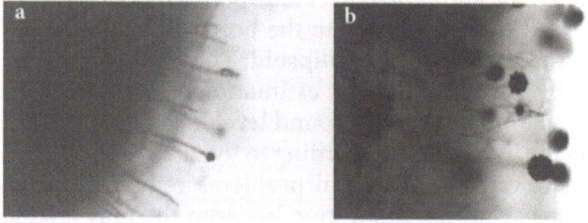

Fig. 20 a, b. *Aspergillus niger* growing on solid substrate (sugar-beet)

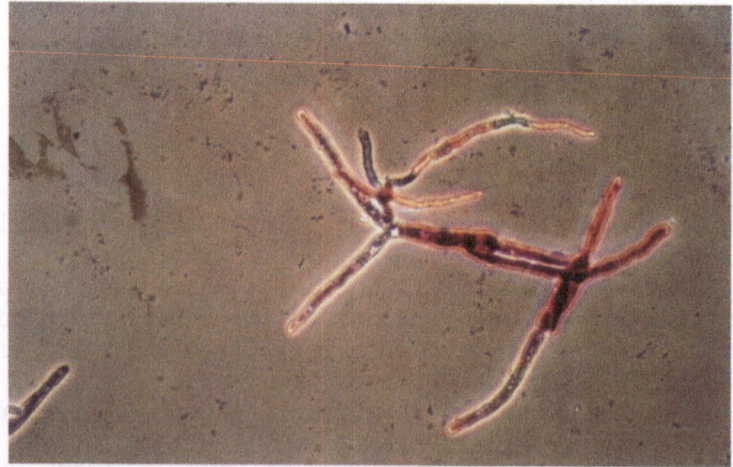

Fig. 21. Color image of a stained *Penicillium chrysogenum* filament (differential staining)

created, without taking into account their physiological state. Each area is characterized by a set of attributes: the mean values of V, B/V and R/V over the area. After a comparison of their attributes, similar neighboring areas are merged into larger ones, the attributes of the resulting areas being the barycenter of the attributes of neighbors. This iterative process is stopped when no similarity could be detected between adjacent areas. After merging is completed, a fourth parameter, the mean of R on the filament centerline is computed;
- classification of zones into the six available classes: a fuzzy C-mean classification procedure was selected to create a partition of the merged zones into six clusters represented by their center. They were then allocated to a physiological state by comparison with a reference, a set of images for which the classification was made manually by a group of experts.

The analysis of 40 images per sample proved to be a good compromise between the precision (20% for all physiological states) and the time required for image capture with respect to the stability of the staining procedure. The agreement between automated and visual classification was in the range 80 to 90%, with a total processing time of 4–6 min per image (after severe optimization of the PC memory management!).

Figure 22 presents the kinetics of a fed-batch penicillin production run in a 20-l fermenter. Rapid growth occurred until 50 h, but growing zones represented only 10% of the total mycelium, showing that differentiation was taking place very early during the fermentation time course. This is lower than the fraction determined by Paul et al. (30%) [68] with a simple neutral red staining, but this may be due to the behavior of the particular strain used. The proportion of growing material was not constant during the production phase: a sudden increase of the growing apices fraction could be observed at 60 h. This phenomenon was transitory and is probably linked to the recycling of material released by lysed fragments. It should be noted that this high proportion of growing material was not directly related to an increase of dry cell weight, because compensation can take place through lysis. Fluctuations, yet of smaller amplitude, could also be noticed for zones 2 and 3.

Because this strain has small vacuoles, their full characterization in terms of shape and size was impossible but a clear distinction could be made between dead and vacuolized segments: it appears that the proportion of degenerated segments (zones 5 and 6) can be very high and reach up to 50% of the mycelium. In particular, the proportion of dead zones increased steadily during the fermentation. Although it was not yet possible to determine statistically any clear link between the physiological stages of biomass and penicillin production, such links should exist and image analysis techniques offer a way to quantify them. The next step would be to develop comprehensive structured kinetic models taking into account the physiology of the microorganism and offering ways to improve the antibiotic production [70].

Acridine Orange is a well-known metachromatic stain which fluoresces orange when bound to single-stranded nucleic acids and green when bounded to double-stranded nucleic acids. The color, which ranges from orange to yellow and green, can give an indication of the relative proportions of RNA (mostly single-stranded) and of DNA (mostly double-stranded). When RNA:DNA is large, such as in fast growing cells, the color is orange and tends to green in slowly growing cells. Acridine Orange would therefore be a good candidate for physiology quantification in filamentous species, as in other types of cells [71, 72]. However color image analysis is required to distinguish the subtle differences in the staining: the pellet shown in Fig. 23 exhibits a color range from yellow for the core to orange and green for the hair filaments. These color changes are not visible on a gray-level image which takes into account only the intensity. The other difficulty is the halo around the cells, which should be minimized by a careful selection of the dye concentration.

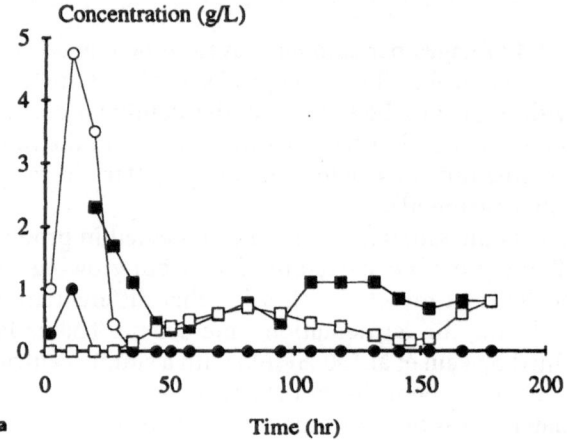

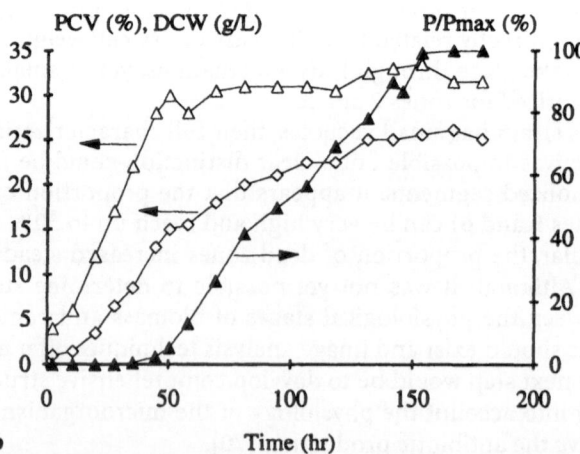

Fig. 22 a, b. Fed-batch kinetics of *Penicillium chrysogenum:* **a** (○) glucose, (●) sucrose, (■) ammonium and (□) phenyl acetic acid; **b** (▲) penicillin, (△) packed cell volume and (◇) dry cell weight

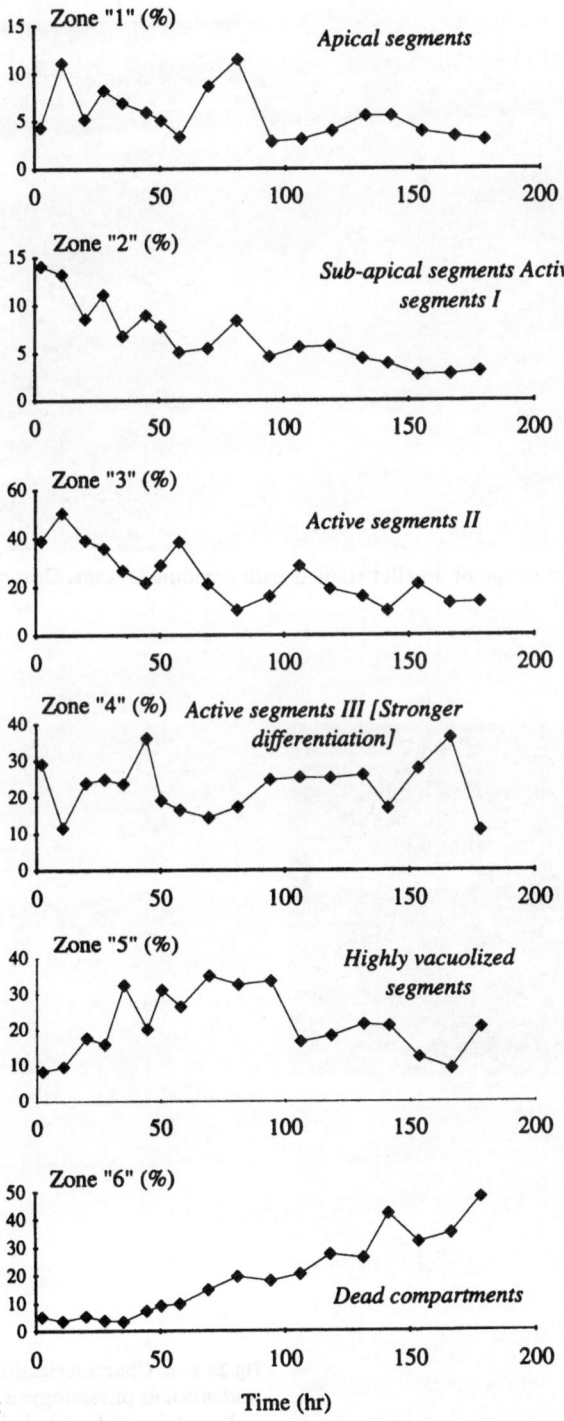

Fig. 22 c. c physiological zones kinetics

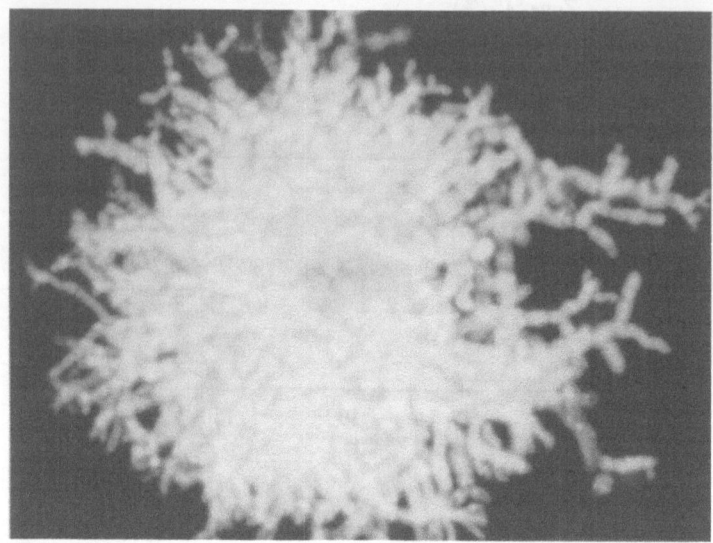

Fig. 23. Grey-level image of a pellet stained with Acridine Orange. Observation by epifluorescence

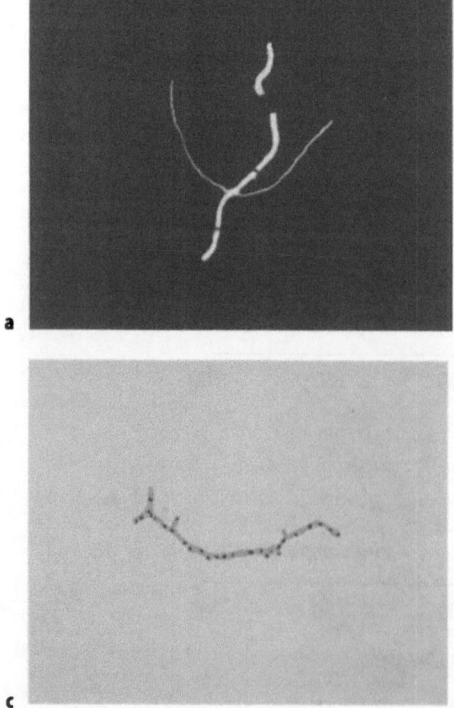

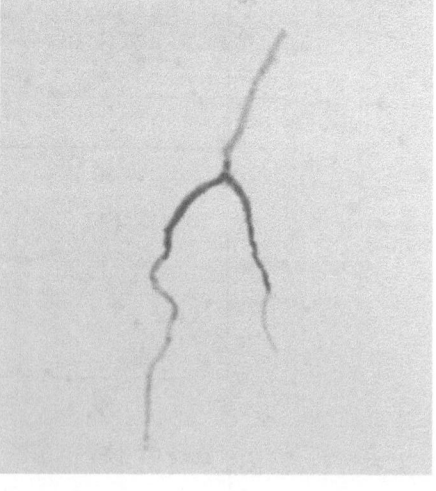

Fig. 24 a–c. Characterization of *Streptomyces ambofaciens* physiology: **a** propidium iodide staining; **b** carbol violet staining; **c** INT staining

Not all the fluorescent dyes require color image analysis. Drouin et al. [73] developed a technique allowing the quantification of *S. ambofaciens* thin hyphal parts and septa using propidium iodide, which fluoresces red (Fig. 24a). The fluorescence signal is low as the dye concentration has to be optimized to reduce the halo and make septa visible. The images are captured by means of a steerable frame integration time camera. The assessment of the physiological state is completed by a carbol violet staining, which reveals the empty zones of the filament (Fig. 24b), and by INT = 2 (*p*-iodophenyl)-3-(*p*-nitrophenyl)-5-phenyl-tetragolum chloride staining with methylene blue counterstaining for respiratory activity quantification (Fig. 24c).

8
Going On-Line

The ultimate goal of size and morphology assessment by image analysis techniques would be to provide an on-line monitoring of these characteristics, which could then be used in the control strategy of the bioprocesses. Frankly speaking we are still today far away from that point. To operate on-line there are a few hurdles to overcome:

- sampling
- sample dilution
- sample staining
- visualization
- image treatment
- statistical analysis

As in many bioreactor monitoring systems for biochemical variables, an approach is the side-loop, by pumping the broth to a flow chamber mounted on the microscope (Fig. 25). It will usually be too concentrated and an automated dilution step would be required. This could be combined with a staining step: technologies used in air-segmented flow analyzers or more recently in flow-injection analyzers would be useful to solve these problems.

Visualization would be the next step: a "good" image, with sufficient contrast between the cells, well in focus, with the background being a prerequisite for a correct final characterization. Automated focus and brightness controls are available on modern light microscopes. However the lenses are mechanically driven, which introduces some lag in the visualization loop. Lichtfield et al. [74] proposes microglass channels: the channel diameter should be adapted to the microorganism size as the sample should not be affected by the flow through too narrow a channel. For *Bacillus thuringiensis* cells and spores a 20 μm thick channel gave the best results. However the detection of inclusion bodies in cells was difficult at a magnification of x1000 because the floating cells were continuously moving in and out of the focus plane.

Maruhashi et al. [75] has monitored the concentration and viability of animal cells without any sample pretreatment such as dilution or staining, using a flow chamber mounted on a microscope. Viable cells are assumed to be larger than dead cells. Flow cytometry has been widely used for the characterization of large

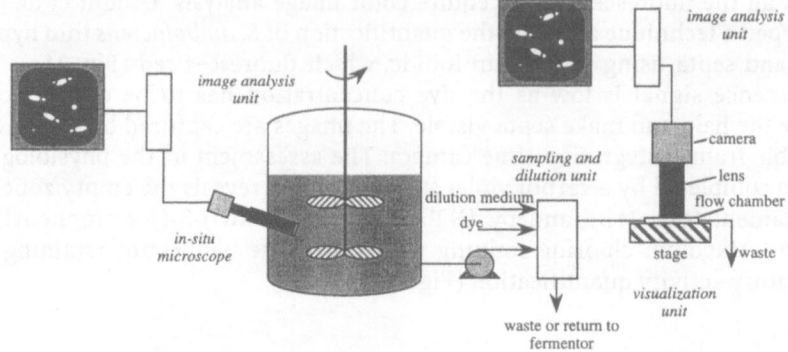

Fig. 25. On-line image analysis on bioreactors

populations of cells, and especially mammalian cells, in a short amount of time. But it is, from the point of view of image analysis, a "blind" technique as it cannot eliminate unwanted cell categories. This sorting could be done by image analysis, which is still considered too slow to give statistically valuable data within an acceptable delay. However some attempts are being made to combine both techniques [76].

To avoid the problem of focus, Zalewski et al. [20] proposed a stop flow system, with an inverted microscope: the flow in the visualization chamber is stopped while the *Saccharomyces cerevisiae* cells are settling on the lower face of the chamber.

The ultimate sophistication would be to have an in-situ microscope (Fig. 25) such as the one developed by Suhr et al. [77]. A front-end sensor head is inserted directly in one of the fermenter ports and mounted on an epifluorescence microscope. Still images generated using pulse illumination are captured by a

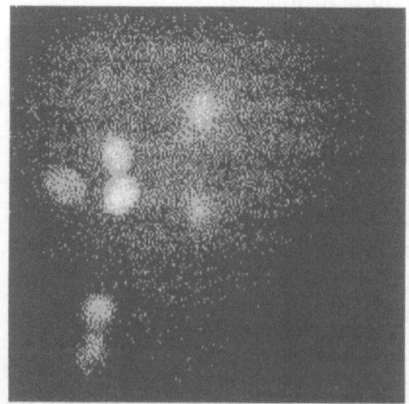
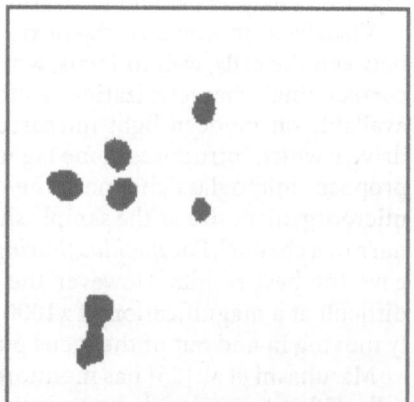

Fig. 26 a, b. In-situ images of *Saccharomyces cerevisiae*: **a** gray-level image; **b** corresponding binary image (with authorization of T. Scheper)

silicon-intensified target camera. The cost for going on-line can be seen by comparing Fig. 26a and Fig. 3a (still image of *Saccharomyces* cells under an off-line microscope): the on-line image is highly blurred as the cells are in different planes of focus. A sophisticated image treatment based on a depth-from-focus algorithm has been developed [78] (Fig. 26b), but is still too slow (1 image/min) for real on-line applications: a typical image contains 5–20 cells, depending upon biomass concentration and it may be necessary to examine at least 200 cells to get valuable data, from the point of view of statistics. Physiology assessment can be combined with sizing, as NADH fluorescence is modulated by metabolic changes such as glucose uptake (Fig. 27) or aeration shutdown (Fig. 28).

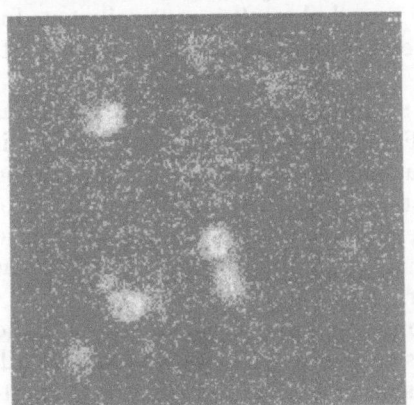
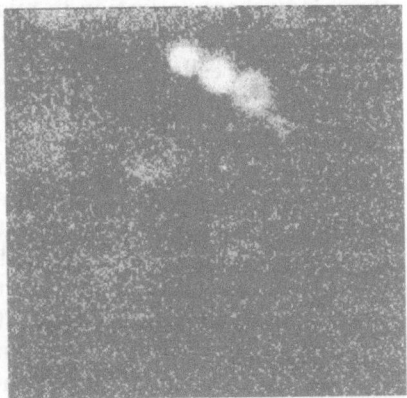

a   b

Fig. 27 a,b. In-situ images of *Saccharomyces cerevisiae*: **a** before; **b** after a pulse of glucose (with authorization of T. Scheper)

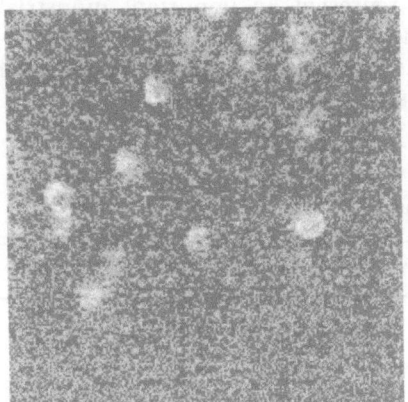
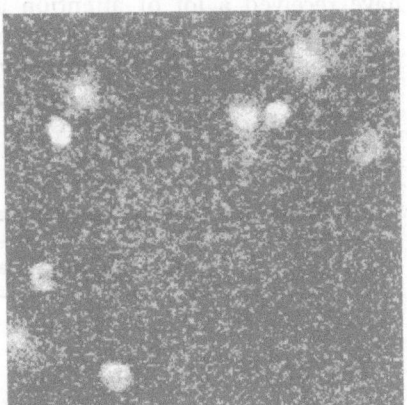

a   b

Fig. 28 a,b. In-situ images of *Saccharomyces cerevisiae*: **a** before; **b** after aeration shutdown (with authorization of T. Scheper)

9
Conclusions

As bioengineers, our aim is to improve the performance of the bioprocesses. In that context what are the prospects of image analysis? It will depend on the goals which we set.

Image analysis is the foremost technique to examine the morphology of microorganisms, as well as of particulate matter in general. Often "size" measurement has some meaning only in relation to the morphology: we can speak of the "length" of rod-like bacteria or of the "diameter" of quasi-spherical yeast cells only because we have some hints of their shapes. The return of classical staining techniques (methylene blue, carbol violet, neutral red, etc.) and the development of new dyes, mostly fluorescent, have opened new horizons for physiology characterization in relation to morphology. It is especially true for filamentous species, for which flow cytometry is more restricted than for yeasts or mammalian cells.

As in any data acquisition problem, the sampling period should be chosen as a function of the process time constant. Furthermore, data processing time is not negligible: it depends on the image quality, on the quantity of information to be extracted, and on the computer system capacity. The statistical point of view should be also kept in mind: it will be difficult, probably for a long time, to examine within a few seconds with an image analysis system as many cells as with a Coulter counter or a flow cytometer.

Just considering the biological time constants given in Table 3, in-situ and real-time image analysis of bacteria seems very difficult. Optical problems will probably make it impossible but on-line vision with a side-loop may be attempted [74]. Yeasts are just at the limit and the work of Suhr et al. [77] show how sophisticated the system should be.

One of the aims of this book was to show how the information gained on morphology could be used to improve our knowledge of microbial physiology and how this information could be integrated in models. Filamentous species have received a lot of attention as they represent an important industrial challenge. Some of the experience has already been transferred to other microorganisms such as plant cells, which also present a large morphological spectrum, and, to a lesser extent, yeasts cells.

Table 3. Some typical time constants for various cells

Cell	Specific growth rate (h^{-1})	Generation period
Bacteria	1–2	20–40 min
Yeast	0.45	90 min
Filament. fungus	0.23	180 min
Plant cell	0.029	≈ 1 day
CHO	0.025 [55]	≈ 1 day
BHK	0.029–0.015 [55]	≈ 1–2 days

As in other applications of image analysis, a compromise should be reached between the number of microorganisms examined, the time required for their analysis, and the richness of the information provided (counts, size, morphology, physiology). This trade-off will be the basis for the decision between an off-line application, mainly directed toward the acquisition of new data, the development of knowledge on the cell physiology, the reactor performance, etc., the bioprocess modeling, and an on-line application, directed mostly toward monitoring and hopefully control.

10
References

1. Schröder D, Krambeck C, Krambeck HJ (1991) Acta Stereol 10:123
2. Bianchi A, Giuliano L (1996) Appl Environ Microbiol 62:174–177
3. de Roissart H, Luquet FM (1988) Les bactéries Lactiques: aspects fondamentaux et technologiques. Lavoisier, Paris
4. Meijer BC, Kootstra GJ, Wilkinson MHF (1990) Binary 2:21
5. Singh A, Pyle BH, McFeters GA (1989) J. Microbiol Methods 10:91
6. Dubuisson MP, Jain AK, Jain MK (1994) J. Microbiol. Methods 19:279
7. Monot F, Martin JR, Petitdemange H, Gay R (1982) Appl Environ Microbiol 44:1318
8. Coello M (1990) PhD Thesis, UTC, Compiègne, France
9. Brandis JW, Ditullio DF, Lee JF, Armiger WB (1989) Process controlled temperature induction during batch fermentations for recombinant DNA products. In: Computers applications in fermentation technology. Elsevier, London, p 235
10. Tsai AM, Betenbaugh MJ, Shiloach J (1995) Biotechnol Bioeng 48:715
11. Lee SY (1996) Biotechnol Bioeng 49:1
12. Fowler JD, Robertson CR (1991) Appl Environ Microbiol 57:102
13. Yeh, TY, Godschalk JR, Olson GJ, Kelly RM (1987) Biotechnol Bioeng 30:138
14. Bärtels CC, Chatzitheodorou, Rodriguez-Leiva M, Tributsch H (1989) Biotechnol. Bioeng 33:1196
15. Monroy Fernandez MG, Mustin G, de Donato P, Barres O, Marion P, Berthelin J (1995) Biotechnol Bioeng 46:13–21
16. Berner JL, Gervais P (1994) Biotechnol. Bioeng 43:165
17. Hirano T (1990) ASBC J 48:79
18. Huls PG, Nanninga N, van Spronsen EA, Valkenburg JAC, Vischer NOE, Woldringh CL (1992) Biotechnol Bioeng 39:343
19. Pons MN, Vivier H, Rémy JF, Dodds JA (1993) Biotechnol Bioeng 42:1352
20. Zalewski K, Götz P, Buchholz R (1994) On-line estimation of yeast growth rate using morphological data from image analysis. In: Galindo E, Ramirez OT (eds) Advances in bioprocess engineering. Kluwer Academic Publishers, Dordrecht, p 191
21. Walsh PK, Isdell FV, Noone SM, O'Donovan MG, Malone DM (1996) Enz Microb Technol 18:366–372
22. O'Shea DG, Walsh PK (1996) Biotechnol Bioeng 51:679–690
23. Shabtai Y, Ronen M, Mukmenev I, Guterman H (1996) Computers Chem Engng 20: S321
24. Heald PJ, Kristiansen B (1985) Biotechnol Bioeng 27:1516–1519
25. Catley BJ (1980) J Gen Microbiol 120:265–268
26. Wecker A, Onken U (1991) Biotechnol Lett 13:155–160
27. Guterman H, Shabtai Y (1996) Biotechnol Bioeng 51:501
28. Huang LC, Chi CM, Vits H, Staba EJ, Cooke IJ Hu WS (1993) Biotechnol Bioeng 41:811–818
29. Grand d'Esnon A, Chee R, Harrell RC, Cantliffe DJ (1989), Biofutur 26:3
30. Cazzulino D, Pedersen H, Chin CK (1991) Bioreactors and image analysis for scale-up and plant propagation. In: Cell culture and somatic cell genetics of plants. Academic Press, New York

31. Rolland T, Pons MN, Vivier H, Dodds JA, Thomas A (1996) Récents Prog. Génie Procédés 10 (45):111–116
32. Uozomi N, Yoshino T, Shiotani S, Suehara KI, Arai F, Fukuda T, Kobayashi T (1993) J Ferment Bioeng 76:505
33. Hämäläinen JJ, Hurten U, Kauppinen V (1993) Biotechnol Bioeng 41:35
34. Chi CM, Vits H, Staba EJ, Cooke TJ, Hu WS (1994) Biotechnol Bioeng 44:368
35. Chi CM, Zhang C, Staba J, Cooke T, Hu WS (1996) J. Ferment. Bioeng. 81:445–452
36. Chi CM, Zhang C, Staba EJ, Cooke TJ, Hu WS (1996) Biotechnol Bioeng 50:65
37. Kieran PM, O'Donnel HJ, Malone DM, MacLoughlin PF (1995) Biotechnol Bioeng 45: 415–425
38. Smith MAL, Reid JF, Hansen AC, Li Z, Madhari DL (1995) J Biotechnol 40:1–11
39. Folkman J, Moscona A (1978) Nature 243:345
40. Renner WA, Lee KH, Hatzimanikatis V, Bailey JE, Eppenberger HM (1995) Biotechnol Bioeng 47:476
41. Gutsche AT, Zurlo J, Deyesu E, Leong KW (1996) Biotechnol Bioeng 49:259
42. Wierzba A, Reichl U, Turner RFB, Warren RAJ, Kilburn DG (1994) Biotechnol Bioeng 46: 185–193
43. Young JC, DiGiusto D, Backer MP (1996) Biotechnol Bioeng 50:465
44. Buettner HM (1996) AIChE J. 42:1127
45. Hu WS, Meier J, Wang DIC (1985) Biotechnol Bioeng 27:585–595
46. Miller SJO, Henrotte M, Miller AOA (1986) Biotechnol Bioeng 28:1466
47. Pichon DR, Vivier HL, Pons MN (1992) Acta Stereol 11/Suppl 1:243
48. Pichon D, Vivier H, Pons MN (1992) Growth monitoring of mammalian cells on microcarriers by image analysis. In: Modeling and Control of Biotechnical Processes. IFAC, Colorado, p 311
49. Pichon D (1993) Analyse d'images en biotechnologie: quantification de la morphologie de microorganismes filamenteux au cours de fermentations et suivi de la croissance de cellules animales sur microporteurs au cours de cultures. INPL, Nancy, France
50. Pons MN, Wagner A, Vivier H, Marc A (1992) Biotechnol Bioeng 40:187
51. Ng YC, Berry JM, Butler M (1996) Biotechnol Bioeng 50:627
52. Nikolai TJ, Peshwa MV, Goetghebeur S, Hu WS (1991) Cytotechnol 5:141
53. Bancel S, Hu WS (1996a) J Ferment Bioeng 81:437
54. Bancel S, Hu WS (1996b) Biotechnol Prog 12:398
55. Spier RE, Griffiths JB, MacDonald C (1992) Animal cell technology: developments, processes and products: sect. 15: cell aggregate cultures. Butterworth-Heinemann, Oxford, p 409
56. Chevalot I, Visvikis A, Nabet P, Engasser JM, Marc A (1994) Cytotechnology 16:121–129
57. Wu FJ, Friend JR, Hsiao CC, Zilliox MJ, Ko WJ, Cerra FB, Hu WS (1996) Biotechnol Bioeng 50:404
58. Dai W, Saltzman WM (1996) Biotechnol Bioeng 50:349
59. Moreira JL, Alves PM, Aunins JG, Carrondo MJT (1995) Biotechnol Bioeng 46:351–360
60. Tucker KG, Chalder S, Al-Rubeai M, Thomas CR (1994) Enz Microb Technol 16:29–35
61. Al-Rubeai M, Singh RP, Goldman MH, Emery AN (1995) Biotechnol Bioeng 45:463–472
62. Al-Rubeai M, Singh RP, Emery AN, Zhang Z (1995) Biotechnol Bioeng 46:88–92
63. Coppen SR, Newsam R, Bull AT, Baines AJ (1995) Biotechnol Bioeng 46:147–158
64. Gorsky G, Guilbert P, Valenta E (1989) Mar Ecol Prog Ser 58:133
65. Vynalek V (1991) Acta Stereol. 8:27
66. Rorrer GL, Zhi C, Polne-Fuller M (1996) Biotechnol. Bioeng 49:559
67. Packer HL, Keshawarz-Moore E, Lilly MD, Thomas CR (1992) Biotechnol Bioeng 39: 384–391
68. Paul GC, Kent CA, Thomas CR (1992) Trans I Chem E 70:13–20
69. Vanhoutte B, Pons MN, Thomas CR, Louvel L, Vivier H (1995) Biotechnol Bioeng 48:1
70. Paul GC, Thomas CR (1996) Biotechnol Bioeng 51:558
71. Wentland EJ, Stewart PS, Huang CT, McFeters GA (1996) Biotechnol. Prog 12:316
72. Wirtanen G, Nissinen V, Tikkanen L, Mattila-Sandholm T (1995) Int J Food Sci Technol 30: 523–533

73. Drouin JF, Maus P, Vivier H, Pons MN, Germain P, Lebrihi A, Louvel L, Vanhoutte B (1995) Cellular differentiation study of *Streptomyces ambofaciens* in submerged culture by image analysis. 7th European Congress on Biotechnology, Nice, France
74. Lichtfield JB, Reid JF, Richburg BA (1992) Machine vision microscopy for on line sampling, analysis and control. In: Modeling and control of biotechnical processes. IFAC, Colorado, p 275
75. Maruhashi F, Murakami S, Baba K (1994) Cytotechnology 15:281–289
76. Boudry C, Herlin P, Coster M, Chermant JL, Sola B, Henry-Amar M (1995) Acta Stereol. 14: 121–128
77. Suhr H, Wehnert G, Schneider K, Bittner C, Scholz T, Geissler P, Jähne B, Scheper T (1995) Biotechnol Bioeng 47:106–116
78. Scholz T (1995) Ein depth-from-focus Bildanalyseverfahren zur Online-In-Situ Bestimmung von Zellkonzentrationen in Fermentern. Ph.D. thesis, Heidelberg, Germany

Received December 1996

Mathematical Modelling of the Morphology of Streptomyces Species

Rudibert King

Institut für Prozeß- und Anlagentechnik, Fachgebiet Meß- und Regelungstechnik,
Technische Universität Berlin, Budapester Str. 48, 10787 Berlin, Germany
E-mail: king0630@mailszrz.zrz.tu-berlin.de

List of Symbols and Abbreviations

a_i coefficient of the lumped pellet model
A   area
c_i concentration of substance i

Advances in Biochemical Engineering /
Biotechnology, Vol. 60
Managing Editor: Th. Scheper
© Springer-Verlag Berlin Heidelberg 1998

c_{Hy}	hyphal concentration
c_{Tip}	concentration of tips
D	diffusion coefficient
f	number density function
F	total number of particles per volume
g	concentration based on cell's volume
HGU	hyphal growth unit
k	rate constant
l_i	coordinate of i-th section, i = 1, 2, 3
l_0	length of a hypha associated with one nucleotide, or length of a spore
L_i	length of i-th section
M	number of substrates
N	number of tips
p	density of the breaking probability
q	specific volumetric flow
Q	volumetric flow
r	reaction rate
$r_{Hy,Tip}$	synthesis rate of hyphae, tips
$r_{HyD,TipD}$	degradation rate of hyphae, tips
$r_{HyB,TipB}$	breakage rate of hyphae, tips
s	characteristic property
t	time
u	growth rate along a characteristic property
V	volume
Y	yield coefficient
z	spatial coordinate
z_R	radius of a pellet

Greek symbols

α	apical growth velocity
β	branching rate
γ	constant
λ	constant
μ	specifc rate
σ	source or sink
ψ_{Hy}	activity of branching
ψ_{Tip}	activity of apical growth

1
Introduction

About 70% of all antibiotics known today are produced by *Actinomycetales*, about 90% of which are related to *Streptomyces* species [1]. Other metabolites such as vitamin B12 and industrial enzymes, e.g. glucose isomerase, are also produced by streptomycetes [1].

Despite their industrial importance, knowledge about the primary metabolism [2, 3] and morphology of streptomycetes is rather limited. In recent decades much effort has focussed on the discovery of new antibiotics and on the improvement of the respective production methods. However, when problems of up-scaling of a cultivation on a rational basis, or design of modern model-based concepts of control engineering, are addressed, more information about the fundamental growth behaviour is needed with respect to morphology, primary and secondary metabolism.

In this contribution the main emphasis will be put on a morphological description of *Streptomyces* cultures. The models developed can be used for a better interpretation of basic phenomena observed in such processes. It will be shown in the following that morphological models:

(a) provide a tool to better understand the complex fundamental growth behaviour, which is especially important when correlations between morphology and productivity are proposed [4];

(b) provide a means to test biochemical hypotheses in a dynamic context;

(c) form a basis on which more complex models including a detailed description of metabolism [5, 6] can be build up in the future; and

(d) can be used to solve dedicated technical problems.

The modelling of the morphology of *Streptomyces* species will be done from different points of view (see Fig. 1). In the first part a microscopic approach is chosen to address more fundamental questions of morphology. The second part is devoted to a macroscopic description of single pellets which reveals the influences of mass transport on growth and production. In the last part the behaviour of a cultivation with pellets of different sizes is explained. Although the models have been developed for *Streptomyces* species in the first place, a major part of the models can also be used for fungal cultivations.

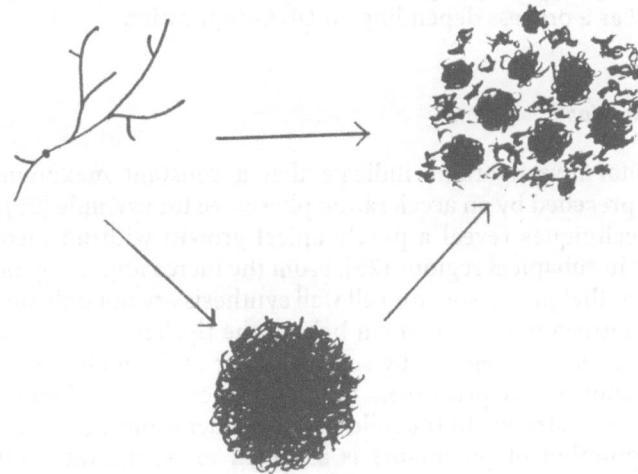

Fig. 1. Different aspects of morphology

2
Microscopic Morphology

Biological details about the processes leading to hyphal elongation and branching in streptomycetes are only known to a very limited extent. A summary of the available information shows that for the development of a dynamic model many questions have even not been stated. In many biologically orientated publications it is evident, however, that elongation and branching cannot be explained satisfactorily if a consideration of dynamic processes is omitted. Hence, only mathematical models can help.

Up to now only a few attempts to describe microscopic growth have been made even when models for fungi are included in this review. A large group of models considers lumped parameters only, such as the number of tips, the total length of all hyphae or producing and non-producing hyphae [7–14]. These models do not give a correct geometric picture of the mycelium and do not contribute to a better understanding of the complex morphological growth patterns. More sophisticated approaches have to be adopted. Apart from some models proposed for fungi [15–17] mainly one research group is active in this area [18–21].

A detailed description of morphology is interesting, for example if a correlation between morphology and productivity is found. Unfortunately, only very few data are available in this respect [4, 10, 22, 23]. However, a quantitative experimental determination of morphology and therefore a systematic approach to solve this problem became possible only a few years ago with modern image analysis systems, see Paul and Thomas in this volume, or [24].

2.1
A model of Microscopic Growth

As a first step, hyphal elongation will be considered. Branching will then be formulated as a process depending on DNA-duplication.

2.1.1
Hyphal Elongation

Experimental investigations indicate that a constant maximum elongation velocity is preceded by an acceleration phase; see for example [25]. Radioactive labelling techniques reveal a purely apical growth with no incorporation of precursors in subapical regions [25]. From the increasing elongation velocity it is concluded that precursors for cell wall synthesis are not only produced at the tip but in a growing compartment behind the tip. This compartment is separated from subapical regions by septa. Prosser and Tough [25] assume a diffusive transport of the precursors to the tip since a cytoskeleton is not known in streptomyces strains. In the following considerations production, transport and incorporation of precursors is assumed to be the rate limiting step in hyphal elongation. To derive a model, a schematic part of a mycelium is sketched in Fig. 2. It consists of a mother compartment (2, 3) of constant length

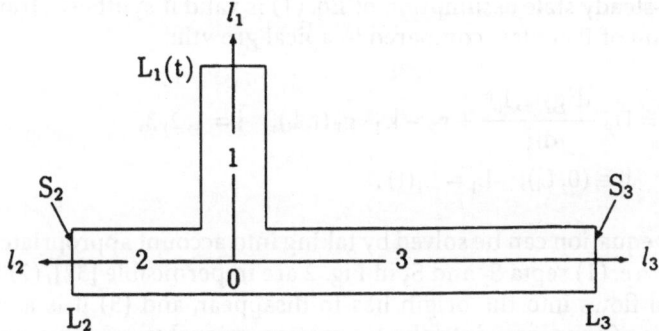

Fig. 2. Schematic representation of a section of a mycelium. S_i – septum at the border of the i-th section of a mother compartment

which is bounded by two inpermeable septa (S_2, S_3) and an apically growing hypha (1).

The model formulation follows the lines shown in [19, 20]. It is assumed that a limiting precursor P is produced in all parts (1, 2, 3) with a velocity $r_P(t, L_i)$. The precursor is degraded by a reaction which is assumed to be first order ($k_d \cdot g_P$) for simplicity. P is transported to the tip by diffusion only. For a constant cross-sectional area A of a hypha a balance yields

$$\frac{\partial g_P(t, l_i)}{\partial t} = D_P \frac{\partial^2 g_P(t, l_i)}{\partial l_i^2} + r_P(t, l_i) - k_d \cdot g_P(t, l_i), \quad i = 1, 2, 3$$

$$t > 0, \quad l_i \in (0, L_i), \quad L_1 = L_1(t) \tag{1}$$

where l_1, l_2 and l_3 are the co-ordinates of the three parts of the mycelium shown in Fig. 2.

In general, the synthesis velocity $r_P(t, l_i)$ is a function of the physiological state of a hypha at $l_i = \xi$, and a function of the local extracellular medium composition. For small mycelia external gradients can be neglected. Therefore, a homogeneous medium composition is assumed. Although a first approach to describe the effects of a time-variant medium composition can be found in [17, 26, 27] a constant composition is assumed here additionally for simplicity.

Little information is available on the spatial distribution of the physiological state in a hypha. It was found for example [28, 29] that DNA- and RNA-synthesis did not show a profound distribution along a hypha. Protein biosynthesis, in contrast, was not constant when mycelia were broken down in 10 μm sections [30]. However, in these investigations it was not known which sections were obtained from apical or subapical regions. Therefore, a constant synthesis velocity $r_P(t, l_i) = r_P$ is assumed in the following model. This assumption is supported by experiments with S. tendae in which branching was influenced by glucose [24] or ammonium [4, 31] but not apical elongation.

A quasi-steady state assumption of Eq. (1) is valid if synthesis, transport and consumption of P are fast compared to apical growth:

$$0 = D_P \frac{d^2 g_P(t, l_i)}{dl_i^2} + r_P - k_d \cdot g_P(t, l_i), \quad i = 1, 2, 3,$$

$$l_i \in (0, L_i), \quad L_1 = L_1(t).$$ (2)

This linear equation can be solved by taking into account appropriate boundary conditions, i.e. (1) septa S_2 and S_3 in Fig. 2 are impermeable [32], (2) the sum of all material flows into the origin has to disappear, and (3) it is assumed that the the flow of precursor into the tip is proportional to an increase in hyphal volume. The apical growth velocity is then given by [33]

$$\frac{dL_1(t)}{dt} = \alpha_m \cdot \frac{e^{2\omega L_1(t)} - \delta}{e^{2\omega L_1(t)} + \delta},$$ (3)

with

$$\omega = \frac{\mu_{Hy}}{\alpha_m}, \quad \delta = \frac{3 + e^{2 \cdot \omega \cdot L_2} + e^{2 \cdot \omega \cdot L_3} - e^{2 \cdot \omega \cdot L_2} \cdot e^{2 \cdot \omega \cdot L_3}}{-1 + e^{2 \cdot \omega \cdot L_2} + e^{2 \cdot \omega \cdot L_3} + 3 \cdot e^{2 \cdot \omega \cdot L_2} \cdot e^{2 \cdot \omega \cdot L_3}},$$ (4)

in which $\mu_{Hy} = \text{const} \cdot r_P$ is the exponential growth rate seen at the beginning of an experiment and $\alpha_m = \text{const} \cdot \sqrt{D_P/k_d} \cdot r_P$ is the constant apical growth velocity found for large $L_1(t)$. Obviously, the exponential growth rate μ_{Hy} is determined by the synthesis rate of the precursors. The maximal growth velocity α_m, however, is also a function of the ratio of degradation to diffusive transport. If degradation increases compared to transport, α_m is slowed down.

Very importantly, the only model parameters μ_{Hy} and α_m of Eqs. (3, 4) can be determined very easily in experiments. Most investigations in this respect have been performed on solid agars, although technical processes are run under submersed conditions. An exception can be found in the work of Reichl et al. [24, 34, 35]. Here growth is monitored and many parameters are determined by image analysis in a growth chamber which is continuously supplied by fresh medium. By analyzing a very large number of growing mycelia as a function of medium composition, statistically reliable data are obtained.

A first comparison between measured data [34] and a numerical integration of Eq. (3) is given in Fig. 3. It can be seen that the observed growth behaviour of a germ tube of S. tendae can be described very well. The growth rate, Eq. (3), derived here from a mechanistic point of view, differs significantly from the assumed growth rate for fungi; see Eq. (16) of the contribution of Krabben and Nielsen in this volume.

More examples, including fungi with impermeable septa, are given in [19, 20], where mycelial compartments with two growing tips are also described.

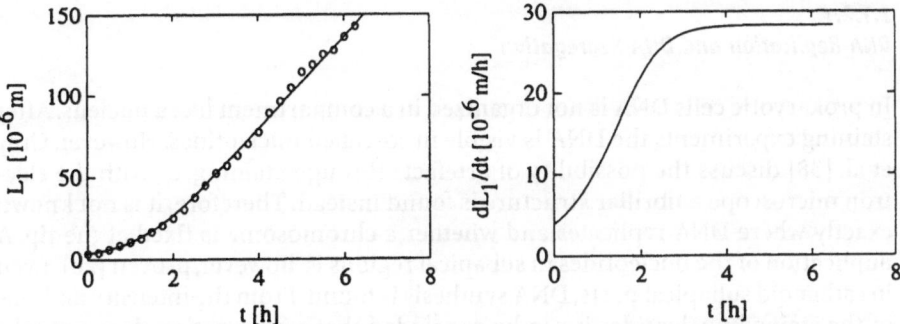

Fig. 3. Comparison of simulated (–) and measured (o) development of a germ tube of *S. tendae* with $\mu_{Hy} = 1.081/h$, $\alpha_m = 28.3\ \mu m/h$, $L_1(0) = L_2(0) = 0$, $L_3(0) = 3.6\ \mu m$ and calculated growth velocity $dL_1(t)/dt$. The measurements are taken from [34]

2.1.2
Branching

When a mycelium grows the total length of all hyphae increases exponentially. This can only be explained by an exponentially increasing number of tips due to branching since individual hypha reach a maximum growth velocity. Therefore, a branching rate has to be described. To yield a geometrical picture of a mycelium which is similar to microscopic observations the branching points have to be determined, too. They seem to be related to the sites where septa are built [29, 32]. However, the sequence of septation and branching has not be revealed up to now, because stains which make septa visible without disturbing growth are not known [29, 32].

Unicellular organisms are considered first, to develop a model assumption for the septation and branching processes of mycelial organisms. A unicellular organism grows by increasing all of the cell's constituents, partitioning of the chromosomes, and then dividing the whole cell into two daughter cells. The time course of division is mainly determined by the doubling of the DNA. It is obvious [36] that the division of a unicellular organism resembles the process of septation in hyphae. Therefore, it is assumed in the following that septation is coupled to DNA-duplication and that there exists a correlation between DNA-doubling and the time and site of branching. A similar idea has been proposed already for fungi [37]. A sequence of septation and branching makes sense, since it allows the subapical part further growth using its still existing metabolic activity for synthesis all of the cell's constituents. Otherwise, the metabolism would have to be cut down drastically to nongrowth conditions. The sequence, however, assumes that branching only takes place in subapical regions. This was found in most experimental observations [32]. However, in some cases the opposite was true.

From this discussion it is clear that a mathematical description of branching has to include DNA-replication, DNA-segregation, septation and initiation of a new tip.

2.1.2.1
DNA-Replication and DNA-Segregation

In prokaryotic cells DNA is not organized in a compartment like a nucleus. After staining experiments the DNA is visible in so-called nucleotides. However, Gray et al. [38] discuss the possibility of artefacts through staining, as with the electron microscope a fibrillar structure is found instead. Therefore, it is not known exactly where DNA replicates and whether a chromosome is fixed at the tip. A duplication of the nucleotides in subapical regions is, however, proven [28]. Even in rather old subapical parts, DNA synthesis is found. From the intensity and size of the stained nucleotides it can be concluded that a segregation does not take place when a certain nucleotide concentration is exceeded. In an apically growing hypha a gradient of nucleotide concentration towards the tip is found. Mechanisms of segregation are not known as the replicon model [39], which is under discussion anyway, cannot be used for apical growth.

It can be summarized that, due to the lack of exact knowledge in the field of DNA-replication and DNA-segregation, only hypotheses can be formulated. Here mathematical modelling provides a powerful tool to compare the outcome of simulations based on such hypotheses with available measurements. An example is given in Fig. 4 in which the nucleotide pattern of a growing apical hypha is calculated based on different hypotheses [33]. The hypothesis H1 b) resembles the asymmetrical picture obtained in staining experiments. It is obtained by a very simple mathematical model [19]. The replication rate decreases according to

$$\mu_D(t,l) = k_D \cdot (g_{Dm} - g_D(t,l)), \quad g_{Dm} = \text{const.,} \tag{5}$$

where g_{Dm} represents a maximum DNA-concentration. With this expression the effect of a regulation mechanism is taken into account. DNA-replication is stopped when segregation fails, i.e. when the DNA-concentration exceeds a certain limit, g_{Dm}. A segregation to a l_0-neighbourhood takes place whenever no other nucleotides are present in this neighbourhood. The considered unit length l_0 can be defined for example as the length or section which is associated to one nucleotide.

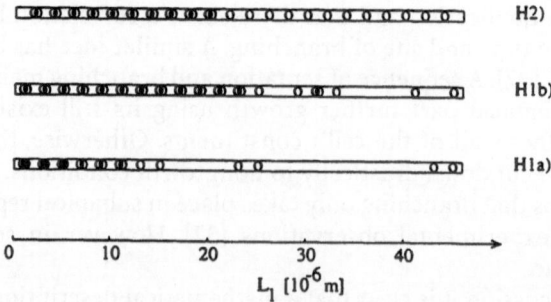

Fig. 4. Nucleotide density in a growing hypha based on different hypotheses. Number of nucleotides: ○ – one; ∞ – two, not segregated; ● – four, not segregated

2.1.2.2
Septation

Staining experiments show that the first septum behind a tip is usually nearer to the second septum than to the tip. A ratio of 1/3 to 2/3 is often found. In subapical hyphae septa are placed in a middle position between two existing septa. As the staining experiments do not reveal when a septum was formed, different possible hypotheses could be discussed. In the following a simple model assumption will be used instead. When the number of nucleotides between two septa or between a first septum and the respective tip is doubled (e.g. 4 → 8) a new septum will be formed. The site of the septation is chosen in a way that on both sides the same number of nucleotides is found. With this model assumption and the saturation kinetics of Eq. (5), the above-mentioned asymmetrical and symmetrical septum distributions in apical and subapical regions are obtained, respectively. Hence, the timing and location of septation is known.

A similar approach was chosen from Prosser and Trinci [16] in a completely discretized model for the growth of fungi. For the numerical simulation of the model proposed here, Yang [19] introduces a hyphal discretization with the already mentioned unit length l_0. In the simulation, nucleotides are modelled as discrete elements. For every unit length the processes of replication and segregation have to be considered.

2.1.2.3
Branching

All processes considered up to now are described in a deterministic way. From experiments it is known, however, that the appearance of mycelia shows stochastic influences. To include these effects in a first approach, branching is modelled as a normally distributed random process. The actual sites of branch initiation and the time delay between septation and branch initiation are represented by random variables. For the first stochastic process experimental data for the mean and the variance are available with the method shown in [32]. For the latter moments have to be chosen in the simulation.

2.1.2.4
Growth and Branching Directions

To obtain a geometrical picture of hyphal growth it has to be considered that hyphae do not grow and branch in a fixed direction. Statistical investigations [21] show a normal distribution of both directions with different moments, respectively. Hence, for a simulation of the development of the unit length sections (l_0) the growth and branching directions are modelled as random processes.

2.2
Simulation and Comparison with Experimental Results

Figure 5 shows a simulation of the growth of a mycelium starting from a spore. A comparison of this 2-dimensional projection of a 3-dimensional simulation

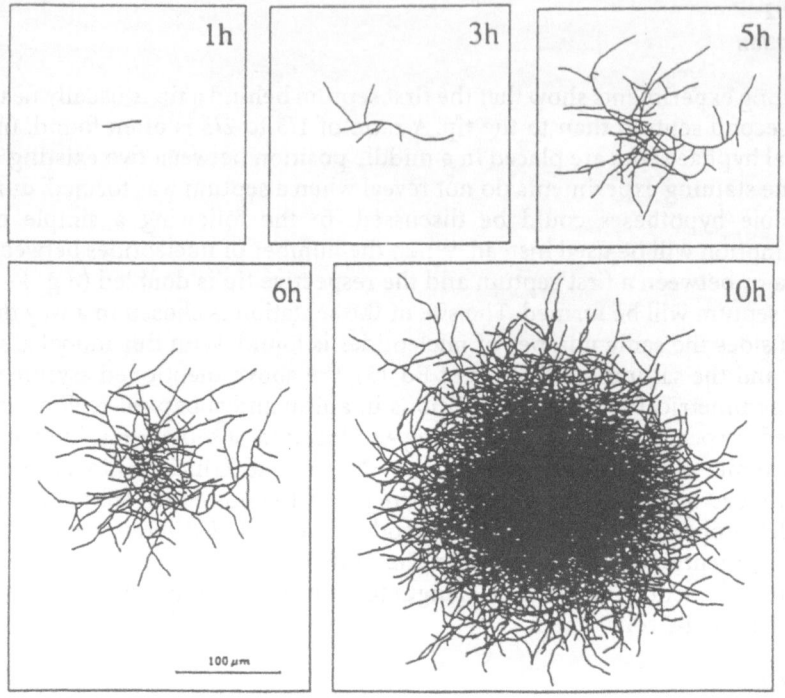

Fig. 5. A 2-dimensional projection of a 3-dimensional simulation of the growth of *S. tendae* for different time instants; from Yang et al. [20]

with real microscopic data shows a very good visual representation of the morphology. This is even enhanced when the real microscopic picture is digitized by means of an image analysis system and the thicknesses of all hyphae are then reduced to one pixel (skeletonized; see contribution of Paul and Thomas in this volume).

The quality of the simulation also becomes obvious in Fig. 6 where the measured total length of all hyphae, L_t, the number of tips, N, and the calculated 'hyphal growth unit', L_t/N, are compared to simulation results. While the evaluation with the image analysis system has to stop after about 6.25 h because of multiple crossings of hyphae and hyphae growing out of focus, the simulation can be continued until a pellet is formed. The simulation should then be stopped, too, since the assumption of a homogeneous medium supply is violated. This problem will be addressed in the next section.

Yang [26] showed with a rather ad-hoc model extension that in principle it is possible to include the influences of mechanical stress due to a stirrer as well. The breakage probability of a single hypha is modelled as a function of the history of this hypha. Microkinetic data in this respect are very rare, of course. The simulation shown in Fig. 7 displays the well known microscopic picture from samples out of a fermenter. Some more examples for the application of this model can be found in [19, 20, 26].

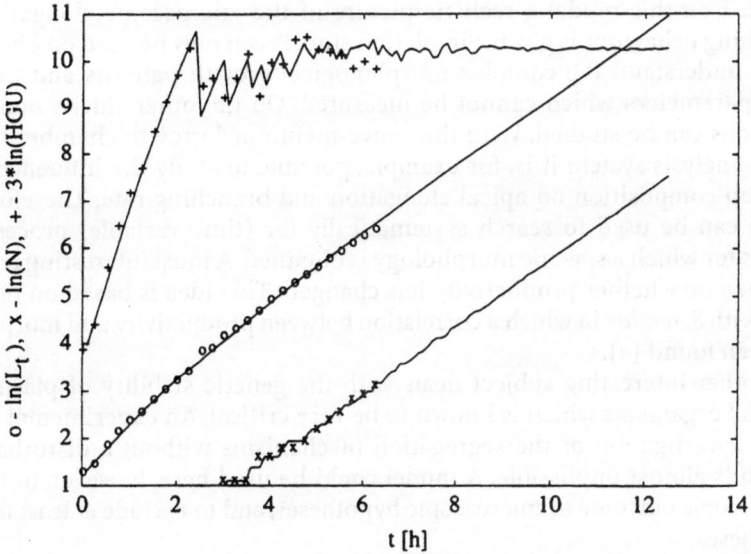

Fig. 6. Simulated and measured total hyphal length L_t (o), number of tips N (×) and hyphal growth unit HGU (+) of a mycelium of *S. tendae* from Yang et al. [20]. *Solid lines* indicate simulations

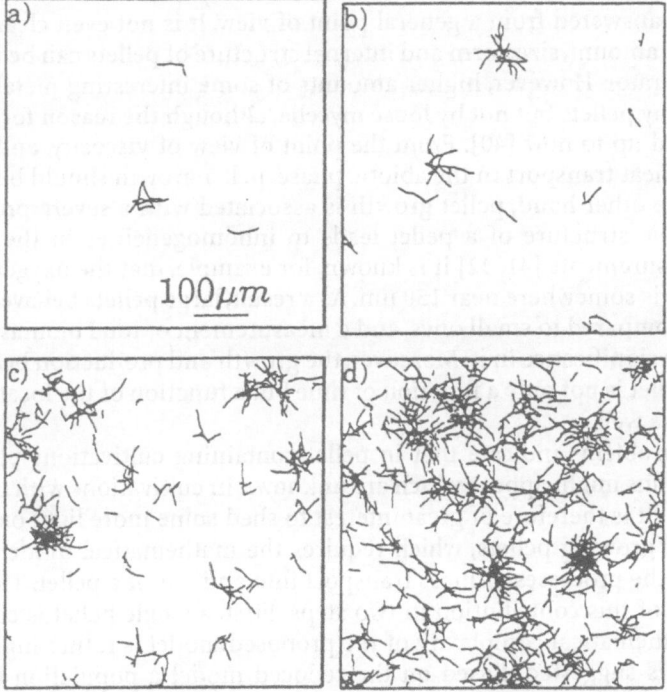

Fig. 7a–d. Growth in a fermenter: **a** t = 10 h; **b** t = 15 h; **c** t = 20 h; **d** t = 25 h; from Yang [26]

Based on this model a realistic picture of the microscopic elongation and branching behaviour is given. Simulation studies can now be used on one side to better understand the complex morphological growth patterns and to determine parameters which cannot be measured. On the other side, a number of questions can be studied. With the above-mentioned growth chamber and the image analysis system it is, for example, possible to study the influence of the medium composition on apical elongation and branching rate. Therefore, this model can be used to search systematically for (time variable) process conditions for which a specific morphology is obtained. A most interesting question will then be whether productivity has changed. This idea is based on observations with S. tendae in which a correlation between productivity and morphology has been found [4].

Another interesting subject deals with the genetic stability of plasmids in mycelial organisms which is known to be very critical. An experimental microscopic investigation of the segregation mechanisms without a disturbance of growth is almost impossible. A model could be used here, however, to test the macroscopic outcome of microscopic hypotheses, and to exclude at least unlikely hypotheses.

3
Morphology of a Single Pellet

Whether a cultivation should be run in such a way that there are many pellets cannot be answered from a general point of view. It is not even clear by which factors the amount, size, form and internal structure of pellets can be influenced by the operator. However, higher amounts of some interesting metabolites are produced by pellets but not by loose mycelia, although the reason for this is not understood up to now [40]. From the point of view of viscosity, and therefore mass and heat transport in the abiotic phase, pellet growth should be favoured, too. On the other hand, pellet growth is associated with a severe problem: the macroscopic structure of a pellet leads to inhomogeneities in the biophase. From measurements [41, 42] it is known, for example, that the oxygen penetration depth is somewhere near 150 μm. As a result, large pellets behave in a different way compared to small ones, and a measurement of total biomass does not bear much significance in it. Moreover, the growth and production behaviour of a single pellet is not only a function of time but a function of the location inside the pellet as well.

These problems indicate that in pellet containing cultivations phenomena and questions might appear which are unknown in cultivations with unicellular organisms. It is therefore of great interest to shed some more light onto the behaviour of growing pellets, which requires the mathematical model to be extended by the processes of mass transport into and out of a pellet. This is done in the rest of this contribution in two steps. First, a single pellet is considered. As the mathematical complexity of the proposed model is rather high a model reduction is suggested. Based on the reduced model a population balance is formulated in the second step, which describes the behaviour of a cultivation consisting of pellets of different sizes. The emphasis will again be put on morphological details. Although the influence of the complex metabolism on growth behaviour is only taken into account by a simple, unstructured approach, basic problems of such cultivations can be discussed.

3.1
A Model of Pellet Growth

As in the case of microscopic growth, only very few approaches are known to model pellet growth. Apart from some very simple formulae to describe growth as a function of time [43 – 45], most of the work was done to predict the oxygen distribution in pellets with a constant radius and constant density [46 – 50]. Similar approaches are found in models for biofilms [51, 52]. However, the simulations shown in the last section as well as histograms [41] reveal that the biomass density in a pellet is not at all constant. A space dependent biomass density is considered only by some authors [17, 19, 27, 41, 42, 53, 54]. Wittler et al. [41] do not include growth in their model, and the profiles of biomass density are somewhat arbitrary and do not reflect the above-mentioned results. As the microscopic model introduced above is already very complex, another approach is chosen here according to Buschulte et al. [42, 53, 54], describing growth and production from a more macroscopic point of view. The model starts from the observation that pellets are usually very dense. Therefore, a description with continuous variables, such as the local concentrations of hyphae and tips, seems meaningful. As in the model for microscopic growth, the macroscopic approach also includes apical elongation and subapical branching as the main growth related processes.

3.1.1
Hyphal Elongation

As new branches are usually formed from rather long mother compartments $(L_2 + L_3)$ an almost constant apical growth velocity follows from Eq. (3). Moreover, Yang [19] showed that in existing hyphae dL_1/dt does not change significantly, irrespective of the formation of new septa in subapical regions. The reason is the still rather long apical part of the hypha in which the supply of precursors at the tip is limited by the diffusive transport to the tip but not by the volume for production behind the tip. Therefore, an apical elongation velocity is considered which only depends on the extracellular composition of the medium, $\underline{c}_S = (c_{S1}, \ldots, c_{SM})$, but not on the microscopic morphology. From a macroscopic point of view the local elongation velocity r_{Hy} of all hyphae is therefore proportional to the local concentration of tips, c_{Tip}:

$$r_{Hy}(t, \underline{z}) = \bar{\alpha} \cdot \psi_{Tip}(\underline{c}_S(t, \underline{z})) \cdot c_{Tip}(t, \underline{z}), \quad \underline{z} \in R^3. \tag{6}$$

The apical growth rate is a function of a mean value of dL_1/dt, i.e. $\bar{\alpha}$, which is observed under ideal nutrient supply, and a function of the local growth activity $\psi_{Tip}(\underline{c}_S(t, \underline{z}))$. $\bar{\alpha}$ is usually smaller than α_m from Eq. (3) due to the assumption of constant elongation velocity. For ψ_{tip} Michaelis-Menten type kinetics can be used.

Colonies on solid agars and thin sections of pellets show a symmetric appearance. Therefore, a reduction to $\underline{z} \in R^1$ is possible. The radius of a pellet is then given by z_R.

To derive a balance equation for the local hyphal concentration a differential spherical shell is considered. Hyphae located in this shell do not leave it as they do not possess any means for motion due to the lack of subapical growth. However, they can be degraded, $r_{HyD}(t, z)$, or broken apart, $r_{HyB}(t, z)$:

$$\frac{\partial c_{Hy}(t, z)}{\partial t} = \bar{\alpha} \cdot \psi_{Tip}(\underline{c}_S(t, \underline{z})) \cdot c_{Tip}(t, \underline{z}) - r_{HyD}(t, z) - r_{HyB}(t, z). \tag{7}$$

The breakage rate is usually not only a function of the local conditions but of the history of the respective hypha as well [55].

3.1.2
Tip Formation

In microscopic investigations the hyphal growth unit, HGU, is a very well known value to characterise a mycelium. It represents the ratio of the total hyphal length to the number of tips. The HGU can be interpreted as the mean length available for every growth centre. In experiments and in the simulations shown above the HGU very soon becomes constant. Hence, if the total hyphal length grows, the number of tips has to grow accordingly, for example with

$$r_{Tip}(t, z) = \bar{\beta} \cdot \psi_{Hy}(\underline{c}_S(t, z)) \cdot c_{Hy}(t, z), \tag{8}$$

where $\bar{\beta}$, ψ_{Hy} and c_{Hy} are the mean branching rate, the activity of branching and the local concentration of hyphae, respectively.

In contrast to hyphae, tips do 'move' or are 'transported' due to apical growth. For a nondirectional growth this movement is proportional to the local gradient of hyphal elongation velocity, i.e. $\partial r_{Hy}/\partial z$, showing a diffusion-like behaviour. Again, degradation and breakage are considered, too. In radial coordinates the balance reads

$$\frac{\partial c_{Tip}(t, z)}{\partial t} = \frac{1}{z^2} \cdot \frac{\partial}{\partial z} \left(k_{Tip} \cdot \bar{\alpha} \cdot z^2 \cdot \frac{\partial(\psi_{Tip} \cdot c_{Tip})}{\partial z} \right)$$

$$+ \bar{\beta} \cdot \psi_{Hy}(\underline{c}_S(t, z)) \cdot c_{Hy}(t, z) - r_{TipD}(t, z) - r_{TipB}(t, z). \tag{9}$$

For all equations appropriate initial and boundary equations have to be formulated, e.g.

$$\left. \frac{\partial(\psi_{Tip} \cdot c_{Tip})}{\partial z} \right|_{z=0} = 0 \tag{10}$$

$$\left(k_{Tip} \cdot \bar{\alpha} \cdot \frac{\partial \psi_{Tip} \cdot c_{Tip})}{\partial z}\right)\Bigg|_{z = z_R} = \frac{dz_R(t)}{dt} \cdot c_{Tip}(t, z_R) \tag{11}$$

$$c_{Tip}(t_0, z) = c_{Tip0}(z). \tag{12}$$

In the boundary equation (11) it is assumed that tips moving over the pellet radius $z = z_R$ imply an increase in pellet radius. The importance of the formulation of a moving boundary problem should not be underestimated. There is no physical argument why only growth inside a pellet but not on the boundary should be considered. Hence, an assumption of nongrowing pellets, as mentioned above, cannot lead to meaningful results. This was confirmed during the development of the model when other definitions of the moving boundary led to different profiles of the concentrations of hyphae and tips. Therefore, a correct boundary formulation is crucial.

3.1.3
Pellet Radius

The change in pellet radius is defined by the motion of that tip which has the largest distance to the centre:

$$\frac{dz_R(t)}{dt} = \gamma \cdot \bar{\alpha} \cdot \psi_{Tip}(c_S(t, z_R)). \tag{13}$$

As hyphae do not grow straight ahead the factor γ has to be introduced, which is smaller than 1. It can be determined from microscopic experiments or from simulations with the microscopic model.

3.1.4
Substrates

The balance equations for the substrates inside a pellet can be easily derived:

$$\frac{\partial c_{Si}(t, z)}{\partial t} = \frac{1}{z^2} \cdot \frac{\partial}{\partial z}\left(D_{Si,eff} \cdot z^2 \cdot \frac{\partial c_{Si}(t, z)}{\partial z}\right) - r_{Si}(t, z), \quad i = 1, ..., M \tag{14}$$

$$\frac{\partial c_{Si}(t, z)}{\partial z}\Bigg|_{z = 0} = 0, \quad i = 1, ..., M \tag{15}$$

$$D_{Si,eff} \cdot \frac{\partial c_{Si}(t, z)}{\partial}\Bigg|_{z = z_R} = k_{Si} \cdot (c_{SiL}(t) - c_{Si}(t, z_R)), \quad i = 1, ..., M \tag{16}$$

$$c_{Si}(t_0, z) = c_{Si0}(z), \quad i = 1, ..., M. \tag{17}$$

$D_{Si,eff}$, r_{Si} and c_{SiL} represent the effective diffusion coefficient of substrate i based on the void volume in a pellet, all substrate consuming reactions, and the concentrations of the substrates in the bulk, respectively. Equations for the substrate concentrations in the bulk can be derived easily. For a detailed discussion of all assumptions included in this model see [53].

3.2
Simulation and Application Studies

Although the model is outlined for all substrates, only oxygen is considered here due to its low solubility. In most cases the other substrates give rise to an approximately uniform behaviour inside a pellet. This does not apply to continuous or to some fed-batch processes in which a depletion of other substrates inside a pellet can occur. However, these cases are not considered here for simplicity.

Figure 8 shows the simulated development [33] of the concentrations of hyphae, tips and oxygen of a growing pellet as a function of spatial location inside the pellet without degradation and breakage. The radii for different time instants can be best estimated from the span of the oxygen concentration profiles. Initial profiles c_{Hy0} and c_{Tip0} are taken from a simulation with the microscopic model. The oxygen concentration is assumed to be uniform over the pellet radius for $t = 0$. The simulation results show that after an initial increase in hyphal length and therefore biomass in the centre of the pellet ($z = 0$), growth is restricted to the outer part after some 10 h.

In [42, 53] it is shown that this model describes experimentally determined oxygen profiles very well.

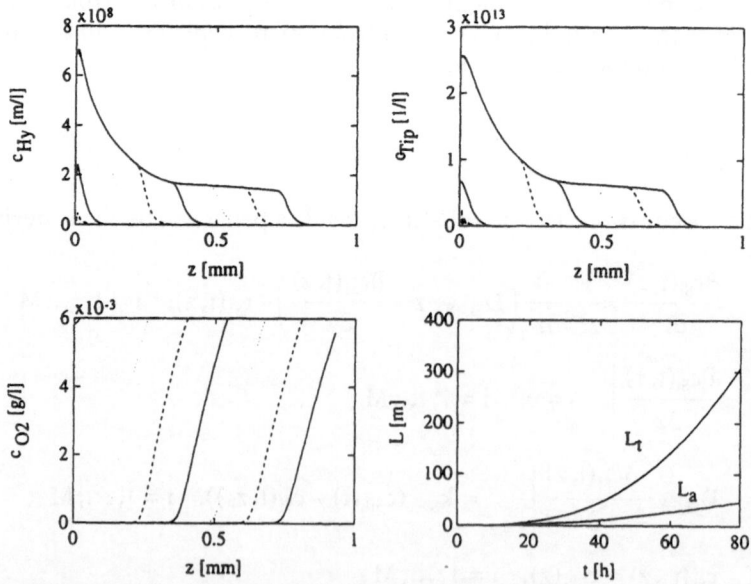

Fig. 8. Simulation of the concentrations of hyphae, tips and oxygen as a function of z for $t = 0$, 6, 18, 30, 42, 54, 66, 78 h and simulated development of the total hyphal length of a pellet, L_t, and of that part of a pellet which is supplied by oxygen, L_a (active length)

In the center of the pellet a growing zone without oxygen can be observed in which neither growth nor production takes place. As a result hyphae (and tips) are degraded, leaving a pellet with an empty core. This phenomenon can be included in the model by appropriate expressions for r_{HyD} and r_{TipD} (not done here). Instead, a simulation with the microscopic model [26] is given in Fig. 9, including an approximate approach for mass transport as well as degradation of hyphae some time after depletion of oxygen. The simulation results in Fig. 9 resemble quite well pictures known from microscopy.

Returning to Fig. 8, two additional variables are shown as functions of time: the total hyphal length (L_t) and that part of the total hyphal length which is supplied with oxygen, called active length here (L_a). These variables will be important in a model reduction explained below. They can be obtained by integration over the whole pellet and the active part of the pellet, respectively. It should be noted that the concentrations of hyphae and tips are formulated as length and number of tips per volume, respectively. The results in Fig. 8 show that an exponential growth at the beginning is followed by an increase which is approximately proportional to t^3. This behaviour is known as the 'cube root law' [45].

3.2.1
Production

To show an application of this model, the production of the antibiotic Nikkomycin by *Streptomyces tendae* is considered in a fluidized-bed reactor. In

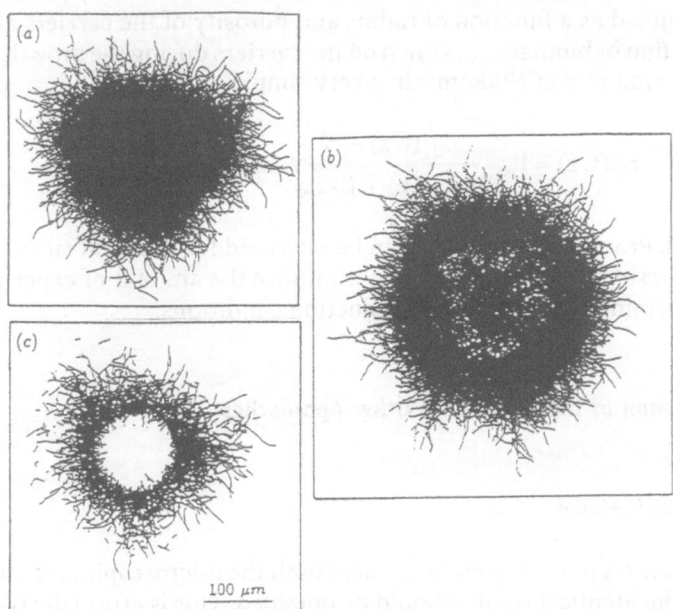

Fig. 9 a–c. Simulation with the microscopic model including oxygen transport into the pellet and degradation of hyphae, from Yang [26]: **a** 7 h; **b** 20 h; **c** simulated 2 μm thin section for t = 17 h

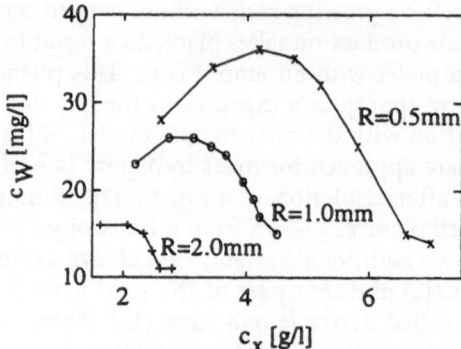

Fig. 10. Stationary concentration of Nikkomycin, c_w, as a function of biomass, c_x, determined during the growth phase, and of radius R of the carriers. The porosity amounts to 0.5

this application the organisms are grown on carriers, which are retained in the reactor. As a result, it is very easy to apply different optimal media for the growth and the production phase by a complete medium exchange [56]. However, the optimal values for the amount of biomass grown on the carriers, as well as the radius and the porosity of the carriers used, have to be determined in time consuming experiments. To get a first idea, the above stated model is extended to describe growth into and on a carrier. Figure 10 shows results of simulations in which the concentration of the antibiotic in the stationary production phase is determined as a function of radius and porosity of the carriers and of the concentration of biomass, c_x, grown on the carriers during the growth phase. For the production rate of Nikkomycin a very simple expression

$$r_W(t, z) = k_m \cdot \frac{c_{O_2}(t, z) - c_{O_2}^*}{c_{O_2}(t, z) + K_{O_2 W}} \cdot c_{Hy}(t, z) \tag{18}$$

is used. Pronounced maxima can be observed for different sizes of the carriers. These results can now be used to minimize the amount of experiments needed to determine finally optimal production conditions.

3.3
Comparison of the Model with Other Approaches

3.3.1
Microscopic Model

For those regions of operation where both the microscopic and the pellet model are valid, identical results should be obtained. This is especially true when there is no limitation in oxygen supply; see Fig. 11. Here, the initial profiles c_{Hy0} and c_{Tip0} of the pellet model for $t = t_0 = 16$ h are taken from a simulation of the microscopic model. By adaptation of the diffusion parameter k_{Tip}, see Eq. (9), it

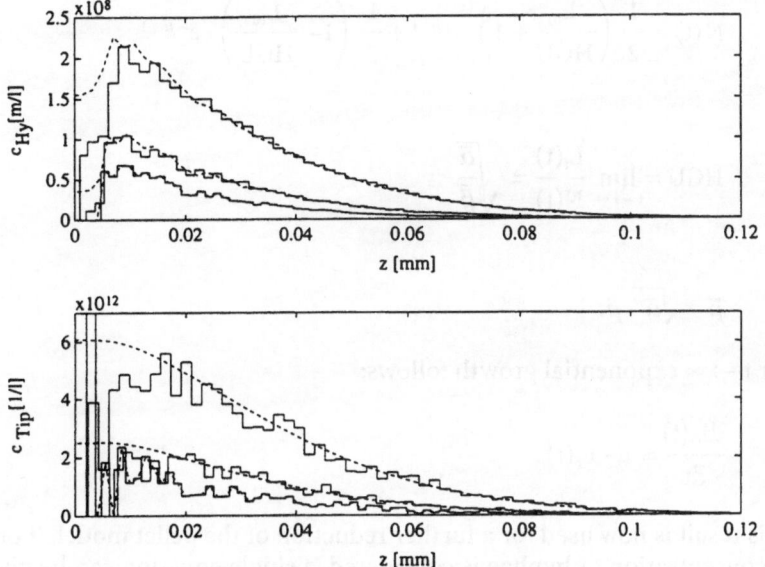

Fig. 11. Comparison of simulations obtained with the microscopic model (—) and pellet model (---) for t = 16 h, 17.6 and 20 h; from Buschulte [53]

is possible to predict the same development using the very different models (Fig. 11). Hence, it is shown that the consideration of local concentrations of hyphae and tips which is adopted in the pellet model leads to meaningful results.

3.3.2
Pellet Model Without Tips

Equations (6) and (8) are based on a model of Schuhmann and Bergter who describe growth of a small mycelium [57]. In this model the total hyphal length L_t increases proportionally to the number of tips N

$$\frac{dL_t(t)}{dt} = \bar{\alpha} \cdot N(t) \tag{19}$$

and the number of tips increases proportionally to the total hyphal length

$$\frac{dN(t)}{dt} = \bar{\beta} \cdot L_t(t). \tag{20}$$

With the initial conditions $L_t(0) = L_{t0}$ and $N(0) = 1$ integration yields

$$L_t(t) = \frac{1}{2} \left(\frac{\bar{\alpha}}{\bar{\mu}} + L_{t0} \right) \cdot e^{\bar{\mu} \cdot t} + \frac{1}{2} \left(L_{t0} - \frac{\bar{\alpha}}{\bar{\mu}} \right) \cdot e^{-\bar{\mu} \cdot t} \tag{21}$$

$$N(t) = \frac{1}{2}\left(\frac{L_{t0}}{HGU} + 1\right) \cdot e^{\bar{\mu} \cdot t} + \frac{1}{2}\left(1 - \frac{L_{t0}}{HGU}\right) \cdot e^{-\bar{\mu} \cdot t}, \tag{22}$$

with

$$HGU = \lim_{t \to \infty} \frac{L_t(t)}{N(t)} = \sqrt{\frac{\bar{\alpha}}{\bar{\beta}}} \tag{23}$$

and

$$\bar{\mu} = \sqrt{\bar{\alpha} \cdot \bar{\beta}}. \tag{24}$$

For $t \to \infty$ exponential growth follows:

$$\frac{dL_t(t)}{dt} \approx \bar{\mu} \cdot L_t(t). \tag{25}$$

This result is now used for a further reduction of the pellet model. If only the local concentration of hyphae is considered, a single equation can be given for the biotic phase in which a formal diffusion of hyphae is introduced:

$$\frac{\partial c_{Hy}(t, z)}{\partial t} = \frac{1}{z^2} \cdot \frac{\partial}{\partial z}\left(k_{Hy} \cdot \bar{\mu} \cdot z^2 \cdot \frac{\partial(\psi_{Hy} \cdot c_{Hy})}{\partial z}\right)$$
$$+ \bar{\mu} \cdot \psi_{Hy} \cdot c_{Hy} - r_{HyD}(t, z) - r_{HyB}(t, z). \tag{26}$$

The development of the total length L_t of hyphae calculated on this basis gives the same results as with the pellet model considered so far. A comparison of the results in Fig. 8 and Fig. 12 reveals, however minor differences in the concentration profiles as a function of radius.

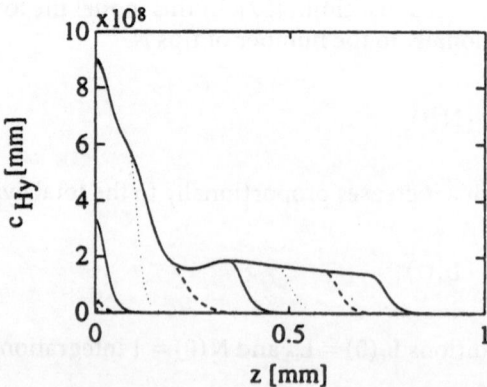

Fig. 12. Simulation of the local hyphal concentration as a function of the radius for $t = 0, 6, 18, 30, 42, 54, 66, 78$ h with model (26)

3.3.3
Lumped Pellet Model

A comparison of different simulations based on the original pellet model indicates that the development of total hyphal length, $L_t(t)$, and therefore biomass, can be explained by a much simpler lumped model. Such a model reduction is necessary when in the next step the behaviour of a population of pellets of different sizes has to be described mathematically. This also requires information about the active length $L_a(t)$, since this quantity determines maintenance demands. However, active length L_a is not a state variable in the sense of system theory. When a large pellet ($L_t > L_a$) is torn up into smaller pellets by the stirrer, all small pellets i might be perfectly supplied with oxygen ($L_{ti} = L_{ai}$). The amount of active biomass is therefore not conserved ($\Sigma L_{ai} \neq L_a$) as would be expected for a state variable. Hence, no differential equation should be used to describe L_a.

A simple black-box model for $L_t(t)$ is given here which is based on numerous simulation studies with the complete pellet model in which combinations of different values of $\bar{\alpha}, \bar{\beta}$ and c_{O_2L} are used to build up a data base for model identification.

As long as total hyphal length is less then a critical value, $L_t < L_{cr}$, the model reads

$$\frac{dL_t(t)}{dt} = a_1 \cdot \mu\,(c_{O_2L}(t)) \cdot L_t(t), \quad L_a(t) = \frac{1}{a_1 \cdot \bar{\mu}} \cdot \frac{dL_t(t)}{dt}, \tag{27}$$

with

$$\mu(c_{O_2L}(t)) = \bar{\mu} \cdot \frac{c_{O_2L}(t)}{c_{O_2L}(t) + K_{O_2L}}, \quad \bar{\mu} = \sqrt{\bar{\alpha} \cdot \bar{\beta}}. \tag{28}$$

For growing pellet sizes ($L_t(t) > L_{cr}$) growth rate decreases according to Eq. (29) from exponential behaviour to smaller values:

$$\frac{dL_t(t)}{dt} = \frac{1 + a_2 \cdot \bar{\beta} \cdot L_{cr}^{\lambda}}{1 + a_2 \cdot \bar{\beta} \cdot L_t^{\lambda}(t)} \cdot a_1 \cdot \mu(c_{O_2L}(t)) \cdot L_t(t) \tag{29}$$

$$L_a(t) = \frac{a_6 \cdot L_t(t)}{1 + a_4 \cdot \bar{\mu}^{-a_5} \cdot L_t x(t)} \cdot \frac{dL_t(t)}{dt}, \quad a_6 = \frac{1 + a_4 \cdot \bar{\mu}^{-a_5} \cdot L_{cr}}{a_1 \cdot \bar{\mu} \cdot L_{cr}}. \tag{30}$$

One set of constant parameters a_i, l_{cr}, λ and K_{O_2L} is identified from the above-mentioned simulations with the complete pellet model (see Table 1).

Figure 13 shows a comparison of the complete and the reduced model. In both cases the oxygen concentration of the bulk, c_{O_2L}, is calculated according to an appropriate balance equation. There is excellent agreement between the complete and the reduced model, indicating that the latter can be used from a macroscopic point of view to describe pellet growth in a population balance.

Table 1. Parameters of the lumped model

L_{cr}	a_1	a_2	a_3	a_4	a_5	λ	K_{O_2L}
0.132 m	1.37	233 $\mu m^{0.622}$ h	0.7 $\dfrac{1}{\mu m^{0.622}}$	2.53 $\dfrac{h^{0.51}}{\mu m}$	0.51	0.378	0.0797 $\dfrac{g}{l}$

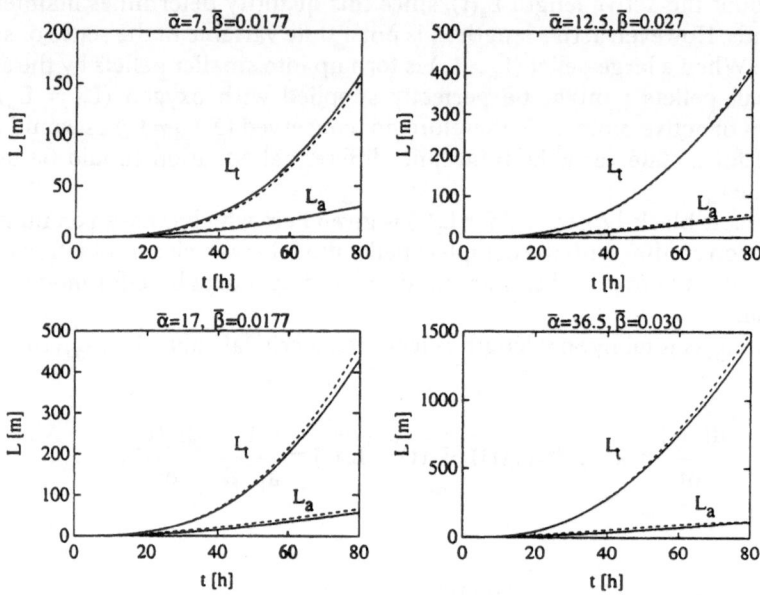

Fig. 13. Comparison of total hyphal length L_t and active hyphal length L_a for different values of $\bar{\alpha}$ ([$\mu m/h$]) and $\bar{\beta}$ ([$1/(\mu m \cdot h)$]). Complete model (——), reduced model (– – –)

For c_{O_2L} = const. and $t \to \infty$ Eq. (29) yields

$$\frac{dL_t(t)}{dt} \approx d_1 \cdot L_t^{1-\lambda}(t) \Rightarrow L_t(t) \approx d_1' \cdot t^{1/\lambda} + d_2. \tag{31}$$

In the identification the exponent $1/\lambda$ according to Eq. (31) is found to be 2.65. A similar result has already been named as the cube root law [45].

4
Morphology in a Submerged Culture

Simulations with the pellet model show that even single pellets are not uniformly supplied with substrates. In cultivations this inhomogeneity is superposed by a pronounced pellet size distribution. Apart from its direct influence on growth and production behaviour in batch, fed-batch or continuous processes this pellet size distribution is of major interest in terms of rheology, heat and mass transfer, e.g. in the context of correlations between these variables. To tackle such kinds of questions a fundamental understanding and a quantitative description of the processes going on in a pellet containing cultivation are necessary. Both experimental and theoretical tools are available today. On one hand image analysis provides an ideal experimental set-up to gain primary data, cf. [58] or Thomas in this volume. Fully automated versions including sampling, sample preparation, image analysis and wash cycles are available [4]. On the other hand, segregated models, especially population balances, are known to describe the influences of processes such as particle birth, growth, breakage and agglomeration on pellet size distribution. In the following section some approaches are summarized.

4.1
Population Balances

Pellet size distribution can be described by a number density function $f(\underline{y}, t)$, in which $\underline{y}$ are characteristic properties, such as total hyphal length, mass of a pellet, age, etc. Only a few attempts have been made so far to describe the development of the pellet size distribution in mycelial cultivations. Nielsen [12] introduces pellet mass and number of tips as characteristic properties; see also Krabben and Nielsen in this volume. After some assumptions they reduce the problem to two ordinary differential equations describing average values. The special influences of mass transport processes inside a pellet on its behaviour are not taken into account as only small mycelia are considered. Mass transport is also not included in Edelstein [59] and Tough [60].

The lumped pellet model discussed above, however, includes these most important effects. Therefore, total hyphal length L_t can be used as a characteristic property, when Eqs. (27–30) are used to account for the influence of oxygen.

As L_t increases exponentially the following population balance is formulated with a normalized characteristic property, i.e. $s = \ln(L_t/L_0)$. L_0 is the length of a spore. To derive the population balance a term $(V(t) \cdot f(s, t) \cdot ds)$ is considered for small values of ds. It represents the number of pellets with a total hyphal length between s and $s + ds$ at time t in volume $V(t)$. This number changes because of pellets growing into and out of the segments $[s, s + ds)$, pellets being washed out of the fermenter and pellets appearing or disappearing due to processes such as

- $\sigma_1(s, t, \ldots)$ birth,
- $\sigma_2(s, t, \ldots)$ death,
- $\sigma_3(s, t, \ldots)$ breakage or
- $\sigma_4(s, t, \ldots)$ agglomeration.

A balance over the number of pellets in a segment $[s, s + ds)$ with $ds \to 0$ yields

$$\frac{\partial}{\partial t}\left[V(t) \cdot f(s, t)\right] = -V(t) \cdot \frac{\partial}{\partial s}\left[u(s, t, \ldots) \cdot f(s, t)\right] + Q_i(t) \cdot f_i(s, t)$$

$$- Q_o(t) \cdot f(s, t) + V(t) \cdot \sum_{j=1}^{4} \sigma_j(s, t, \ldots),$$

$$t \geq t_0, \qquad s \geq s_0, \qquad (32)$$

in which u and $Q_{i,o}$ are the growth velocity of a single pellet and the volumetric flows into and out of the system, respectively. As some σ_j are expressed by integrals the resulting equation is of a non-linear integro-partial differential type. For $Q_i = Q_o = Q, q = Q/V, V = \text{const.}, f_i = 0$, neglecting death and agglomeration processes and introducing a rate of breakage $r_3 = k_z \cdot \exp(\eta \cdot s) \cdot f(s, t)$, the population balance reads [33]

$$\frac{\partial f(s, t)}{\partial t} = -a_1 \cdot \mu(c_{O_2}(t)) \cdot \frac{\partial}{\partial s}\left[\hat{u} \cdot f(s, t)\right] - q(t) \cdot f(s, t) - k_z \cdot e^{\eta \cdot s} \cdot f(s, t)$$

$$+ \int_{\ln(e^s + 1)}^{\infty} 2 \cdot p(s', s, t) \cdot k_z \cdot e^{\eta \cdot s} \cdot f(s', t) \, ds', \qquad (33)$$

with

$$\hat{u} = \begin{cases} 1 & \text{for} \quad s \leq s_{cr} \\ \dfrac{1 + d_3 \cdot e^{\lambda \cdot s_{cr}}}{1 + d_3 \cdot e^{\lambda \cdot s(t)}} & \text{for} \quad s > s_{cr} \end{cases}. \qquad (34)$$

The birth process, σ_1, indicating germination, is included in the boundary condition. The germination rate $r_G(c_{Sp}, \varsigma_S, t)$ is assumed to be a function of the spore concentration $c_{Sp}(t)$ and the substrate concentration $\varsigma_S(t)$:

$$\sigma_1(s, t, c_{Sp}, \varsigma_S) = r_G(c_{Sp}, \varsigma_S, t) \cdot f_G(s, t). \qquad (35)$$

The number density function $f_G(s, t)$ of particles produced by birth is usually known. In the example shown here all particles produced by birth have the same size, i.e. $s^* = s_0 = \ln(1) = 0$. Hence, the smallest particle considered is a spore which just has started germination. The number density function $f_G(s, t)$ of new particles then reduces to a Dirac pulse:

$$f_G(s, t) = \delta(s - s_0). \qquad (36)$$

If $f(s, t), \partial f(s, t)/\partial s$ and $f_i(s, t)$ are bounded for $s \to s_0$ a balance of the element $[s_0, s_0 + ds)$ with $ds \to 0$ yields the missing boundary condition

$$a_1 \cdot \mu(c_{O_2L}(t)) \cdot f(s, t)\big|_{s = s_0} = r_G(c_{Sp}, \varsigma_S, t). \qquad (37)$$

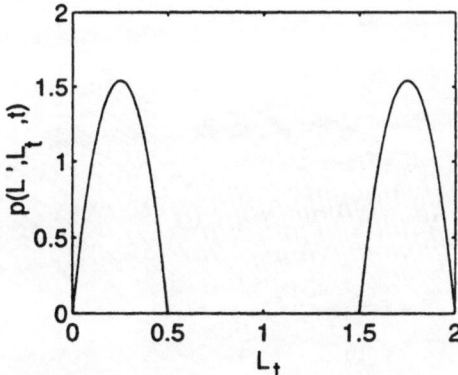

Fig. 14. Density of the breakage probability as a function of L_t

For the germination rate, r_G, a first order rate equation is assumed. The balance for the ungerminated spores reads

$$\frac{dc_{Sp}(t)}{dt} = -r_G(t) - q \cdot c_{Sp}(t)$$
$$= -k_G \cdot c_{Sp}(t) - q \cdot c_{Sp}(t). \tag{38}$$

An example of the breakage probability $p(s', s, t)$ in terms of the original variable L_t is shown in Fig. 14. This kind of presentation is chosen since the symmetry of p is lost upon transformation with $s = \ln(L_t/L_0)$.

All pellets consume oxygen for growth, which is proportional to ds/dt, and for maintenance. The latter is proportional to the amount of active hyphal length $L_a = L_0 \cdot \exp(s_a)$. A balance yields

$$\frac{dc_{O_2L}(t)}{dt} = q(t) \cdot (c_{O_2in}(t) - c_{O_2L}(t)) + k_L a \cdot (c_{O_2}^* - c_{O_2L}(t))$$

$$- L_0 \cdot \int_0^\infty \left(Y_{O_2g} \cdot e^{s(t)} \cdot \left. \frac{ds(t)}{dt} \right|_{Pellet} + Y_{O_2m} \cdot e^{s_a(t)} \right) \cdot f(s, t) \, ds. \tag{39}$$

In Eq. (39) no additional account for mass transfer inside a pellet has to be included, as this behaviour is already part of the reduced pellet model.

4.2
Simulation and Qualitative Comparison with Experimental Results

After a discretization of s, and evaluation of the integrals by trapezoidal rule, a numerical solution of the coupled equation is obtained. Figure 15 shows the result of a simulation. At the beginning, the pellet size distribution moves

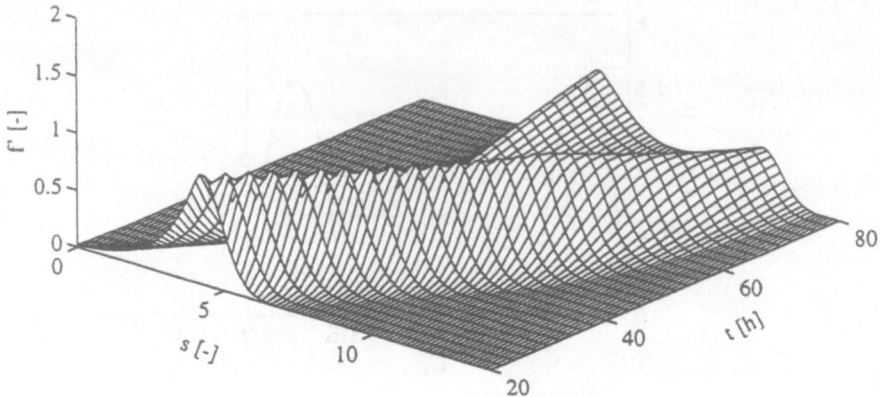

Fig. 15. Development of the normalised number density function $f' = f/f_0$ [61]

towards larger values of s. With growing sizes more and more pellets are broken down into smaller parts. A new subpopulation of smaller pellets emerges.

Figure 16 shows another simulation in which a parameter of the breakage probability is varied. Here, s_{II} relates to a length L_{II} for which the breakage probability becomes zero for the first time (0.5 in Fig. 14).

From the number density function more information can be gained. By integration the total number of pellets

$$F(t) = \int_{s_0}^{\infty} f(s, t) \, ds \tag{40}$$

or total amount of biomass

$$m_t = \sigma_{Hy} \cdot L_0 \cdot \int_{s_0}^{\infty} e^s \cdot f(s, t) \, ds \tag{41}$$

can be obtained.

A very first approach of adapting the model to experimental conditions shows that the behaviour determined in experiments by image analysis [58] can already be described qualitatively (see Fig. 17). When comparing the results it should be noted that in the simulation a distribution is calculated as a continuous function of the normalized hyphal length while the experimental data are given for discrete size classes given as pellet area. To allow at least a rough comparison a relation has to be found between the total hyphal length of a pellet and the pellet area. Therefore, Eqs. (27, and 29) are transformed to the new coordinate s, i.e. ds/dt. Then, Eq. (13)

$$\frac{dz_R(t)}{dt} = \gamma \cdot \bar{\alpha} \cdot \frac{c_{O_2}}{K_{O_2} + c_{O_2}}$$

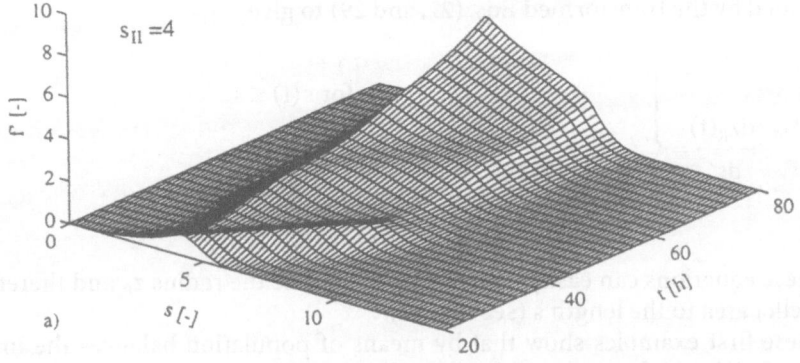

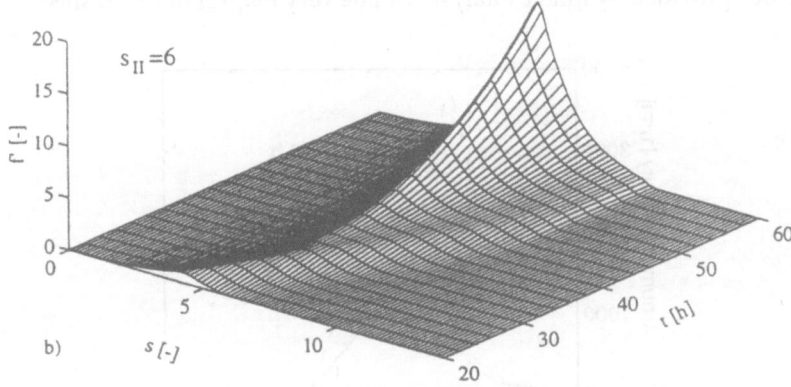

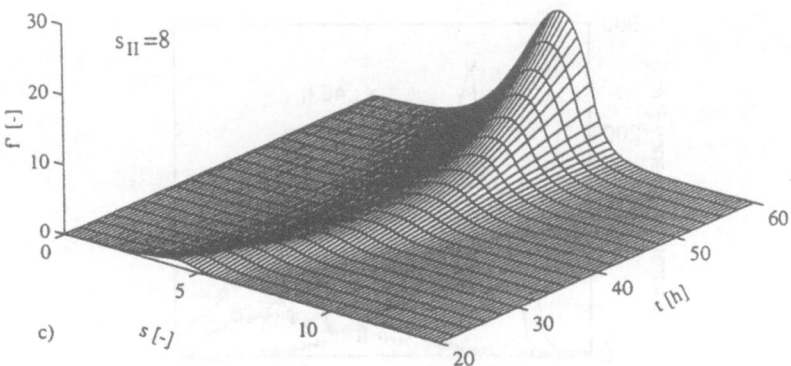

Fig. 16. Development of the normalised number density function $f' = f/f_0$ for different values of s_{II}; see text

is divided by the transformed Eqs. (27, and 29) to give

$$\frac{dz_R(t)}{ds} = \begin{cases} \dfrac{\gamma \cdot \bar{\alpha}}{a_1 \cdot \mu} & \text{for } s(t) \leq s_{cr} \\[3ex] \dfrac{\gamma \cdot \bar{\alpha} \cdot (1 + d_3 \cdot {}^{\lambda \cdot s(t)})}{a_1 \cdot \bar{\mu} \cdot (1 + d_3 \cdot e^{\lambda \cdot s_{cr}})} & \text{for } s(t) > s_{cr} \end{cases}$$

These equations can easily be integrated to relate the radius z_R and therefore the pellet area to the length s (see Fig. 17).

These first examples show that by means of population balances the influences of different factors on pellet size distributions can be studied. Based on such information further questions, e.g. concerning correlations with rheological parameters, can be addressed. However, before this can be done, more experimental data on breakage and agglomeration processes are needed. The methods provided by image analysis will be very helpful in this respect.

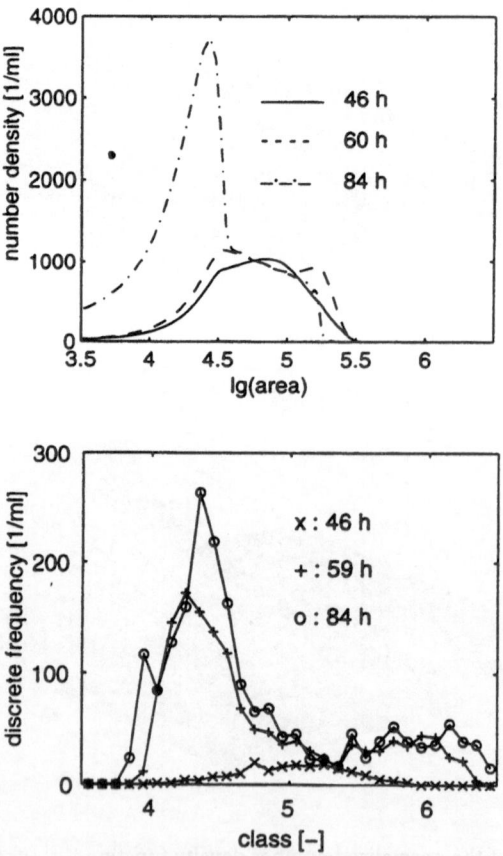

Fig. 17. Qualitative comparison of simulated pellet size distribution with experiments

5
References

1. Bushell ME (1988) Growth, product formation and fermentation technology. In: Goodfellow M, Williams ST, Mordanski M (eds) Actinomycetes in biotechnology. Academic Press, London, p 186
2. Fisher SH (1988) Nitrogen assimilation in streptomycetes. In: Okami Y, Beppu T, Ogawara H (eds) Biology of *Actinomycetes*. Japan Scientific Societies Press, Tokyo, p 47
3. Hirsch CF, McCann-McCormick PA (1985) Biology of streptomyces. In: Demain AL, Solomon NA (eds), Biology of industrial microorganisms. Benjamin/Cummings, Menlo Park, CA, p 291
4. Treskatis SK, Orgeldinger V, Wolf H, Gilles ED (1997) Biotechnol Bioeng 53:191
5. King R (1996) J Biotechnol 52:219
6. King R, Büdenbender Ch (1996) J Biotechnol 52:235
7. Bergter F (1978) Z Allg Mikrobiol 2:143
8. Cagney JW, Chittur VK, Lim HC (1984) Biotechnol Bioeng Symp 14:619
9. Megee RD III, Kinoshita S, Fredrickson AG, Tsuchiya HM (1970) Biotechnol Bioeng 12:771
10. Matsumura M, Imanaka T, Yoshida T, Taguchi H (1981) J Ferment Tech (Japan) 59:115
11. Nestaas E, Wang DIC (1983) Biotechnol Bioeng 25:781
12. Nielsen J (1992) Modelling the growth of filamentous fungi. In: Fiechter A (ed) Adv Biochem Eng Biotechnol 46, Springer, Berlin Heidelberg New York, p 187
13. Patankar DB, Liu TC, Oolman T (1993) Biotechnol Bioeng 42:571
14. Paul GC, Thomas CR (1996) Biotechnol Bioeng 51:558
15. Aynsley M, Ward AC, Wright AR (1990) Biotechnol Bioeng 35:820
16. Prosser JI, Trinci APJ (1979) J of Gen Microbiol 111:153
17. Lejeune R, Baron GV Accepted for Biotechnol Bioeng
18. King R, Buschulte TK, Yang H, Gilles ED (1990) Mathematical models of filamentously growing microorganisms. In: Behrens D (ed) Dechema Biotechnology Conferences, vol. 4. Verlag Chemie, Frankfurt, p 989
19. Yang H (1994) Mathematische Modellierung des Wachstums myzelbildender Mikroorganismen. PhD thesis, Universität Stuttgart
20. Yang H, King R, Reichl U, Gilles ED (1992) Biotechnol Bioeng 39:49
21. Yang H, Reichl U, King R, Gilles ED (1992) Biotechnol Bioeng 39:44
22. Shomura T, Yoshida J, Amano S, Kojima M, Inouye S, Niida T 81979) J Antibiotics 32:427
23. Vitális S, Szabó G, Vályi-Nagy T (1963) Acta Biol Hung 14:1
24. Reichl U, King R, Gilles ED (1991) Biotechnol Bioeng 32:193
25. Prosser JI, Tough AJ (1991) Crit Rev Biotechnol 10:253
26. Yang H (1993) Mathematical model for filamentous growth of mycelial microorganisms in lab chamber and in batch, fed-batch and continuous cultures. In: Alberghina (ed) ECB 6. Elsevier, Amsterdam, p 845
27. Meyerhoff J, Tiller V, Bellgardt KH (1995) Bioproc Eng 12: 305
28. Kummer C, Kretschmer S (1986) J Basic Microbiol 26:27
29. Prosser JI, Gray DI, Gooday GW (1988) Cellular mechanisms for growth and branch formation in streptomycetes. In: Okami Y, Beppu T, Ogawara H (eds) Biology of actinomycetes. Japan Scientific Societies Press, Tokyo, p 316
30. Riesenberg D, Bergter F (1979) Z Allg Mikrobiol 19:415
31. Treskatis S, Reichl U, Orgeldinger V, King R, Gilles ED (1992) Process control of fermentations with filamentous microorganisms by means of digital image processing and pattern recognition. Poster, 6th Europ Conf on Biotechnology, Florence, Italy
32. Reichl U, Yang H, Gilles ED, Wolf H (1990) FEMS Microbiol Letters 67:207
33. King R (1994) Mathematische Modelle myzelförmig wachsender Mikroorganismen. Fortschrittsber 17, 103, VDI, Düsseldorf
34. Reichl U (1991) Einsatz eines Bildverarbeitungssystems zur Erfassung der Morphologie und des Wachstums mzyelbildender Mikroorganismen in submerser Kultur. PhD thesis, Universität Stuttgart

35. Reichl U, Buschulte TK, Gilles ED (1990) J Micros 158:55
36. Kretschmer S (1989) J Basic Microbiol 29:587
37. Fiddy C, Trinci APJ (1976) J Gen Microbiol 97:167
38. Gray DI, Gooday GW, Prosser JI (1990) J Microbiol Meth 12:163
39. Neidhardt FC, Ingraham JL, Schaechter M (1990) Physiology of the bacterial cell: a molecular approach. Sinauer Associates, Sunderland, MA
40. Braun S, Vecht-Lifshitz SE (1991) TIBTECH 9:63
41. Wittler R, Baumgart H, Lübbers DW, Schügerl K (1986) Biotechnol Bioeng 28:1024
42. Buschulte TK, Yang H, King R, Gilles ED (1991) Formation and growth of pellets of streptomyces. In: Reuß M, Chmiel H, Gilles ED, Knackmus HJ (eds) Biochemical Engineering – Stuttgart. G. Fischer, Stuttgart, p 393
43. Chiu ZS, Zajic JE (1976) Biotechnol Bioeng 18:1167
44. Koch AL (1975) J Gen Microbiol 89:209
45. Emerson S (1950) J Bacteriol 60:221
46. Aiba S, Kobayashi K (1971) Biotechnol Bioeng 13:583
47. Kobayashi T, van Dedem G, Moo-Young M (1973) Biotechnol Bioeng 15:27
48. Phillips DH (1966) Biotechnol Bioeng 8:456
49. Pirt SJ (1966) Proc Roy Soc London B166:369
50. Yano T, Kodama T, Yamada K (1961) Agr Biol Chem 25:580
51. Chang HN, Moo-Young M (1988) Appl Microbiol Biotechnol 29:107
52. Skowlund CT (1990) Biotechnol Bioeng 35:502
53. Buschulte TK (1992) Mathematische Modellbildung und Simulation von Zellwachstum, Stofftransport und Stoffwechsel in Pellets aus Streptomyceten. PhD thesis, Universität Stuttgart
54. Buschulte TK, Gilles ED (1990) Modeling and simulation of hyphal growth, metabolism and mass transfer in pellets of streptomyces. In: Christiansen C, Munck L, Villadsen J (eds) Proceedings of the 5th European Congress on Biotechnology, p 279
55. Taguchi H (1971) The nature of fermentation fluids. In: Ghose TK, Fiechter A (eds) Advances in biochemical engineering 1. Springer, Berlin Heidelberg New York, p 1
56. Trück HU, Chmiel H, Hammes WP, Trösch W (1990) Appl Microbiol Biotechnol 33:139
57. Schuhmann E, Bergter F (1976) Z Allg Mikrobiol 16:201
58. Reichl U, King R, Gilles ED (1992) Biotechnol Bioeng 39:164
59. Edelstein L (1983) J Theor Biol 150:427
60. Tough AJ (1989) A theoretical model of Streptomycete growth in submerged culture. PhD thesis, University of Aberdeen
61. King R, Büdenbender Ch, Oswald G (1995) Mathematical models for growth and production of single pellets and pellet populations. In: Munack A, Schügerl K (eds) 6th Int Conf Computer Applications in Biotechnology. Dechema, Frankfurt, p 154

Received December 1996

Modeling the Mycelium Morphology of *Penicillium* Species in Submerged Cultures

P. Krabben · J. Nielsen

Center for Process Biotechnology, Department of Biotechnology, Technical University of Denmark, DK-2800 Lyngby, Denmark

Modeling the mycelium morphology of filamentous fungi is valuable in connection with studies of their growth mechanisms, i.e. tip extension and branching, and in this work a general frame for morphological models is presented. The general frame consist of a population balance equation (PBE) for a two-dimensional density function, which describes the properties, i.e. the number of tips and the total hyphal length, of a population of hyphal elements. From the general PBE, balances for the average properties of the population can be derived. After presentation of the general model frame the kinetics for the different processes influencing the mycelium morphology, i.e. spore germination, growth, and hyphal fragmentation, are reviewed. Thereafter follows an overview of different kinetic models presented in the literature. The models are divided into four groups: single hyphal element/branch models; average property models; population models; and morphological structured models. Models within the first three groups are discussed and presented within the general frame. Finally some solutions to the general PBE are presented and aspects on model verification based on experimental data are discussed.

Advances in Biochemical Engineering/
Biotechnology, Vol. 60
Managing Editor: Th. Scheper
© Springer-Verlag Berlin Heidelberg 1998

List of Abbreviations and Symbols

$B(\alpha, \beta)$	the beta function
c	steepness/skewness parameter in Eq. (14)
c_v	concentration of vesicles at the hyphal tip
d	diameter of the hyphae
d_s	diameter of the stirrer
D	dilution rate
e	the concentration of hyphal elements
e_{spore}	the inoculum concentrations of spores
$f(l_t,n,t)$	number density function
$f_0(l_t,n)$	initial number density function
$f(r)$	absorption rate of vesicles at the hyphal tip at a radius r from the center axis
$g(z,t)$	germination frequency
$h(l_t,n,z,t)$	net rate of formation of hyphal elements with the properties (l_t,n)
$h_f(l_t,n,z,t)$	net rate of formation of hyphal elements with the properties (l_t,n) by fragmentation
$h_g(l_t,n,z,t)$	net rate of formation of hyphal elements with the properties (l_t,n) by germination
k_{bran}	branching parameter
$k_{tensile}$	tensile strength of the hyphal wall
K	allometric coefficient
K_{br}	saturation parameter
K_t	saturation parameter
K_v	saturation parameter in Eq. (25)
k_{tip}	maximum tip extension rate
l_{branch}	the length at which branching starts
$l_{e,av}$	average effective length
$l_{e,max}$	maximum effective length at which fragmentation does not occur
l_t	total length of a hyphal element
$l_{t,av}$	average total length per hyphal element
$l_{t,g}$	length of a newly germinated spore
l_{tip}	constant in Eq. (19)
n	number of active growing tips in a hyphal element
n_{av}	average number of tips per hyphal element
N	stirrer speed
$p((l_t',n'),(l_t,n))$	partitioning function
$p_l(l_t,z,t)$	probability that a newly germinated spore will have the hyphal length l_t
$p_n(n,z,t)$	probability that a newly germinated spore will have n number of tips
P_g	power input at gassed conditions
$q_{bran}(l_t,n,z)$	branching frequency
$q_{frag}(l_t,n,z)$	rate of fragmentation
$q_{tip}(l_t,n,z)$	tip extension rate

r	the radius of the hyphae
r_{abs}	absorption rate of vesicles at the hyphal tip
$r_{abs,max}$	maximum absorption rate of vesicles at the hyphal tip
$\bar{t}$	the normalized germination time
t_c	the circulation time
t_f	the time at which germination stops
t_s	the time at which germination starts
V	the volume of the culture
y_m	state-vector of the hyphal element
y_{viab}	the viability of spores
z	vector of environmental conditions
α	steepness/skewness parameter in Eq. (13)
β	steepness/skewness parameter in Eq. (13)
λ	the time at which half of the viable spores have germinated

1
Introduction

The *Penicillium* species forms a large group of industrially important filamentous fungi (Table 1). Some are or have been used in industrial production, e.g. *P. chrysogenum*, *P. glabrum*, *P. griseofulvin*, and *P. notatum*, and others are used in the food industry, e.g. *P. camberti* and *P. roqueforti*. Finally some species are well known due to their contamination of food stocks, e.g. *P. digitatum* and *P. italicum*. Since no sexual reproduction is observed for most species of *Penicillium* they belong to the group of Deuteromycetes. In those cases where a sexual reproduction is observed they are placed among the *Talaromyces* or in the *Eupenicillium* group of Ascomycetes.

With the industrial importance of these species it is important to elucidate further their growth kinetics and hereby the development of their morphology. The mycelium morphology has been studied most thoroughly for *P. chrysogenum*, which is used for the production of penicillins, the largest group of antibiotics.

Vegetative cells of *P. chrysogenum* form so-called hyphal elements. In submerged cultures the hyphal elements may form either a mycelium, which is a

Table 1. Important *Penicillium* species and their product or effect [1]

Organisms	Product or effect
P. camberti	Fermented chees
P. chrysogenum	Antibiotic/Penicillin
P. digitatum	Deterioration of citrus fruit
P. glabrum	Citric acid
P. griseofulvin	Antibiotic/Griseofulvin
P. italicum	Deterioration of citrus fruit
P. notatum	Antibiotic/Penicillin
P. roqueforti	Blue cheeses

culture of disperse hyphal elements or pellets, which are spherical agglomerates of hyphal elements. The fungal morphology can be divided into two parts, 1) the microscopic morphology which describes the morphology of the hyphal elements and 2) the macroscopic morphology which describes the pellet morphology [2]. Formation and growth of pellets in submerged cultures of *P. chrysogenum* has recently been reviewed by Nielsen and Carlsen [3] and pellet morphology is also the topic of another review in this volume and we therefore concentrate on the mycelium morphology in the present text. The growth of the mycelium takes place at the hyphal tip and new tips can be formed by branching.

In order to facilitate the study and understanding of fungal morphology it is helpful with a general model structure which can be applied to different filamentous microorganisms. A general model structure will facilitate the comparison of different models and hence the understanding of the growth process. We therefore start this review by presenting a general model for the description of the morphological development of the mycelium morphology of filamentous microorganisms, which will be used in our review of previously described mathematical models.

2
Generalized Morphological Model

The intrinsic state of a hyphal element can be described by a state-vector y_m, which includes information on the size, shape etc. However, the state-vector usually only consist of two elements, i.e. the number of tips and the length or mass of the mycelium. The state-vector should at least contain the properties of the hyphal element which are the most important intrinsic variables influencing the growth kinetics, and this will in many cases allow consideration of only the number of tips and the total length of the hyphal element. With these two variables fully characterizing the state of a hyphal element, the overall growth process is a result of two sub-processes: tip extension and formation of new tips by branching.

In submerged cultures there will be a population of hyphal elements. Ideally these will have the same properties, but due to asynchronous spore germination and more or less random fragmentation of the hyphal elements, there will be a distribution of properties for the population. This population distribution may be described with the general population balance equation (PBE) given by Ramkrishna [4] with the number of tips and the total hyphal length as independent variables. In order to simplify the population balance it is assumed that the number of tips can be described as a real number and not as a discrete number as it is by nature. This simplification reduces the number of equations from infinity to 1. The simplified PBE for the two-dimensional number density function $f(l_t,n,t)$ is then given by [5]

$$\frac{\partial f(l_t,n,t)}{\partial t} + \frac{\partial}{\partial l_t} (n \cdot q_{tip} (l_t,n,z) \cdot f(l_t,n,t)) + \frac{\partial}{\partial n} (q_{bran} (l_t,n,z) \cdot f(l_t,n,t))$$
$$= h(l_t,n,z,t) - D \cdot f(l_t,n,t)$$
(1)

where $q_{tip}(l_t, n, z)$ is the average tip extension rate of the hyphal element (μm per tip per h), $q_{bran}(l_t, n, z)$ is the branching frequency (tips per h) which includes apical, subapical branching and formation of extra germ tubes, $h(l_t, n, z, t)$ is the net rate of formation of hyphal elements with the property (l_t, n), and D is the dilution rate. Equation (1) clearly shows that the population distribution is influenced both by the growth kinetics and by the net formation of new hyphal elements, due to spore germination and hyphal fragmentation. Besides being a function of the hyphal element properties all the rates are a function of the environmental condition specified by the vector z, which includes the concentration of the limiting substrate.

The net rate of formation of hyphal elements, $h(l_t, n, z, t)$, can be divided into two terms – one accounting for the net formation of new hyphal elements as a result of spore germination and one accounting for net formation of new hyphal elements as a result of hyphal fragmentation:

$$h(l_t, n, z, t) = h_g(l_t, n, z, t) + h_f(l_t, n, z, t) \tag{2}$$

The net formation of hyphal elements upon germination can be described as

$$h_g(l_t, n, z, t) = g(z, t) \cdot p_l(l_t, z, t) \cdot p_n(n, z, t) \tag{3}$$

where $g(z, t) \cdot dt$ is the number of spores which germinate in the time interval between t and t+dt and $p_l(l_t, z, t)$ is the probability that a spore which germinates at time t will form a hyphal element with length l_t (which is equal to the spore diameter at germination). Finally $p_n(n, z, t)$ is the probability that a spore germinating at time t will form a hyphal element with n number of tips. Normally it can be assumed that a newly formed hyphal element upon germination of a spore will have fixed properties $(l_t, n) = (l_{t,g}, 1)$, whereby Eq. (3) can be specified as

$$h_g(l_t, n, z, t) = g(z, t) \cdot \delta(l_t - l_{t,g}) \cdot \delta(n-1) \tag{4}$$

where δ is the Dirac-delta function.

Generally fragmentation of hyphal element occurs by binary fission and $h_f(l_t, n, z, t)$ is therefore given by

$$h_f(l_t, n, z, t) = 2 \cdot \int_n^\infty \int_{l_t}^\infty q_{frag}(l_t', n', z) \cdot p((l_t', n'), (l_t, n)) \cdot f(l_t', n', t)$$
$$\cdot dl_t' dn' - q_{frag}(l_t, n, z) \cdot f(l_t, n, t) \tag{5}$$

where $q_{frag}(l_t, n, z)$ is the rate of fragmentation of hyphal elements with the properties (l_t, n) and $p((l_t', n'), (l_t, n))$ is the partitioning function which is the probability that an element with the properties (l_t, n) is formed upon fragmentation of an element with the properties (l_t', n') [4, 6].

The boundary conditions of the PBE Eq. (1) includes an initial distribution $f_0(l_t, n)$, i.e.

$$f(l_t, n, 0) = f_0(l_t, n) \tag{6}$$

which in the case of a batch cultivation inoculated with spores is zero [5]. Another set of boundary conditions follows from the assumption of the number of tips being real. This will ensure that both l_t and n start to increase right after formation of a new hyphal element from a spore. We therefore have

$$f(l_{t,g}, n, t) = 0 \quad \text{for} \quad n > 1; f(l_t, 1, t) = 0 \quad \text{for} \quad l_t > l_{t,g} \tag{7}$$

Furthermore, the solution to the PBE should fulfill the condition that there are no hyphal elements with infinite values of either the total hyphal length or the number of tips, i.e.

$$f(l_t, n, t) \to 0 \quad \text{for} \quad l_t \to \infty$$
$$f(l_t, n, t) \to 0 \quad \text{for} \quad n \to \infty \tag{8}$$

This is a result of the fragmentation process, since the rate of fragmentation (Sect. 6) is generally found to be proportional with l_t (or even l_t^2) and it approaches infinity when l_t becomes large. Since n and l_t are coupled the solution should for the same reason not describe the existence of elements with an infinite number of tips. In fact the requirement to the solution can even be strengthened to include

$$l_t \cdot f(l_t, n, t) \to 0 \quad \text{for} \quad l_t \to \infty$$
$$n \cdot f(l_t, n, t) \to 0 \quad \text{for} \quad n \to \infty \tag{9}$$

When evaluating experimental data of the morphology, e.g. average total length of a hyphal element and average number of tips per hyphal element, it is useful to have balances for these properties. Such balances can be derived from Eq. (1) if the growth kinetics follows simple kinetics, i.e. q_{tip} and q_{bran} are described by either 0 or 1. order kinetics with respect to l_t and q_{frag} is 1. order with respect to l_t [5]:

$$\frac{de}{dt} = (q_{frag}(l_{t,av}, \mathbf{z}) - D) \cdot e + g(t) \tag{10}$$

$$\frac{dl_{t,av}}{dt} = (n_{av} \cdot q_{tip}(l_{t,av}, \mathbf{z}) - q_{frag}(l_{t,av}, \mathbf{z}) \cdot l_{t,av} + (l_{t,g} - l_{t,av}) \cdot \frac{g(t)}{e(t)} \tag{11}$$

$$\frac{dn_{av}}{dt} = q_{bran}(l_{t,av}, \mathbf{z}) - q_{frag}(l_{t,av}, \mathbf{z}) \cdot n_{av} + (1 - n_{av}) \cdot \frac{g(t)}{e(t)} \tag{12}$$

From these balances the influence of the four processes – spore germination, tip extension, branching and hyphal fragmentation – on the mycelium morphology is clear, and before we turn to a review of different models we consider these processes separately.

3
Spore Germination

The fungal spore can be regarded as a dormant stage in the fungal life cycle, since it has a low water content and a low metabolic turnover. Normally two types of dormancy are recognized: constitutional dormancy and exogenous dormancy [7]. Spores in exogenous dormancy germinate when placed in a suitable medium while spores in constitutional dormancy need some kind of activation. The nutritional requirement for germination is not necessarily the same as for hyphal growth. Spores of *P. chrysogenum* used for inoculum of submerged cultivations are, however, usually in exogenous dormancy and germinate readily when placed in a growth medium.

During spore germination three phases can be recognized [8]: "spore swelling, germ tube emergence", and "germ tube elongation". Some further subdivide spore swelling into two phases – endogenous and exogenous swelling [9]. Endogenous swelling consists primarily of uptake of water and is independent of the environmental conditions whereas exogenous swelling which takes place after endogenous swelling is highly dependent on the environmental conditions. It has been reported that the uptake of water is proportional to the surface area which yields linear growth of the spore diameter [10]. In the phase of exogenous swelling, formation of all major macromolecules takes place [9, 11, 12], and a germ-tube is formed. For *P. megasporium* one to three additional germ-tubes may be formed after completion of swelling [13], and the number of germ-tubes per spore is most likely dependent on the species.

The germination of spores involves the biosynthesis of active machinery for the production of new cell material. It is therefore expected that there is increased activity of both anaplerotic reactions and the pentose phosphate pathway. Thus the activity of glucose-6-phosphate dehydrogenase which catabolizes the first step in the pentose phosphate pathway has been reported to increase when spores of *P. chrysogenum* germinates [14]. Furthermore, the glyoxylate cycle, which replenishes intermediates of the TCA cycle, is active in germinating spores of *P. oxalicum* [15], and the anaplerotic reaction catalyzed by the enzyme pyruvate carboxylase is thought to be important for the germination of spores [7]. This could explain the fact that carbon dioxide has a drastic effect on the germination [16–19], namely by influencing both the viability of the spores [16] and the duration of the lag phase [17–19]. The lag phase before germination is prolonged both at high and low carbon dioxide concentrations. Low carbon dioxide concentration is thought to result in reduced germination, especially at high spore concentrations [20, 21].

Germination of spores in submerged culture is normally an asynchronous process, but for some *Penicillium* species it can be made synchronous by letting the spores germinate at a temperature at which they are not able to grow [22, 23]. The germination frequency is determined by the spore quality and the environmental conditions at which they germinate. The process of germination has received very little attention with respect to modeling and most quantitative work done so far has been focused on fitting the accumulated germination percentage to different empirical formulas [5, 24, 25] and thereby obtaining the germination frequency as the first time derivative. Thus Nielsen and Krabben [5] showed that the germination frequency, g(t), can in most cases be fitted by the five parameters of Eq. (13): 1) the viability, y_{viab}, of the spores, i.e. the fraction of spores which germinates; 2) the time, t_s, at which germination starts; 3) the time, t_f, at which no more spores germinate and 4, 5) the two parameters α and β of the B-distribution, determining the steepness and skewness respectively of the germination frequency:

$$g(t) = e_{spore} \cdot y_{viab} \cdot \begin{cases} \dfrac{1}{B(\alpha,\beta)} \cdot \bar{t}^{\alpha-1}(1-\bar{t})^{\beta-1}; \bar{t} = \dfrac{t-t_s}{t_f-t_s} & t_s \leq t \leq t_f \\ 0 & \text{else} \end{cases} \tag{13}$$

where e_{spore} is the concentration of spores.

The two form parameters α and β are usually constant for the individual species and for P. chrysogenum they are both found to be 3 (Fig. 1) [5].

In many cases a more simple fit than the one given by Eq. (13) can be used. Lejeune et al. [25] have proposed the use of a log-normal distribution to describe the germination process. In order to get a good fit it is, however, necessary to introduce a lag time before germination starts, and this model therefore has four parameters: 1) the viability, y_{viab}, of the spores; 2) the start time of germination; and 3, 4) the two parameters of the log-normal distribution. This distribution also fits the data of Fig. 1 nicely and with the presently available data it is not possible to discriminate between the two models [25]. The main difference between the log-normal distribution and the B-distribution is that with the B-distribution it is possible to vary the steepness and the skewness independently of each other in contrast to the log-normal distribution where a fixed steepness yields a fixed skewness. Finally, an even more simple three parameter fit has been reported [24]:

$$g(t) = e_{spore} \cdot y_{viab} \cdot \frac{d}{dt}\left(\frac{(t/\lambda)^c}{1+(t/\lambda)^c}\right) \tag{14}$$

In this function c is a steepness/skewness parameter and λ is the time at which 50% of the spores which will germinate have germinated. This simple equation has been shown to be able to fit data of spore germination for five different fungi including Aspergillus and Geotrichum [24].

4
Tip Extension

In contrast to spore growth, where cell wall material is laid down uniformly throughout the surface of the spore, growth of hyphal elements occur only at the tips. This growth phenomenon was already recognized in the 1890s, and at the same time it was correctly speculated that substrate is taken up throughout the hyphal element, processed to cell metabolites, which are sent toward the hyphal tip [26]. Despite extensive research in fungal growth since this early discovery the mechanisms of the tip extension have not been fully elucidated. It has,

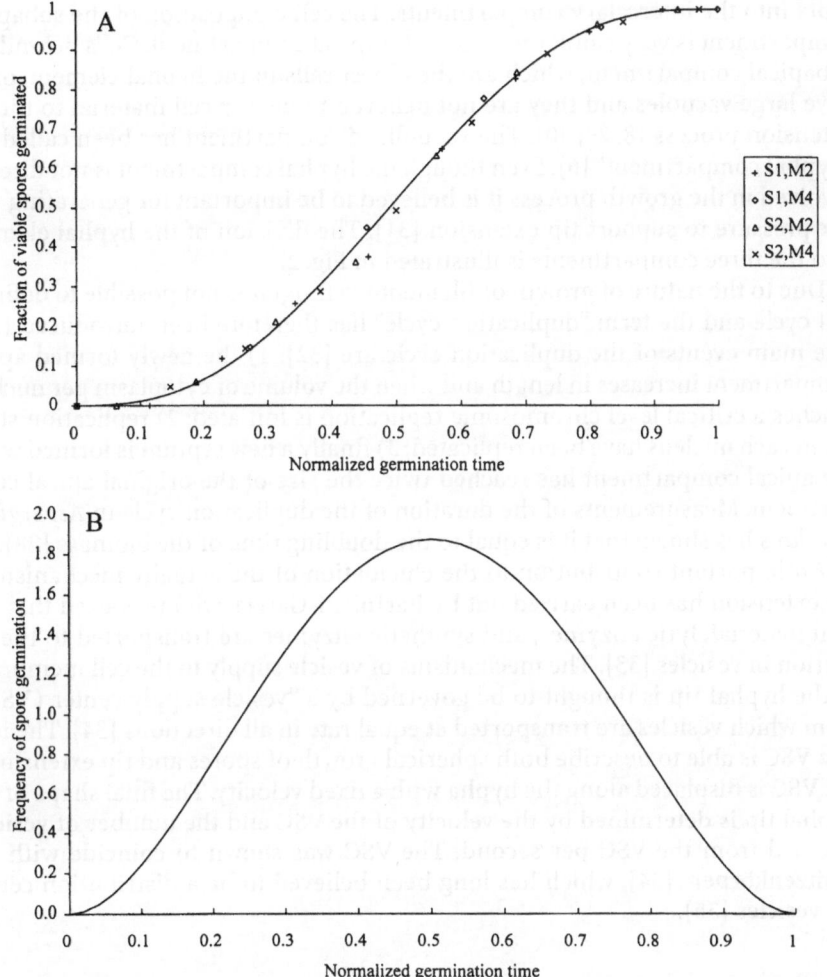

Fig. 1 a, b. Spore germination of *P. chrysogenum* spores in submerged batch cultivations: A the normalized fraction of germinated spores as function of the normalized germination time of Eq. (13). The parameters y_{viab}, t_s, and t_f are listed in Table 2. The data are taken from Paul et al. [8]. S1 refers to fresh spores and S2 refers to old spores, whereas M2 refers to a defined medium and M4 to a complex medium. The curve represents the accumulated frequency function $e(t)/(e_{spore} \cdot y_{viab})$ for spore germination given by Eq. (13) with the parameters $\alpha = \beta = 3$; B the frequency of spore germination $g(t)/(e_{spore} \cdot y_{viab})$

however, been shown that cells distal from the hyphal tip have the same rate of RNA and protein formation as new cells close to the hyphal tip [27]. Furthermore, the hyphal element of *Penicillium* is divided into smaller entities by "septa", which have pores to allow free cytoplastic movement between the compartments. The compartment from the hyphal tip to the first septum is called the "apical compartment" [28]. The cells just behind the apical compartment is called the "sub-apical compartment" and this compartment is further divided by septa into the intercalary compartments. The cell composition of the subapical compartment is very similar to that of the apical compartment. Cells behind the subapical compartment, which are the oldest cells in the hyphal element, often have large vacuoles and they are not believed to deliver cell material to the tip extension process [8, 29, 30]. The vacuolized compartment has been called the "hyphal compartment" [6]. Even though the hyphal compartment is not directly involved in the growth process it is believed to be important for generating turgor pressure to support tip extension [31]. The division of the hyphal element into the three compartments is illustrated in Fig. 2.

Due to the nature of growth of filamentous fungi it is not possible to define a cell cycle and the term "duplication cycle" has therefore been introduced [28]. The main events of the duplication cycle are [32]: 1) the newly formed apical compartment increases in length and when the volume of cytoplasm per nucleus reaches a critical level chromosome replication is initiated; 2) replication stops when each nucleus have been replicated; 3) finally a new septum is formed when the apical compartment has reached twice the size of the original apical compartment. Measurements of the duration of the duplication cycle in *Aspergillus nidulans* has shown that it is equal to the doubling time of the biomass [28].

An important contribution to the elucidation of the actually mechanism of tip extension has been carried out by Bartnicki-Garcia, who proposed that cell wall material, lytic enzymes, and synthetic enzymes are transported to the tip section in vesicles [33]. The mechanisms of vesicle supply to the cell membrane at the hyphal tip is thought to be governed by a "vesicle supply center (VSC)" from which vesicles are transported at equal rate in all directions [34]. The idea of a VSC is able to describe both spherical growth of spores and tip extension if the VSC is displaced along the hypha with a fixed velocity. The final shape of the hyphal tip is determined by the velocity of the VSC and the number of vesicles released from the VSC per second. The VSC was shown to coincide with the "spitzenkörper" [34], which has long been believed to be a distribution center for vesicles [35].

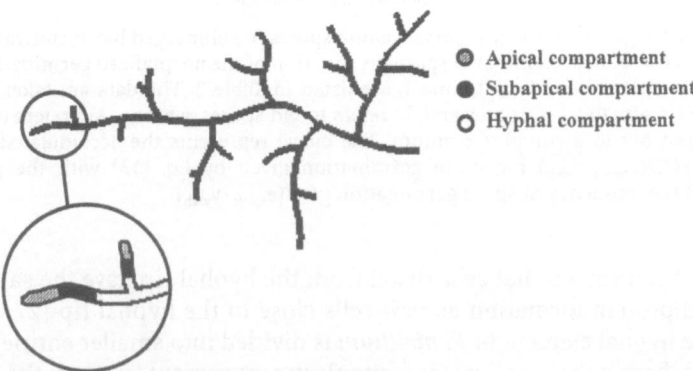

● Apical compartment
● Subapical compartment
○ Hyphal compartment

Fig. 2. The division of a hyphal element into three compartments. The enlargement at the tip shows the different compartments and the septa which separates the apical compartment from the subapical compartment

The limiting factor for tip extension could vary from species to species or even from strain to strain. If the limiting process for tip extension was at the hyphal tip, e.g. vesicle fusion with the cell membrane or the reorganization of vesicles at VSC, then the tip extension rate would be independent of the branch length. However, if biosynthesis of macromolecules is the limiting process for tip extension, then it depends on the length of the branch involved in synthesis of macromolecules used for tip extension. Since there is a maximum tip extension rate this dependence could be expressed by saturation kinetics with respect to the length of the branch, i.e.

$$q_{tip}\,(l_{bran}, z) = k_{tip}\,(z) \cdot \frac{l_{bran}}{l_{bran} + K_{br}} \tag{15}$$

where l_{bran} is the branch length [μm], $k_{tip}(z)$ is the maximum tip extension rate [μm per h] and K_{br} is a saturation constant [μm]. This kind of kinetics has been experimentally confirmed by several species of filamentous fungi among these *Aspergillus nidulans* (Fig. 3) [28], and the hypothesis of the biosynthesis of macromolecules being the limiting factor for tip extension is therefore supported.

In a submerged culture it is not possible to determine the kinetics of the single branches and an average population view is therefore required. From analysis of the average tip extension kinetics of *P. chrysogenum* in submerged cultures the following kinetics has been proposed for the tip extension rate (Fig. 4) [5]:

$$q_{tip}\,(l_{t,av}, z) = k_{tip}\,(z) \cdot \frac{l_{t,av}}{l_{t,av} + K_t} \tag{16}$$

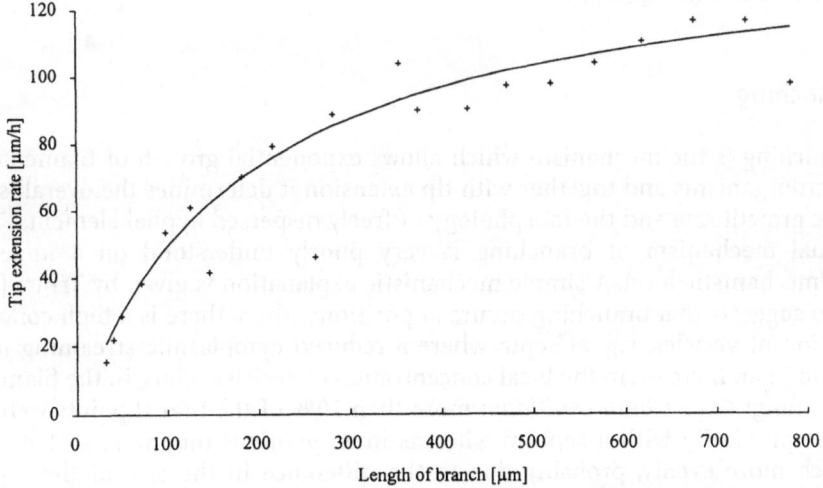

Fig. 3. Tip extension rate for a single branch as function of the branch length of *A. nidulans* grown as a surface culture. The line is given by the kinetics of Eq. (15) with the parameters $k_{tip} = 142$ μm/h and $K_{br} = 188$ μm. The data are taken from Fiddy and Trinci [28]

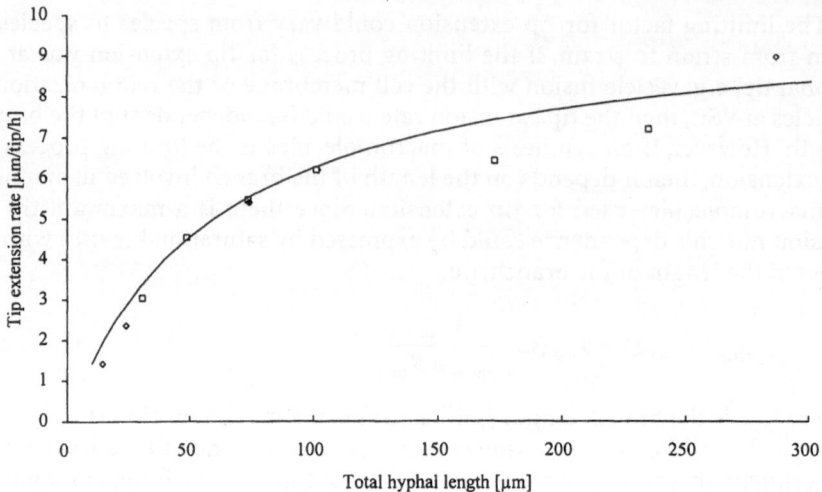

Fig. 4. Tip extension rate for *P. chrysogenum* in two batch cultivations on defined media [5]. The average tip extension rate as function of the average total hyphal length. The line is the kinetics given by Eq. (16) with the parameters $k_{tip} = 10$ μm/tip/h and $K_t = 60$ μm

Equation (16) should be used with caution since it is empirically derived from measurements of average properties in submerged cultures. However, it has been confirmed for several species of filamentous fungi, and it may be useful in average property models. It is, however, important to notice that both parameters are functions of the environmental condition. Thus, both k_{tip} and K_t are different with growth of *P. chrysogenum* on a complex medium compared with a defined medium [5]. Furthermore, k_{tip} has been shown to be a function of the glucose concentration [16].

5
Branching

Branching is the mechanism which allows exponential growth of filamentous microorganisms and together with tip extension it determines the overall specific growth rate and the morphology of freely dispersed hyphal elements. The actual mechanism of branching is very poorly understood on a molecular/mechanistic level. A simple mechanistic explanation is given by Trinci [32] who suggests that branching occurs at positions where there is a high concentration of vesicles, e.g. at septa where a reduced cytoplasmic streaming may result in an increase in the local concentration of vesicles. Thus, in the filamentous fungi *Geotrichum candidum* more than 70% of the branchpoints were in close proximity with a septum whereas in *A. nidulans* they were distributed much more evenly, probably due to the difference in the size of the septal pore [36].

Filamentous fungi are adapted to grow on surfaces where branching is important in order to search for good growth conditions. It should therefore be expected that branching is highly regulated and that at good growth conditions highly branched hyphal elements would appear. This is supported by chemostat cultures of several species where more densely branched hyphal elements are found at high dilution rates [37, 38]. However, for *Fusarium graminearum* the picture is the reverse [39]. Furthermore, when cAMP, which is generally a signal for low glucose concentrations, is added to the medium the branching frequency of this fungus increases [40]. The effect of calcium upon branching has also been examined in some fungi and in *Neurospora crassa* it is proposed that it acts as a branching signal [41, 42].

Since very little is known about the branching mechanisms this process is often viewed as a more or less random process, for which the frequency is proportional to the total hyphal length [5, 25, 43] and we have proposed it be described with [5]

$$q_{bran}(l_t, z) = \begin{cases} 0 & ; \quad l_t < l_{bran} \\ k_{bran}(z) \cdot l_t ; & l_t \geq l_{bran} \end{cases} \tag{17}$$

where q_{bran} is the average branching frequency, k_{bran} is the branching constant [tips per µm per h] and l_{bran} is a parameter [µm] which determines the lag phase for initiation of branching.

6
Fragmentation

At the time when the continuous cultures were introduced it was believed that this cultivation technique could not be used with filamentous microorganisms because they do not grow by fission. However at the start of the 1960s it was recognized that hyphal fragmentation made continuous culture of filamentous micro-organisms possible [44].

Fragmentation of the hyphal element occurs when the local shear rate becomes greater than the tensile strength of the hyphal wall. The tensile strength of the hyphal wall is probably not the same throughout the hyphal element and the hyphal wall is reported to be weaker at the septa [45].This implies that factors influencing the structure of the cell wall should have a strong influence on the fragmentation rate. Based on Kolmoghorov theory for eddies, van Suijdam and

Table 2. Parameters of the spore germination fit Eq. (13) for four different inoculum conditions reported by Paul et al. [8]

Experiment	y_{viab}	t_s [h]	t_f [h]	Comments
S1, M2	0.51	4	39	Fresh spores, defined medium
S1, M4	0.89	2	24	Fresh spores, complex medium
S2, M2	0.33	6	38	Old spores, defined medium
S2, M4	0.63	3	37	Old spores, somplex medium

Metz [46] derived the following relationship between maximum effective length $l_{e,max}$ at which the shearing forces of the eddies is just enough to break the hyphae:

$$l_{e,max} = k_{tensile} \cdot d^{3/8} \cdot \left(\frac{P_g}{V}\right)^{-1/4} \tag{18}$$

The effective length is normally assumed to be the longest hyphal length, i.e. the main hyphae [46], but the real effective length is always smaller than the main hyphae because the hyphal element is curved. However, the length of the main hyphae can be used as long as $k_{tensile}$ is not determined by independent measurements, i.e. it is taken to be an empirical parameter. An approximate length of the main hyphae can be correlated to the number of tips and the total length of the hyphal element [5]

$$l_{e,av} = \frac{n_{av}+1}{2 \cdot n_{av}} \cdot (l_{t,av} - l_{tip}) \tag{19}$$

where the parameter l_{tip} is found to be 10 μm for *P. chrysogenum*.

The rate of fragmentation is a function of how often the hyphal element enters the impeller region with a high shear force and of the residence time distribution of hyphal element in this zone. Based on a simple model for the flow in the bioreactor and the local energy input, van Suijdam and Metz [46] derived the following equation for the overall fragmentation rate, q_{frag}:

$$q_{frag}(l_{t,av}, z) = \begin{cases} 0 & l_{e,av} < l_{e,max} \\ k_{frag}(z) \cdot (l_{e,av}^2 - l_{e,max}^2) & l_{e,av} \geq l_{e,max} \end{cases} \tag{20}$$

where k_{frag} [μm^{-2} h^{-1}] is a function of the power input and hereby the agitation rate, viscosity, and the working volume according to

$$k_{frag}(z) = k(z) \cdot \frac{N^{1.75} \cdot d_s^{0.5}}{d^{3/8}} \tag{21}$$

where N is the agitation rate h^{-1} and d_s is the diameter of the stirrer m. The parameter $k(z)$ is dependent on the tensile strength of the cell wall. From experimental data of Metz [19] we derived an empirical correlation for k_{frag} [5]:

$$k_{frag} = \left(0.125 \cdot \frac{P_g}{V} + 1.5\right) \cdot 10^6 \, m^{-2} \cdot h^{-1} \tag{22}$$

where the power input is given in W per L. This empirical correlation has similarities with Eq. (21) but it is observed that even at very low power input a substantial fragmentation occurs. This may be explained by 1) break-down of the linearity in the low power input area; 2) cell lysis; or 3) the hyphal length measured at low power inputs not representative for the culture due to formation of clumps [47].

The influence of the engineering parameters upon the rate of fragmentation has been examined in terms of the size of the turbulent region in the bioreactor [46, 48–50]. It has been showed by Reuss [48] that the time spent in the impeller shear zone is a determining factor for fragmentation of *Rhizopus nigricans*, whereas the group at UCL [49, 50] has suggested that it is the frequency at which a hyphal element enters the impeller shear zone which determines the fragmentation rate, i.e.

$$k_{frag} \propto \frac{P_g}{V} \cdot \frac{1}{t_c} \quad \text{or} \quad k_{frag} \propto \frac{P_g}{d_s^3} \cdot \frac{1}{t_c} \tag{23}$$

All the above-mentioned studies have focused on quantification of q_{frag}, whereas there have been no reports on the partitioning function of Eq. (5). This is most likely explained by the experimental difficulties in determination of the complete distribution function. However, only when a good expression for the partitioning function has been derived – either empirically or based on some assumed mechanisms – more information on the fundamental mechanisms of fragmentation will be obtained.

7
Morphological Models

The term morphological models is a very wide term and covers many different models which were set up with totally different aims. Morphological models can be grouped according to the scheme shown in Fig. 5. Single hyphal elements or single branch models obviously describes the growth of a single element, whereas population models describe the behavior of populations. Population models can either be totally deterministic [5] or have a stochastic element usually describing branching [25, 43] or fragmentation [5]. Morphological structured

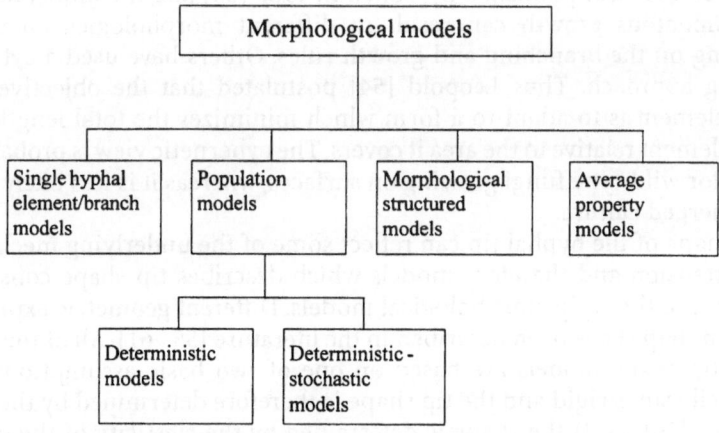

Fig. 5. A grouping of morphological models

models are models which describe the development of different cells/compartments. These models will not be dealt with here since they have been reviewed recently [2, 16]. The last group of models consists of average property models which describes the morphological development in populations but only based on average properties.

The formulation of morphological models is closely linked to the development of image analysis, which may supply the experimental data for model verification. The first generation of morphological models consisted of the single hyphal element/branch models which follow the growth of the single hyphal element, which is easily done from photographs of the developing hyphal element. The experimental basis for setting up average property models requires the measurement of a large number of hyphal elements per sample in order to obtain a good estimate of the average properties in the population. With modern automated image analysis systems it is, however, relatively easy to measure 100 or more elements in each sample, which is sufficient to obtain good measures of the average properties [5]. Even the best image analyzers, however, do not enable routine measurements of the distribution in the population, where of the order of 10 000 or more elements have to be measured. Presently the experimental basis therefore does not exist for verification of population models, but this group of models is important since they may be used to draw some qualitative conclusions on the underlying mechanisms, e. g. for hyphal fragmentation.

8
Single Hyphal Element/Branch Models

The single hyphal element/branch models are generally based on observed mechanisms of the growth process and they have therefore also been referred to as mechanistic models [51]. Due to recent developments in image analysis it has been possible to extent the mechanistic models from single hyphal elements to populations [43, 52]. Most of these models have focused on tip extension, and only a few on the branching process. The first general model for biological branching processes was published by Cohen in 1967 [53] and the model describes how filamentous growth can result in different morphologies on surfaces depending on the branching and growth rules. Others have used a cybernetic modeling approach. Thus Leopold [54] postulated that the objective of the hyphal element is to adapt to a form which minimizes the total length of the hyphal element relative to the area it covers. The cybernetic view is probably reasonable for wild-type fungi growing on surfaces, whereas it is less likely to hold for submerged culture.

The shape of the hyphal tip can reflect some of the underlying mechanisms of tip extension and therefore models which describes tip shape constitute a major part of the early morphological models. Different geometric expressions for the tip shape have been described in the literature [55–61]. All of the mathematical tip shape models are based on one of two basic assumption: 1) the hyphal cell wall is rigid and the tip shape is therefore determined by the supply of vesicles [26], or 2) the shape is determined by the elasticity of the cell wall [62]. The models based on the elasticity of the hyphal cell wall are the so-called surface stress models. The hyphal tip will assume a shape which minimizes the surface stress when the elasticity is given as a function of the distance to the hyphal tip [56, 57]. The cell wall elasticity is very difficult to measure and the models have therefore been used the other way around, namely from a tip shape, to determine how the elasticity varies through the tip extension zone. The surface stress model does not incorporate a mechanism of vesicle supply and thereby indirectly assumes that the hyphal wall instantly assumes the shape with the lowest surface stress when a vesicle fuses with the cell membrane. The models based on vesicle supply as the determining factor for the tip shape were critically analyzed by Saunders and Trinci [56] and they derived a basic relationship between the allometric coefficient K and the vesicle fusion rate f(r), where r is the radius from the center axis of the hyphal tip:

$$f(r) = \text{constant} \cdot \left(1 + \frac{1}{K}\right) \cdot K \cdot r^{K-1} \tag{24}$$

On the basis of this relationship they concluded that a variation in vesicle fusion rate determines the allometric coefficient and not the shape of the tip. This conclusion is however not valid as illustrated by the model by Bartnicki-Garcia et al. [58, 59]. This model is based on the idea of a vesicle supply center (VSC) (see also the section on tip extension). The vesicles are transported towards the hyphal tip where they are organized at the VSC from where vesicles are released with equal probability in all directions. When a vesicle fuses with the membrane the membrane grows with an equal rate in the longitudinal and the circumferential direction and hence the allometric coefficient is 1. The shape of the tip is given as the shape at which f(r) is constant and this shape depends upon the flow of vesicles from the VSC and the displacement velocity of the VSC.

The model by Prosser and Trinci [63] is one of the most important models for a quantitative kinetic description of hyphal growth. This model was probably the first mechanistic model that describes both tip extension and branching, and it has in many cases served as a basis for average property models [52] (Sect. 9) and the population models [43] (Sect. 10). The model by Prosser and Trinci [63] is based on the mechanism of tip extension by vesicle fusion with the plasma membrane at the hyphal tip as proposed by Bartnicki-Garcia [33]. The model is based on a morphological description of the hyphal element by the length of the individual branches, their placement and the placements of septa (Fig. 6).

The vesicles contributing to the growth are formed with a constant rate throughout the hyphal element. The transport rate of the vesicles within the hyphal structure is constant and assumed to be directed towards the hyphal tip. The apex of the hyphal tip is chosen to be the place for regulation of tip extension and the absorption of vesicles is described by saturation kinetics of the Monod-type:

$$r_{abs} = r_{abs,\,max} \cdot \frac{c_v}{c_v + K_v} \tag{25}$$

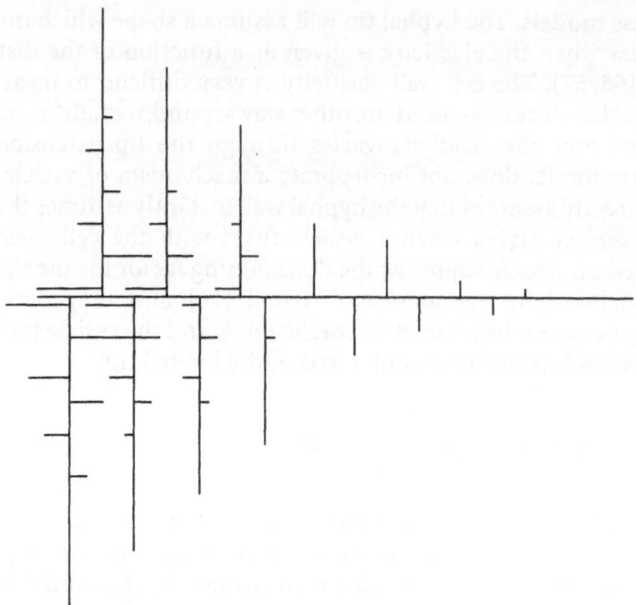

Fig 6. Hyphal element simulated by the Prosser and Trinci model [63]. The formation of branches occur just behind the septa and the placement of branch-points therefore reflects the positioning of septa

where r_{abs} is the rate of vesicle absorption, $r_{abs, max}$ is the maximum rate of vesicle absorption, c_v is the concentration of vesicles at the tip extension zone and K_v is a saturation constant for the absorption process. The elongation of the specific tip is then given as the rate of vesicle absorption multiplied by the increase in the hyphal length per vesicle incorporated into the cell membrane. The duplication cycle concept [32] is also incorporated into the model and it controls indirectly the branching process. Initially the apical compartment gets four nuclei and when the volume per nucleus becomes lower than a critical value, an exponential growth of nuclei begins until eight nuclei exist in the apical compartment. At this time a new septum is formed dividing the apical compartment in the middle and the new apical compartment contains four nuclei. The newly formed septum is "complete" and allows a certain transport of vesicles through the septal pore. The septal pore becomes gradually more and more plugged and the maximum rate of transport through the septal pore decreases at a linear rate. The formation of a branch then occurs when the local concentration of vesicles at a septum attains a critical level.

9
Average Property Models

With the introduction of fully automated image analysis it has become possible to obtain good estimates for the average properties of the mycelial population, i.e. the average hyphal length, the average number of tips per hyphal element,

Table 3. Different average property models.

Model	Mechanistic	Germination	Fragmentation
Aynsley et al. [52]	Yes	n.i.[a]	Yes
Bergter [64]	No	n.i.[a]	n.i.[a]
Viniegra-González et al. [65]	No	Yes	n.i.[a]

[a] n.i.: not included.

and the average length of the hyphal growth unit. The mathematically more complex population models (Sect. 10) may describe the development of the population and they are therefore obviously also capable of describing the changes in the average properties. Average property models reviewed in this section are therefore models which can only describe the average properties and not the distribution of the population. Table 3 summarizes some of the published average property models.

The hyphal growth in a batch or a fed-batch cultivation begins with inoculation either with spores or with growing biomass. The initial conditions have an influence on the development of the average properties. Later in the cultivation fragmentation also starts to become important for the average properties. None of the published average property models includes asynchronous germination in their original formulation but it can easily be included in the model by Bergter [64]. In the full model by Viniegra-Gonzáles et al. [65] it is not possible to include this since synchronous germination is a fundamental assumption in the model. Neither can asynchronous germination easily be included in the model by Aynsley et al. [52].

The purpose of the model by Aynsley et al. [52] is to examine the effect of different substrate and inoculum concentrations on the morphological behavior throughout a fed-batch cultivation of *P. chrysogenum*. This is of industrial interest since penicillin is produced by fed-batch cultivation and the morphology of *P. chrysogenum* during the cultivation is important for control of the process since viscosity of the medium is determined both by the biomass concentration and the morphology. The model is based on growth mechanisms proposed for individual hyphal elements. The mechanisms for tip extension used in this model was originally described by Bartnicki-Garcia [33] and the branching mechanism was originally described by Katz et al. [66]. The model assumes substrate uptake throughout the hyphal element. When the substrate enters the cell it is converted to a growth precursor. The substrate uptake and formation of the growth precursor are combined into one reaction which is described by a Monod-type expression. The formed growth precursor is transported towards the hyphal tip with a constant flux and at the hyphal tip incorporation into the cell wall is described by Michaelis-Menten kinetics. Tip formation can either happen by branching or fragmentation. The branching rate is proportional to the average concentration of the growth precursor and the fragmentation rate is proportional to the biomass concentration and inhibited at high substrate concentration. Despite the mechanistic basis of the model, it is based on a number of assumptions which are required for the derivation of the kinetic expressions for the average properties and they should therefore not be considered as more than empirical correlations which probably have little predictive strength even though it was shown to fit morphological data of P. chrysogenum during a fed-batch cultivation.

The model described by Bergter [64] is the most simple and probably the most robust of the average property models. The model was set up as a tool for analyzing the influence of the environmental conditions on the mycelial growth kinetics. The mycelial growth kinetics is described by the simplest possible kinetics which will give exponential growth in both the average total length and the average number of tips, and with the same specific growth rate:

$$q_{tip}(l_t, n) = k_{tip} \quad \text{and} \quad q_{bran}(l_t, n) = k_{bran} \cdot l_t \tag{26}$$

This simple growth kinetics was observed by examining the growth of single hyphal element on a surface [67]. From a comparison with the tip extension kinetics given in Eq. (16) the constant tip extension rate proposed in the Bergter model could indicate that the saturation constant for tip extension is too low to be measured for the Streptomyces hygroscopicus which was studied, or that the limiting factor for tip extension is at the hyphal tip.

An alternative view on growth of hyphal elements was introduced by Viniegra-Gonzáles et al. [65], who consider hyphal elements as "trees" and apply theories from graph theory to describe the growth of hyphal elements. The objective of the modeling work was to create a link between mycelial growth kinetics and the biomass growth kinetics. In order to determine this link it is assumed that there is synchronous germination and that all the mycelial elements in the population have the same length and number of tips. During most batch cultivations an assumption of synchronous spore germination and growth will not be valid and this reduces the general application of the model.

10
Population Models

The introduction of population models for microbial systems was introduced in 1966 by the group of Fredrickson [68] and the structures of population models were later formalized by Ramkrishna [4]. The major difference between population models and traditional average property models is that population models are based on the growth kinetics of single hyphal elements while average property models are based on the kinetics of the population. Most of the previously published population models describe bacteria and yeast cultures where it is possible to obtain the essential morphological data, for example using flow cytometry, for model verification. Since similar data are not available for populations of hyphal elements there have only been a few examples of population models, and in this section we will review these models and consider some solutions to the generalized PBE of Sect. 2.

Modeling the growth in submerged culture involves a large number of hyphal elements and it is therefore not necessary to use a stochastic description [4]. However, a stochastic element might be introduced as a result of the growth process, e.g. branching, and/or the cultivation condition, e.g. taking fragmentation as a stochastic process. We will therefore divide our review of population models into two parts, one covering completely deterministic models and one covering models with some stochastic element included.

10.1
Deterministic Models

Fully deterministic models of a mycelial population are only possible in cases where the morphology is not influenced by stochastic processes, e.g. fragmentation. A typically submerged cultivation situation where fragmentation is not significant is the early stages of batch cultivations, and it is therefore the only situation where fully deterministic population models can be used. We have described a deterministic population model of this type [5,69], and the solution is described here in further detail.

When fragmentation can be ignored the complexity of the PBE is drastically reduced (see Eq. (27)) and it is possible to solve it analytically for different kinds of growth kinetics, i.e. tip extension rate and branching frequency:

$$\frac{\partial f(l_t, n, t)}{\partial t} + \frac{\partial}{\partial l_t} (n \cdot q_{tip} (l_t, n) \cdot f(l_t, n, t))$$

$$+ \frac{\partial}{\partial n} (q_{bran} (l_t, n) \cdot f(l_t, n, t)) = g(t) \cdot \delta (l_t - l_{t,g}) \cdot \delta (n-1) \quad (27)$$

The solution is found by finding the growth curves, i.e. $l_t(t)$ and $n(t)$, for a hyphal element and thereafter convoluting them with the germination fre-

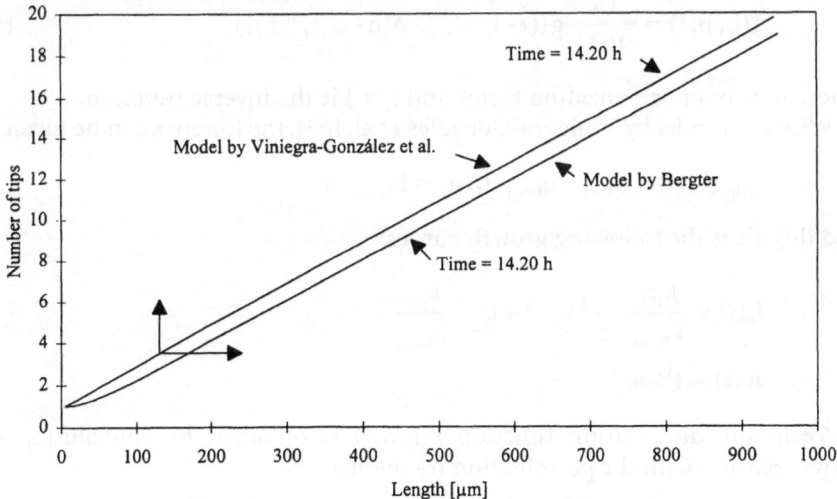

Fig. 7. Phase-plane plot of (l_t, n) during a batch cultivation. The growth curves are simulated using the models by Viniegra-Gozáles et al. [65] and Bergter [64]. The kinetics parameters are chosen so that both have the same maximum specific growth rate of the total hyphal length and number of tips and also the same tip extension rate. The velocity of a given hyphal element is given by $q_{bran}(l_t, n, z)$ in the tip direction and by $n \cdot q_{tip}(l_t, n, z)$ in the length direction

quency. In order to find the growth curves the growth kinetics is described as a system of two ordinary differential equations:

$$\frac{dl_t(t)}{dt} = n \cdot q_{tip}(l_t, n), \quad l_t(0) = l_g$$

(28)

$$\frac{dn(t)}{dt} = q_{bran}(l_t, n), \quad n(0) = 1$$

The growth curves are both monotonous increasing with time which means that all the hyphal elements at a given length have the same number of tips, and Fig. 7 shows a phase plane plot of the growth curve for two models.

With the model by Bergter [64] (see Sect. 9) the PBE can be solved by finding the growth curves for one hyphal element as a function of time and thereafter convoluting it with the germination frequency. The growth curves are found to be

$$l_t(t) = \frac{1}{2} \cdot \left(\sqrt{\frac{k_{tip}}{k_{bran}}} + l_{t,g} \right) \cdot e^{\sqrt{k_{tip} \cdot k_{bran}} \cdot t} + \frac{1}{2} \cdot \left(l_{t,g} - \sqrt{\frac{k_{tip}}{k_{bran}}} \right) \cdot e^{-\sqrt{k_{tip} \cdot k_{bran}} \cdot t}$$

(29)

$$n(t) = \frac{1}{2} \cdot \left(1 + l_{t,g} \cdot \sqrt{\frac{k_{bran}}{k_{tip}}} \right) \cdot e^{\sqrt{k_{tip} \cdot k_{bran}} \cdot t} + \frac{1}{2} \cdot \left(1 - l_{t,g} \cdot \sqrt{\frac{k_{bran}}{k_{tip}}} \right) \cdot e^{-\sqrt{k_{tip} \cdot k_{bran}} \cdot t}$$

and the distribution function $f(l_t, n, t)$ is obtained by convoluting the above solution with the germination frequency

$$f(l_t, n, t) = \frac{1}{N(t)} \cdot g((t - l_t^{-1}(l_t)) \cdot \delta(n - n(l_t^{-1}(l_t)))$$

(30)

where $N(t)$ is a normalization factor and $l_t^{-1}(\ )$ is the inverse function.

With the model by Viniegra-Gonzáles et al. [65], the kinetics can be given as

$$q_{tip}(l_t, n) = k_{tip}; \quad q_{bran}(l_t, n) = k_{bran} \cdot n$$

(31)

and this gives the following growth curves:

$$l_t(t) = \frac{k_{tip}}{k_{bran}} \cdot e^{k_{bran} \cdot t} + l_{t,g} - \frac{k_{tip}}{k_{bran}}$$

(32)

$$n(t) = e^{k_{bran} \cdot t}$$

Again the distribution function $f(l_t, n, t)$ is obtained by convoluting the above solution with the germination frequency

$$f(l_t, n, t) = \frac{1}{N(t)} \cdot g\left(t - \frac{1}{k_{bran}} \cdot \ln\left(1 + \frac{k_{bran}}{k_{tip}} \cdot (l_t - l_{t,g}) \right) \right) \cdot \delta\left(n - \frac{k_{bran}}{k_{tip}} \cdot (l_t - l_{t,g}) - 1 \right)$$

(33)

where $N(t)$ is the normalization factor.

When the growth curves of the two models are convoluted with a germination frequency distributions of the morphological properties can be obtained. From Fig. 8 it can be seen that the distributions from the two models resemble each other very much and an experimental discrimination of two models on the basis of the distribution is impossible unless a very large number of hyphal elements are measured. However, from the phase plane plot in Fig. 7 it may be possible to discriminate the two models on basis of the lag phase predicted by the Bergter model.

It is interesting to note that even though both models gives a reasonable prediction of the one-dimensional distribution they fail to give a correct description of the two-dimensional distribution since the completely deterministic description of the growth process for all hyphal elements implies that at a given length they have the same number of tips. The use of a continuous $p_l(l_t, z, t)$ instead of a Dirac-function would give a continuous density function, but even with a continuous function for $p_l(l_t, z, t)$ it will not be possible to obtain a wide density function as observed experimentally because the growth curves are rather insensitive to the start conditions. This indicates that there must be a stochastic element in the growth kinetics, i.e. in branching and tip extension. A mechanistic explanation for this can be that there is a distribution of properties in the spore population, e.g. some spores may have a composition that enables rapid branching of the hyphal element developing from these spores, whereas other spores have different compositions that result in slow branching of the resulting hyphal elements. Furthermore, it is likely that there is a stochastic element in the branching process, even for hyphal elements having the same properties. Finally, the assumption of no fragmentation in the initial phase may

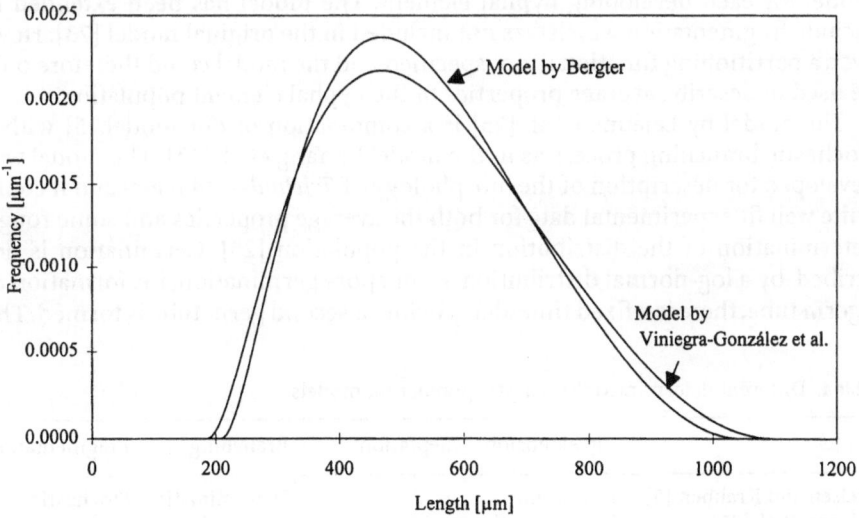

Fig. 8. Distributions of the hyphal length. The curves are obtained by convolution of the growth curves with the germination frequency. The germination frequency is given as a B-distribution with the parameters $t_s = 0$ h, $t_f = 8$ h and $\alpha = \beta = 3$

be wrong – even some fragmentation will change the resulting distribution dramatically.

10.2
Deterministic/Stochastic Models

A stochastic element has been introduced into the deterministic models either because the mechanism of branching is still unknown or because the process of fragmentation is stochastic. The deterministic/stochastic models described in literature are summarized in Table 4 together with some of their main characteristics.

The model by Yang et al. [43, 70] was primarily developed to simulate the growth of one hyphal element in three dimensions and thereby describe how it is possible to generate a pellet from one spore. The model has been modified further to include fragmentation at the pellet surface and mass transfer and lysis within the pellet [71]. The model is deterministic with regard to the growth of the length, but stochastic branching is included and three-dimensional morphology of the hyphal element is determined by stochastic processes – namely a stochastic placement of the branches, a stochastic branching angle and a stochastic tip extension direction. The basis of the model is the work of Prosser and Trinci [63] with one major modification. The transport of building blocks towards the tip is thought to be governed by diffusion. This mechanism may be correct in *Streptomyces* for which the model was developed, but it would not be able to describe the increase in the vesicle concentration towards the tip as has been measured in fungi [72]. The model cannot be directly formulated within the general PBE of Sect. 2 and it is therefore practically incapable of describing the formation of a realistic population of hyphal elements if it is combined with a germination-frequency, since it requires simulation of the rather complex model for each developing hyphal element. The model has been extended to include fragmentation which was not included in the original model [73]. However, a partitioning function is not specified and the model could therefore only be used to describe average properties of the hyphal element population.

The model by Lejeune et al. [25] is a combination of our model [5] with a stochastic branching process as in the model by Yang et al. [43]. The model was developed for description of the morphology of *Trichoderma reesei* and it could quite well fit experimental data for both the average properties and some rough determination of the distribution in the population [25]. Germination is described by a log-normal distribution. After spore germination, i. e. formation of a germ tube, there is a fixed time-delay before a second germ tube is formed. The

Table 4. Different deterministic/stochastic population models

Model	Tip extension	Septation	Branching	Fragmentation
Nielsen and Krabben [5]	Deterministic	n. i.[a]	Deterministic	Stochastic
Lejeune et al. [25]	Deterministic	n. i.[a]	Stochastic	n. i.[a]
Yang et al. [43, 70]	Deterministic	Deterministic	Stochastic	Included

[a] n. i.: not included.

growth of the individual branch is described by saturation kinetics with an offset (Eq. (34)), and the probability of branching is related to the total hyphal length (Eq. (35)). The model parameters were determined by fitting the model to measurements of average properties, and the model was then shown to give a fairly good prediction of the distribution. The model results in exponential growth of both the number of tips and the total hyphal length of the average properties. The specific growth rates for the individual hyphal elements converges but only very slowly towards the average specific growth rate (Fig. 9). This

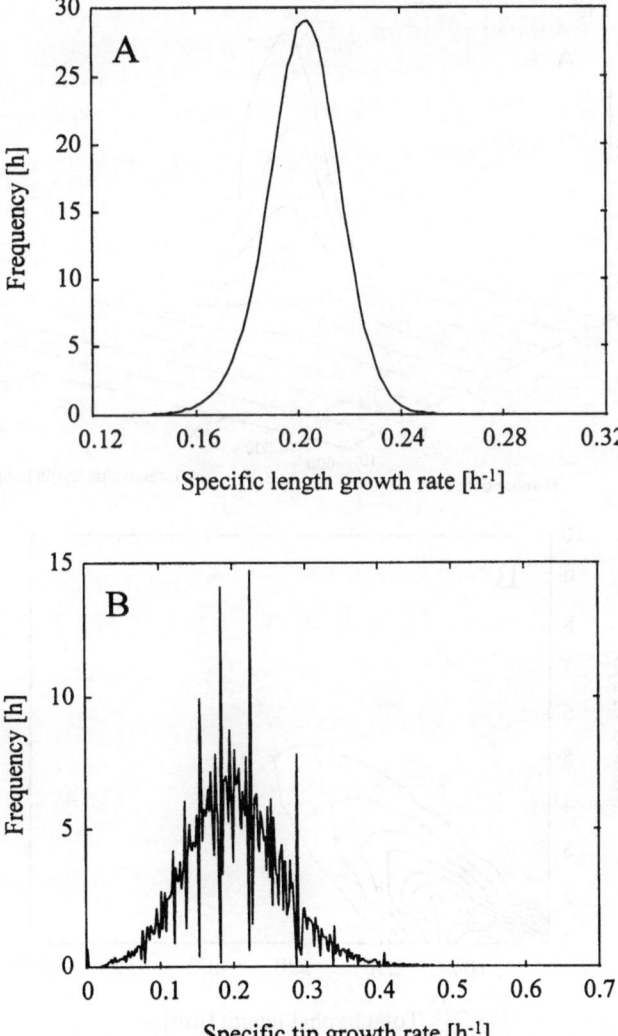

Fig 9 A, B. The distribution of the individual hyphal element growth rates after 18 h of growth. The data are simulated using the kinetic parameters for experiment TRI4 in Lejeune et al. [25]

model structure is probably the best model to describe the developing morphology in the early phase of batch-cultivations due to the stochastic element involved in branch formation (Fig. 10):

$$q_{tip}(l_{bran}) = k_{tip,1} + k_{tip,2} \cdot \frac{l_{bran}}{l_{bran} + K_{br}} \tag{34}$$

$$q_{bran}(l_t) = \begin{cases} k_{bran} \cdot l_t & l_t \geq l_{branch} \\ 0 & l_t < l_{branch} \end{cases} \tag{35}$$

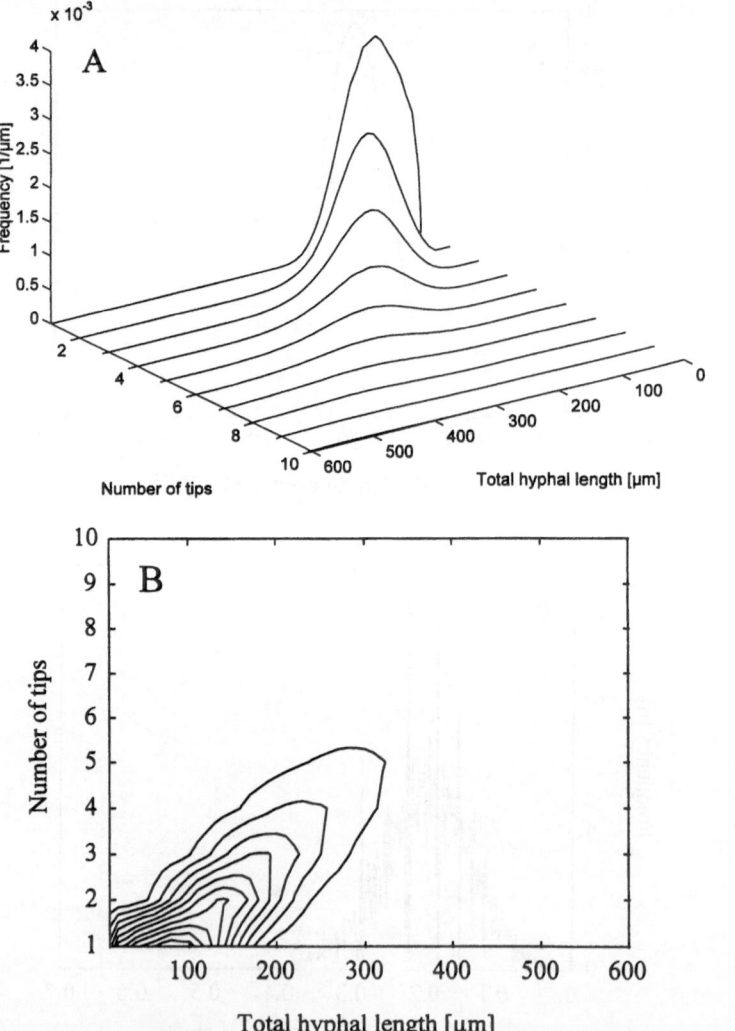

Fig 10 A, B. The hyphal element distribution at 18 h for a population growing with the kinetic parameters given for experiment TRI4 in Lejeune et al. [25]. A three-dimensional plot of the hyphal element distribution; B contour-plot of the hyphal element distribution

From our analysis of distribution functions in this and the previous section it is quite obvious that even if a model gives a good description of the one-dimensional distribution it may not necessarily be correct. It is therefore necessary to consider the two-dimensional distribution, even though it is (presently) practically impossible to obtain this experimentally. By the use of contour plots it is, however, possible to do a rough comparison of model simulations with experimental data, e.g. by checking whether different fractions of the experimental data lies within the corresponding contour lines.

11
References

1. Gravesen S, Frisvad JC, Samson RA (1994) Microfungi. Munksgaard, Copenhagen
2. Nielsen J (1992) Adv Biochem Eng/Biotechnol 46:187
3. Nielsen J, Carlsen M (1996) Fungal pellets. In: Willaert RG, Baron, GV, De Backer (eds) Immobilised living cell systems: modelling and experimental methods. John Wiley. (in press)
4. Ramkrishna D (1979) Adv Biochem Eng/Biotechnol 11:1
5. Nielsen J, Krabben P (1995) Biotech Bioeng 46:588
6. Nielsen J (1993) Biotech Bioeng 41:715
7. Sussman (1966): The fungi, vol 2, p 733
8. Paul GC, Kent CA, Thomas CR (1992) Biotechnol Bioeng 42:11
9. Yanagita T (1957) Arch Mikrobiol 26:329
10. Ekundayo JA, Carlile MJ (1964) J Gen Microbiol 35:261
11. Ohmori K, Gottlieb, D (1965) Phytopathol 55:1328
12. Gottlieb D, van Etten, JL (1964) J Bacteriol 88:114
13. Fletcher J (1969) Trans Br Mycol Soc 53:425
14. Kornfield JM, Knight SG (1960) Bact Proc 19:151
15. Gottlieb D, Ramachandran S (1960) Mycologia 52:599
16. Nielsen J (1997) Physiological engineering aspects of *Penicillium chrysogenum*. World Science Publishing Co., Singapore
17. Nyeri L, Lengyel ZL (1965) Biotechnol Bioeng 7:343
18. Nyeri L (1967) Z Allg Mikrobiol 7:107
19. Metz B (1976) PhD thesis, Technical University of Delft, Delft
20. Trinci APJ, Whittaker C (1968) Trans Br Mycol Soc 51:594
21. Fletcher J, Morton AG (1970) Trans Br Mycol Soc 54:65
22. Sekiguchi J, Gaucher GM, Costerton JW (1975) Can J Microbiol 21:2048
23. Sekiguchi J, Gaucher GM, Costerton JW (1975) Can J Microbiol 21:2059
24. Bosch A, Maronna RA, Yantorno OM (1995) Process Biochem 30:599
25. Lejeune R, Nielsen J, Baron G (1995) Biotechnol Bioeng 47:609
26. Reinhardt MO (1892) Jahrb Wiss Bot 23:479
27. Zalokar M (1959) Am J Bot 46:602
28. Fiddy C, Trinci APJ (1976) J Gen Microbiol 97:169
29. Paul GC (1992) PhD thesis, University of Birmingham, Birmingham
30. Paul GC, Kent CA, Thomas CR (1993) Trans I Chem E 70:13
31. Money NP (1994) Osmotic adjustment and the role of turgor in mycelial fungi. In: Wessels JGH, Meinhardt F (eds) The mycota, vol I: growth, differentiation and sexuality. Springer, Berlin Heidelberg New York
32. Trinci APJ (1978) The duplication cycle and vegatative development in moulds, p 134–163. In: Smith JE, Berry DR (eds) The filamentous fungi, vol III. Edward Arnold, London
33. Bartnicki-Garcia S (1973) Symp Soc Gen Microbiol 23:245

34. Bartnicki-Garcia S (1993) Role of vesicles in apical growth and a new mathematical model of hyphal morphological model of hyphal morphogenesis, pp 211–232. In: Heath IB (ed.) Tip growth in plant and fungal cells. Academic Press, San Diego, Calif
35. Grove SN (1978) The cytology of hyphal tip growth, pp 28–50. In: Smith JE, Berry DR (eds.) The filamentous fungi, vol III. Edward Arnold, London
36. Trinci APJ (1984) Regulation of hyphal branching and hyphal orientation. pp 23–52. In: Jennigs DH, Rayner ADM (eds) The ecology and physiology of the fungal mycelium. Cambridge University Press, Cambridge
37. Morrison KB, Righelato RC (1974) J Gen Microbiol 81:193
38. Robinson PM, Smith JM (1979) Proc Br Mycol Soc 72:39
39. Robson GD, Wiebe MG, Trinci APJ (1991) J Gen Microbiol 137:963
40. Wiebe MG, Trinci APJ (1991) Biotechnol Bioeng 38:75
41. Dicker JW, Turian G (1990) J Gen Microbiol 136:1413
42. Reissig JL, Kinney SG (1983) J Bacteriol 154:1397
43. Yang H, Reichl U, King R, Gilles ED (1992) Biotechnol Bioeng 39:49
44. Pirt SJ, Callow, DS (1960) J Appl Bact 23:87
45. Savage GM, van der Brook MJ (1946) J Bacteriol 52:385
46. van Suijdam JC, Metz B (1981) Biotechnol Bioeng 23:111
47. Tucker KG, Kelly T, Delgrazia P, Thomas CR (1992) Biotechnol Prog 8:353
48. Reuß M (1988) Chem Eng Technol 11:178
49. Ayazi Shanlou P, Makagiansar HY, Ison AP, Lilly MD, Thomas CR (1994) Chem Eng Sci 49:2621
50. Smith JJ, Lilly MD, Fox RI (1990) Biotechnol Bioeng 35:1011
51. Prosser JI (1991) Mathematical modeling of vegative growth of filamentous fungi. In: Arora DK, Rai B, Mukerji KG, Knudsen GR (eds) Handbook of applied mycology, vol 1. Marcel Dekker, New York, p 591
52. Aynsley M, Ward AC, Wright AR (1990) Biotechnol Bioeng 35:820
53. Cohen D (1967) Nature 216:246
54. Leopold LB (1971) J Theor Biol 31:339
55. Trinci APJ, Saunders PT (1977) J Gen Microbiol 103:243
56. Saunders PT, Trinci APJ (1979) J Gen Microbiol 110:469
57. Koch AL (1982) J Gen Microbiol 128:947
58. Bartnicki-Garcia S, Hergert F, Gierz G (1990) A novel computer model for generating cell shape: application to fungal morphogenesis. In: Kuhn PJ, Trinci APJ, Jung MJ, Goosey MW, Copping LG (eds). Springer, Berlin Heidelberg New York
59. Bartnicki-Garcia S, Hergert F, Gierz G (1989) Protoplasma 153:46
60. Green PB (1969) Annu Rev Plant Physiol 20:365
61. Riva Ricci D da, Kendrick B (1972) Can J Bot 50:2455
62. Thompson D'Arcy W (1952) On growth and form. Cambridge University Press, Cambridge
63. Prosser JI, Trinci APJ (1979) J Gen Microbiol 111:153
64. Bergter F (1978) Z Allg Mikrobiol 18:143
65. Viniegra-Gonzáles G, Saucedo-Castañeda G, López-Isunza F, Favela-Torres E (1993) Biotech Bioeng 42:1
66. Katz D, Goldstein D, Rosenberger RF (1972) J Bacteriol 109:1097
67. Schuhmann E. Bergter F (1976) Z Allg Mikrobiol 16:201
68. Eakman JM, Fredrickson, AG, Tsuchiya HM (1966) Chem Eng Prog Symp Ser 69:37
69. Krabben P, Nielsen J (1996) MS in preparation
70. Yang H, Reichl U, King R, Gilles ED (1992) Biotechnol Bioeng 39:44
71. Meyerhoff J, Tiller V, Bellgardt K-H (1995) Bioproc Eng 12:305
72. Collinge AJ, Trinci APJ (1974) Arch Microbiol 99:353
73. Yang, H. (1994) Mathematical model for filamentous growth of mycelial microorganisms in lab chambers and in batch, fed-batch and continuous cultures. In: Alberghina L, Frontali L, Sensi P (eds). Proceedings of the 6th European Congress on Biotechnology 13–17 June 1993. Elsevier Amsterdam, Vol II, p 845

Received December 1996

Process Models for Production of β-Lactam Antibiotics

K.-H. Bellgardt

Institute for Chemical Engineering, University of Hannover, D-30176 Hannover, Germany

Great progress has been made in the modelling of biotechnical processes using filamentous microorganisms. This paper deals with cultivations of *Penicillium chrysogenum* for the production of Penicillin and of *Acremonium chrysogenum* for the production of Cephalosporin C. The properties of the processes and the existing models are reviewed. Models are presented for both processes that consider aspects which are important for industrial cultivation. The process model for Penicillin production is based on a detailed morphological description of growth of hyphal filaments and pellets. The model allows for simulation of the production process including the preculture and considering the inhomogenous pellet population. It opens new possibilities for understanding the complex kinetics of the process and improvement of its control. The structured segregated model for Cephalosporin C production considers soy oil as second carbon source besides sugar. The application of the model for dynamic optimization of feeding strategies by Iterative Dynamic Programming is demonstrated. As an alternative approach, modelling of the Cephalosporin production by an artificial neural network is discussed.

Advances in Biochemical Engineering/
Biotechnology, Vol. 60
Managing Editor: Th. Scheper
© Springer-Verlag Berlin Heidelberg 1998

List of Symbols and Abbreviations

a	Model parameter
A	Mean culture age, h
b	Model parameter
C	Concentration, $g\,l^{-1}$
CPR	Carbon dioxide production rate, $g\,l^{-1}\,h^{-1}$
d_h	Diameter of hyphae, μm
D	Dilution rate, h^{-1}
D_i	Diffusion coefficient for component i, $cm^2\,s^{-1}$
E_2	intrinsic enzyme concentration
f_{12}	Model parameter
F	Flow rate, $l\,h^{-1}$
F_{Oil}	Feeding rate of soy oil, $g\,h^{-1}$
hgu	hyphal growth unit, μm
I	Performance criterion
k	Model parameters
K	Limitation constants, $g\,l^{-1}$
K_I	Inhibition constant, $g\,l^{-1}$
l_{ap}	Length of apical region, μm
L	Length of hyphae, μm
m	specific maintenance coefficient, h^{-1}
m_i	Mass of substance i, g
M	Dry syrup concentration, $g\,l^{-1}$
n	number of tips
Oil	Soy oil concentration, $g\,l^{-1}$
P	Product concentration, $g\,l^{-1}$
P_{Br}	Breakage probability
PM	Pharma medium concentration, $g\,l^{-1}$
q	Specific reaction rate, h^{-1}
r	radius, μm
R	Radial position of fictive layer, μm
S	Substrate concentration, $g\,l^{-1}$
t	time, h
t_G	Germination time, h
u_{max}	Model parameter
V_L	Liquid volume, l
V_R	Volume of fictive layer at radius R, μm^3
X	Dry biomass concentration, $g\,l^{-1}$
xCO_2^A	Mole fraction of CO_2 in the exhaust gas
Y	Yield coefficient, $g\,g^{-1}$
α	Linear hyphal extension rate, $\mu m\,h^{-1}$
β	Model variable
γ	Model variable
μ	Specific growth rate, h^{-1}
κ	Model variable
φ	Normalized limitation kinetics

π   Specific production rate, h^{-1}
ρ   Density, $g\, l^{-1}$
λ_{Shear} Shear stress parameter.

1
Introduction

In the production of secondary metabolites, such as the antibiotics Penicillin or Cephalosporin C by the fungi *Penicillium chrysogenum* and *Acremonium chrysogenum (Cephalosporium acremonium)*, many biological, chemical and technical factors influence the growth of the cells and product formation. Small alterations in preculture, media, and operating conditions change the productivity in a seldom fully predictable way. The actual research is, therefore, directed towards broader understanding of the processes. On the basis of new and improved analytical techniques, mathematical models can help to get an integral view of the entire process by putting together the knowledge of the different levels from biology over cultivation to process control.

Synthesis of β-lactam-antibiotics is a complex reaction network coupled with catabolism and anabolism and undergoes various kinds of metabolic regulation. Precursors for the antibiotics come from the metabolism of amino acids [1–5]. High productivity is only possible in the presence of sufficient amounts of carbon source and additional precursors in the medium. It is known that antibiotics synthesis is at several points of the pathway subject to catabolite repression by hexoses such as glucose [2, 6, 7]. High concentrations of easily convertible sugars lead to high growth rates but reduced product synthesis by catabolite repression. Therefore, the kinds of C- and N-sources and their amounts present in the medium have a strong influence on growth and productivity [8–10]. This aspect of the medium composition is discussed by Wittler and Schügerl [11] in detail. Another, and often limiting, factor for productivity is the oxygen supply to the culture because it sets bounds on the maximum attainable growth rate and cell density. Furthermore, the cyclization reaction of the tripeptide is directly oxygen-dependent [2, 12, 13]. Even short oxygen limitations can have lasting negative effects on the product synthesis. Morphology of the cells, which is influenced by the conditions during preculture and cultivation, is another important parameter for antibiotics production. The morphology is closely related to the fungal life-cycle and to aging effects, as result of the growth history of the cells, namely the properties of the preculture and the cultivation conditions. It is a general view that only non-growing morphological states contribute to product synthesis.

Therefore, the antibiotics production processes can be characterized by a diversity of influencing factors originating from medium, growth conditions, and biological properties of the strains. The difficulties in quantifying these effects and relating them to the morphological heterogeneity of the culture and to resulting variations in the productivity are the cause for the so-called bad reproducibility of the process. Mathematical models can help to throw some light onto the kinetics and mechanisms on the cellular level and connect them to the observed macrokinetic effects.

Mathematical modelling of antibiotics production is a wide field. Depending on the modelling target, different approaches are used for the description of certain aspects of the cultivations. Known models cover all process levels beginning with the intracellular reaction network, growth of hyphae in space and time, over kinetics of substrate uptake and product formation, up to segregated or distributed models for population dynamics, aging, and morphological heterogeneity [14]. These are accompanied by efforts to describe inhomogeneities in the bioreactors as a result of transport processes and hydrodynamic effects, in connection with the rheological properties of the mycelia suspensions, which are not reported here [15, 16]. The biological models can be divided into a few categories [17, 18].

Unstructured models have the highest degree of simplicity. The biomass is viewed as homogeneous and constant over time in its properties. The only characterising variables for the process are cell mass, and kinetics for substrate uptake, growth and production. This type of model is less suited for the complex phenomena observed in antibiotics production.

Structured models account for variations in the growth properties of the cells over time due to slow metabolic regulation or other intracellular processes. The structural elements of these models are intracellular concentrations of metabolites, enzymes or cell constituents.

Simple segregated models distinguish a few types of cells with different properties, which can be of genetic, age-related or morphological origin. These models are already a better approximation to the reality of antibiotics production. With some mathematical care they can be combined with structured models and are a suitable approach for description of antibiotics processes.

Complex segregated distribution models are able to describe a continuous variation of a cellular property, which may be related to individual cells, e.g. age, size or intrinsic concentration of a certain cellular component. The mathematical and computational effort for such models is rather high. Therefore, for practical reasons, they are mostly complemented by only unstructured growth models, which on the other hand restricts their applicability.

Morphologically structured models may be taken as a special case of simple segregated models, where the cell classes represent the different morphological cell types of hyphae. Their advantage is that the model variables can be directly interpreted in biological terms and relative easily be compared to experimental data.

Detailed morphological models for growth of single hyphae and their extensions for three-dimensional description of mycelia flocs and pellets have the advantage that no simplifying assumption about the lumping of cells of different morphological states into a few classes have to be met. On the other hand, the effort for considering each individual cell of mycelia and pellets is tremendous. Therefore, the application of such models for process simulation requires simple biological models and further simplifications on the population level. The advantage is that the models conserve the inner dependence from the microkinetics of hyphal growth up to the process level.

In the following, some new models for the production of Penicillin and Cephalosporin are presented, which try to cover further aspects of the cultivations, such as consideration of additional complex substrates, age and morphological state of the culture, or the development of mycelia pellet distributions. In a first step, these models can improve our understanding of the mechanisms of observed macrokinetic phenomena during the cultivations, and of the interrelation of preculture, operating conditions, growth, and production. In a second step, the models can be used to evaluate control strategies and optimize the production process.

2
Penicillin Process

The biochemical and morphological aspects of cultivations of filamentous fungi and their growth mechanisms are discussed in several papers and reviews [18–21, 108]. The morphology is influenced by the strain properties, preculture, and cultivation conditions. *Penicillium chrysogenum* can grow in the form of mycelia filaments or more dense pellets [22–26]. The morphology has a strong influence on the viscosity of the cultivation broth, and indirectly on the oxygen supply. In contrast to pellets, free filaments increase viscosity of the cultivation broth and hinder oxygen transfer, which then limits cell density and productivity. On the other hand, substrates such as nitrogen or carbon source, and oxygen can enter the inner parts of pellets only diffusively. Experimental results show that pellets up to a size of 200–400 μm are optimum and have no negative effect on the productivity, because the viscosity still remains low and there is no serious transport limitation into the pellets as is observed for larger pellets. The pellet form itself seems to have no negative effect on the productivity. The models presented in this article deal preferentially with the aspects of the production process.

Pellets may be formed by agglomerates of spores, hyphae and solid particles from the medium constituents. Ideally, a pellet evolves from only one spore. Small pellets often have a decreasing density over the radius [19, 20], but generally the density can pass through a maximum over the radius [11]. The pellet size is strongly influenced by shear forces in the reactor [27, 28]. High energy input supports small, smooth, and compact pellets, while under low shear stress the pellets are larger and more hairy. Large (> 400 μm), old pellets are often hollow in the center as a result of autolysis. This is caused by strong limitation of substrate and mainly oxygen into the inner zones [11, 19, 29]. The measured oxygen profiles by microprobes clearly support the autolysis hypothesis. The mechanisms of the intra pellet transport are still under discussion. It was found [11, 26] that molecular diffusional, turbulent and directed convective flows are in the same order of magnitude. Diffusion seems to be the main mechanism in the center of the pellets while turbulence penetrates only the outer parts.

For maximum production, the cells need sufficient amounts of carbon source and precursor, phenylacetic acid and phenoxyacetic acid, which are incorporated into the penicillin molecule in the last synthesis step. Sugars such as glucose or lactose, starch hydrolyzates, and plant oil may serve as substrate for the cultivation. The disadvantages of oil are the increased oxygen demand and the hindrance of the oxygen transfer by bubble coalescence. Addition of protein to the substrates improves the productivity by freeing resources for product formation in the amino-acid pathways.

Models for industrial-like cultivations should consider as many as possible of these aspects to be able to give a proper description of measured process variables. The model can then be used to analyze the cultivations and to draw conclusions about the course of non-measured variables, or of the influence of control variables on the cultivation, to guide the process optimization. In the next paragraph, existing models for penicillin production are summarized and then two models are refereed in more detail, which put emphasis on the above-mentioned aspects.

2.1
Models for Penicillin Production

The complexity of models for penicillin production has increased significantly over the years. The early models were purely unstructured [30–33], and tried only to give a formal description of the distinction of growth and production phase. Based on the early experimental results [22, 34], Righelato [35], Bull and Trinci [36], and van Suijdam and Metz [19] proposed models for the development of pellet populations in the average. It is assumed that hyphae are only growing at the free tips. The important parameter is the hyphal growth unit as the ratio of the average hyphal length per free tip. The average hyphal length, the number of free tips and the hyphal extension rate were found to be proportional to the specific growth rate.

Calam and Russel [32] used a model with extremely long lag-phase to explain the delayed product formation. Hegewald et al. [37] incorporated not only the dependence of the kinetics on carbon source and dissolved oxygen, but also the effects of the nitrogen source in the model (ammonia and amino nitrogen). A mechanistic model based on the Contois-kinetics for carbon source and oxygen limitation was developed by Bajpai and Reuss [38]. Yield variations are considered by a maintenance term in the balance equations. The unstructured model included a substrate inhibition term for repression of product synthesis during the growth phase. Due to its simplicity, the model served as a basis for a number of further investigations, not only to improve the validity of the model, but also to use it for extensive process optimization studies. Montague et al. [39] extended this model for carbon dioxide production and then used it for on-line state estimation and parameter-adaptive control purposes. A further extension and improved parameter estimation of the Baijpal-Reuss model was given by Nicolai et al. [40], who also included some ideas of the balancing approach of Heijnen et al. [41]. To overcome model inconsistencies during batch and fed-batch simulations, the authors introduced a variation in the maintenance metabolism related to the substrate concentration. Genon and Saracco [42] extended the Baijpal-Reuss model by economic balances to evaluate the profitability of fed-batch, repeated fed-batch, and continuous cultivations in one- and two-stage reactor systems. It was found that with this simple unstructured model the last system performed best when neglecting the known operational difficulties for continuous cultivations. Repeated batch with a cycle period of

96 h and a draw-off ratio of >95% of the reactor operating volume was the second choice.

Another simple model – which correlates penicillin production, maintenance metabolism and growth with the CO_2 production – was given by Calam and Ismail [43]. The model is applied to control of substrate feeding in a fed-batch process. Similar problems were attacked by several authors [44–47] with balancing methods.

In some models an age dependence of the product formation is assumed rather than a correlation to the specific growth rate, e.g. Calam [48], and Calam and Ismail [43, 44]. An age distribution model is used by Germain et al. [49] to explain the stagnation of the product at the end of longer cultivations. Calam [50] interprets this aging effect somewhat differently as a decrease in the concentration of active cells. In contrast, Heijnen et al. [41] refer the slowdown of the production to a simple dilution effect due to the substrate feeding and as a consequence of product hydrolysis. They propose a direct coupling of growth and production.

The morphologically structured model of Nestaas and Wang [51] also postulates coupling of growth and production assuming a production rate dependent on the specific growth rate. The cell mass is segregated into three types: growing tips, non-growing but producing, and degenerated non-producing inactive cells. Beside endogenous metabolism of the fraction of active biomass, a cell lysis mechanism is also used to explain variations in the yield. A fictive inner enzyme is assumed to be responsible for product synthesis after derepression. The proposed mechanisms cannot account for substrate limitation, which is an important property for application of the model to fed-batch cultivations. Although the model is not predictive, it was used by Guthke and Knorre [52] to investigate the optimal control of repeated fed-batch cultivations under singular conditions. They found an optimal cycle period of 48 h and a draw-off ratio of about 80%. This is lower than the result of [42] but still unusually high. Nevertheless, the tendency of the predicted optimal control policy was in agreement with experimental data.

The growth of fungal pellets with consideration of transport limitation was modeled by van Suijdam et al. [53] and van Suijdam and Metz [19, 21]. They assume a constant biomass density within the pellets and pellets of uniform size, which is not found in reality. The aspect of intra-pellet substrate diffusion and the total substrate turn-over of a pellet were under practical and theoretical investigation by Metz and Kossen [23]. Pellet breakage and disruption due to mechanical sheer stress as a consequence of high energy input for agitation of the cultivation medium was looked at by van Suijdam and Metz [21] and Bhavaraju and Blanch [54]. The formation of hollow pellets and lysis effects during substrate limitation was described by Wittler [29] and Wittler and Schügerl [11] in a mathematically simple form. The last paper proposes an equation for the development of the average pellet diameter as a function of the inoculation density and the actual concentration of the pellets.

Kluge et al. [55] presented an extension of the model of Heijnen et al. [41] which uses a first-order delay system for induction of the penicillin synthesis after derepression of the glucose. The segregated model considers active and inactive biomass as well as a temperature dependency of some model parameters, and thus can be used for simulation of experiments with temperature shifts or for temperature profile optimization.

A detailed hyphae model was proposed by Ainsley et al. [56] as an extension of the ideas of Prosser and Trinci [57]. The hyphae are considered as self extending tubular reactors, where substrate diffusion into the cells take place all over the hyphae. The substrate is converted into vesicles, which are transported to the tip. The flow of vesicles determines the tip extension rate, and their level the branching frequency. Fragmentation was assumed to occur mainly at substrate limitation. The model could be fitted to experimental data of total biomass and to the hyphal growth unit.

A morphologically structured model was combined with a population balance model by Nielsen [58] to analyze morphological data from different batch and continuous cultivations of filamentous microorganisms. The model discriminates three compartments, growing apical cells, subapical cells which still participate directly in the tip extension process, and hyphal cells that are inactive with respect to the growth process, but provide material for the growth of apical cells by contributing to the stream of vesicles. Branching points are assumed to be located at the subapical region. The morphological model is complemented by balance equations for the average number of hyphal elements and actively growing tips as well as the hyphal mass. These equations are derived from a distribution model. It is an important point for the description of submerged cultures that the model discriminates between active and inactive tips since fragmentation is a frequent event under agitated conditions. If one assumes that practically every mycelia floc was subject to fragmentation, it must contain exactly two inactive tips which add to the measurable total number of tips. The model was in good agreement with experimental data of continuous cultures of *Penicillium chrysogenum*. It was also found that the specific rate of fragmentation is linearly correlated with the energy input, regardless of batch or continuous operation.

Using the basic ideas of Nestaas and Wang [51], Paul and Thomas [59] presented an interesting structured-segregated model for submerged growth of *Penicillium chrysogenum* filaments with far-reaching experimental verification by data from image analysis. In this model, the vacuole formation is considered as an important physiological process during growth and aging of hyphae. The model divides the biomass into basically three distinct regions according to the activities and structure of hyphal compartments: these are actively growing tips (A_0), nongrowing Penicillin-producing regions, and degenerated or metabolic inactive regions (A_3) that can be subject to fragmentation and autolysis. The nongrowing region is considered to consist of cytoplasm (A_1) and vacuoles (A_2). Growing tips are transformed by septation into nongrowing type-A_1-cells that initially contain no vacuoles. Under substrate-limited conditions vacuole formation is initiated, which then grow in size until they fill the entire cell volume and finally lead to the degenerated A_3-form. The vacuole size distribution in the hyphae is described by a population balance equations. Penicillin production is assumed to take place in the cytoplasm at low substrate concentrations. The differentiation phenomena are represented by simple kinetic equations of the Monod-type. Unknown model parameters were estimated from experimental data for the development of cell-dry-mass equivalents of types A_0-A_3 during fed-batch cultivations. This quantitative information was obtained by methods of digital image analysis. The model was in good agreement with the experimental data and had far reaching predictive capabilities for a number of fed-batch cultivations with varying sugar-feeding strategies. The highest product concentration was obtained by a feeding profile with gradually decreasing glucose-feed during the production phase.

2.2
A Segregated Model for Near-Production Cultivations in the Presence of Pellets

None of the above-mentioned investigations considered an effect of the macroscopic fungal morphology on the overall model structure. But for cultivations in the presence of pellets one must expect variations on the macrokinetics due to inhomogenous biomass distribution in the pellets, inhomogenous substrate supply by transport limitation, and alterations in the size and age distribution of the pellet population [23]. Growing cells are located mainly in the outer layers of the pellets, while production is assumed to take place preferably by the non-growing hyphae in the inner regions of the pellets under substrate limitation. Tiller et al. [60] presented a segregated model that was aimed at cultivations of a highly-producing strain of *Penicillium chrysogenum* with pronounced pellet formation. A segregated model for the antibiotics production can be taken as a first approximation to the inhomogeneities in the population and the effects of morphological differentiation and aging of cells. The most developed models so far of Montague et al. [39], Bajpai and Reuss [38, 61] and Heijnen et al. [41] showed no agreement with the experimental data under these conditions.

The model presented here was aimed at the description of cultivations with a high-yielding strain of *Penicillium chrysogenum* in a 20-l reactor. In a first batch growth phase, a mixture of 20 g/l Pharma-medium and glucose was used as a substrate. In the growth phase up to 90% of the total biomass was found as pellets with a diameter of > 250 μm. During the fed-batch production phase, glucose was continuously added to the culture to avoid catabolite repression of the production. Figure 1 shows a plot of the development of total biomass and of pellet fractions with small and large diameters during the cultivation. After the exponential growth phase (up to 60 h), the formation of bigger pellets stagnates for about 40 h. Then a pronounced fragmentation and lysis of these large pellets can be observed, leading to a further increase in the sieve fraction with small diameter.

The characteristics of these cultivations were a significant variation in the yield from the batch to the fed-batch phase. This can only be referred in part to the Pharma medium which is used in the batch phase. A detailed analysis showed that morphological factors contribute to this effect since the yield variation is accompanied by a significant change in the pellet size distribution and a decrease in the specific metabolic activity in the second process phase. This can be explained by aging effects and mechanical damage of the cells. On the contrary, the maintenance turnover increases significantly, so one may conclude

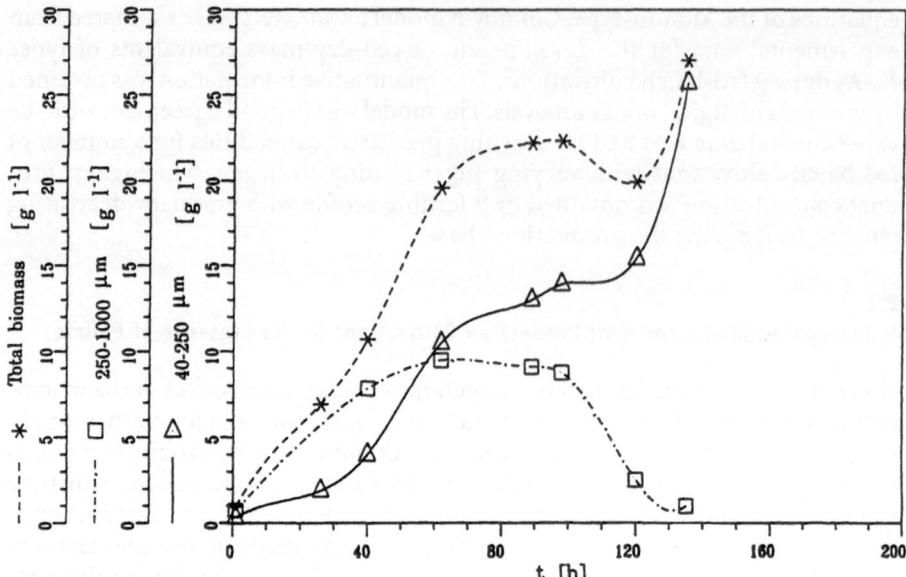

Fig. 1. Biomass sieve fractions of a *Penicillium chrysogenum* cultivation containing pellets [60]. The growth phase ends at about 50 h. The data points are connected by spline interpolation. Reprinted from Tiller et al. (1994) Segregated mathematical model for the fed-batch cultivation of a high-producing strain of *Penicillium chrysogenum*. J Biotechnol 34:119–131, with kind permission of Elsevier Science, Sara Burgerhartstraat 25, 1055 KV Amsterdam, The Netherlands

that cells within pellets have lower maintenance requirements compared to free mycelia.

The model subdivides the entire biomass, X,

$$X = X_1 + X_2 \tag{1}$$

into two fractions to take the morphological change into account, a growing and producing fraction, X_1, and a non-growing but producing fraction, X_2. The balance equations for fed-batch cultivations are

$$\frac{dX_1}{dt} = (\mu_s + \mu_{PM}) X_1 - (D + k_{12}) X_1 \tag{2}$$

$$\frac{dX_2}{dt} = k_{12}X_1 - (D + k_{ly}) X_2 \tag{3}$$

where the specific growth rates on glucose and Pharma medium are taken as Monod-kinetics:

$$\mu_s = \mu_{S, max} \frac{S}{K_S + S} \tag{4}$$

$$\mu_{PM} = \mu_{PM,max} \frac{PM}{K_{PM} + PM} \tag{5}$$

The balance equations for the substrates then become

$$\frac{dS}{dt} = -\frac{\mu_s}{Y_{XS}} X_1 - \left(\frac{\pi}{Y_{PS}} + m\right) X + F - DS \tag{6}$$

$$\frac{dPM}{dt} = -\frac{\mu_{PM}}{Y_{XPM}} X_1 - DPM + k_{ly} X_2 \tag{7}$$

The third term in the balance, Eq. (7), of Pharma medium accounts for reformation of complex media constituents by cell lysis. The amino acid and protein constituents of Pharma medium are an important source of nitrogen during the cultivation. The rate of morphological differentiation, k_{12}, of lysis, k_{ly}, and of the maintenance requirements, m, are introduced as age-dependent according to

$$k_{ly} = a_{ly} + b_{ly} A \tag{8}$$

$$k_{12} = f_{12} A \tag{9}$$

$$m = b_m + a_m A \tag{10}$$

where the mean culture age can be calculated by the simplified equation

$$A(t) = \frac{1}{X(t)} \int_0^t X(\tau) d\tau \tag{11}$$

which assumes that the initial age is zero. The error made by this assumption is small because of the low inoculation concentration of cells. Since both types of cells are assumed to produce, the balance equation for penicillin becomes

$$\frac{dP}{dt} = \pi X - (D + k_h)P \tag{12}$$

This also considers a product decay by a first order mechanism. The specific penicillin production during the cultivations was not consistent with the assumption of a substrate inhibition mechanism as proposed by Baijpal and Reuss [38]. Instead, growth related production according to the function given in Fig. 2 was used, where the specific production rate is maximum for regions of moderate growth rates. At strict limitation or above a critical level, $2\mu_{P2}$, no product is synthesized. Such a mechanism is close to earlier studies [41, 62].

The model was successfully applied for simulation of fed-batch cultures. The model parameters are given in Table 1. Here, only the results for substrate, cell mass and product are reported in Fig. 3. Figure 4 shows some of the control

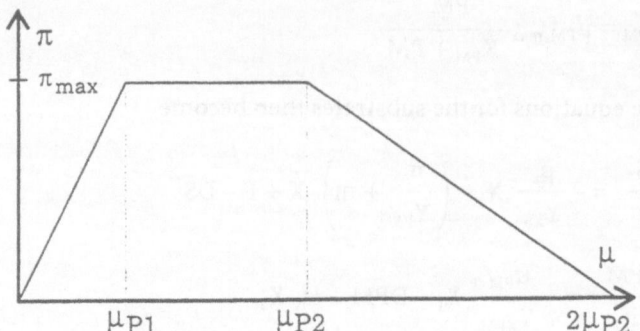

Fig. 2. Schematic plot of the specific product formation rate vs specific growth rate. Reprinted from Tiller et al. (1994) Segregated mathematical model for the fed-batch cultivation of a high-producing strain of *Penicillium chrysogenum*. J Biotechnol 34:119–131, with kind permission of Elsevier Science, Sara Burgerhartstraat 25, 1055 KV Amsterdam, The Netherlands

Table 1. Parameters of the model of Tiller et al. [60]

Parameter	Unit	This model (Fig. 3)	Bajpai and Reuß [61]	Nestaas and Wang [51]	Heijnen et al. [41]	Genon and Saracco [42]
$\mu_{S,max}$	h^{-1}	0.06	0.11		0.18	0.11
π_{max}	h^{-1}	0.0046	0.004	0.006		0.004
K_S	$g\,l^{-1}$	0.07	0.12[a]		1.0	
$\mu_{PM,max}$	h^{-1}	0.03				
K_{PM}	$g\,l^{-1}$	2.0				
Y_{XS}	$g\,g^{-1}$	0.47	0.47	0.6	0.5–0.78	0.47
Y_{XPM}	$g\,g^{-1}$	0.51				
Y_{PS}	$g\,g^{-1}$	1.2	1.2		0.85–1.17	1.2
Y_{XO}	$g\,g^{-1}$	1.25				
Y_{PO}	$g\,g^{-1}$	6.25	6.25			
Y_{XC}	$g\,g^{-1}$	0.9			0.34	
Y_{PC}	$g\,g^{-1}$	0.2				
k	h^{-1}	0.0006	0.04	0.003	0.002	0.01
μ_{P1}	h^{-1}	0.003				
μ_{P2}	h^{-1}	0.014		0.0115	0.01	
f_{12}	h^{-2}	0.00046				
a_m	h^{-2}	0.00015				
b_m	h^{-1}	0.001				
m_O	$g\,l^{-1}\,h^{-1}$	0.03				
m_C	$g\,l^{-1}\,h^{-1}$	0.022				
a_{ly}	h^{-1}	−0.0008				
b_{ly}	h^{-1}	$3.0 \cdot 10^{-6}$				

[a] Modified model from [60].

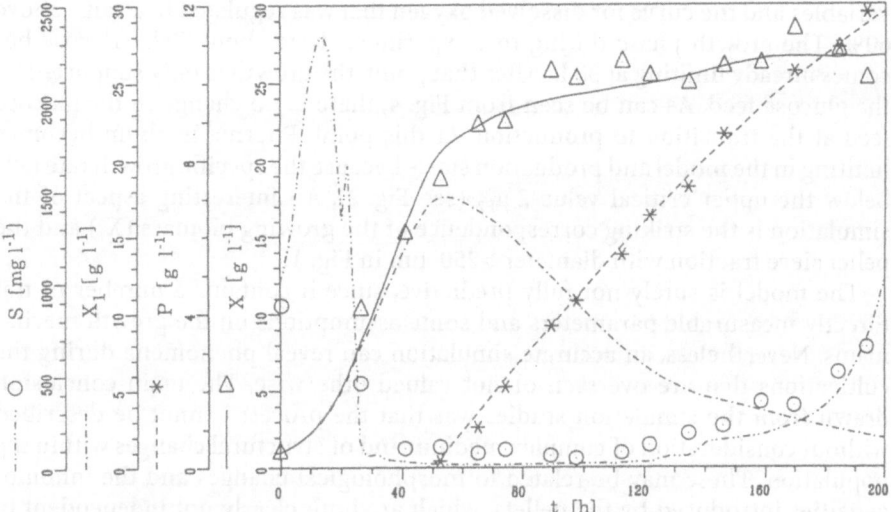

Fig. 3. Experimental data (*symbols*) and simulation results with the model of Tiller et. al. [60] (*lines*) for a fed-batch experiment of *Penicillium chrysogenum*: Total biomass (X), growing biomass (X₁), glucose concentration (S) and product concentration (P). Reprinted from Tiller et al. (1994) Segregated mathematical model for the fed-batch cultivation of a high-producing strain of *Penicillium chrysogenum*. J Biotechnol 34: 119–131, with kind permission of Elsevier Science, Sara Burgerhartstraat 25, 1055 KV Amsterdam, The Netherlands

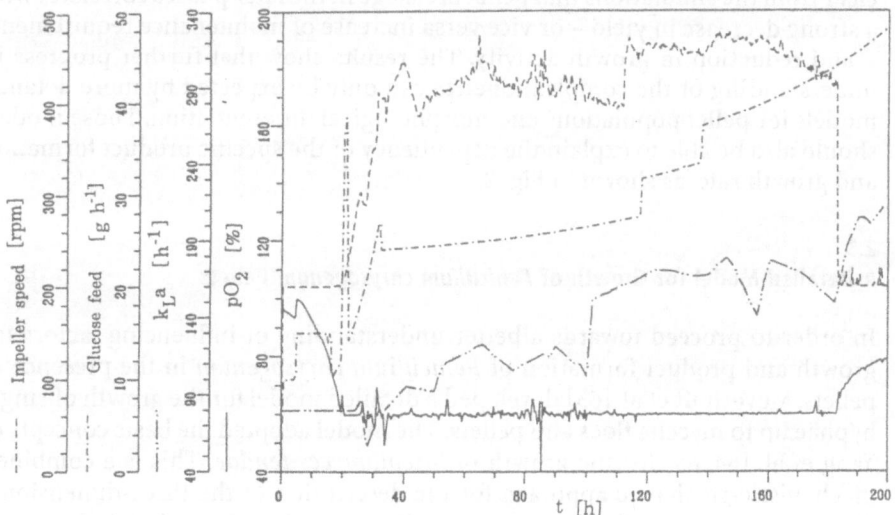

Fig. 4. Glucose feed, impeller speed, specific mass transfer coefficient for oxygen, k$_L$a, and dissolved oxygen concentration, pO$_2$, for the experiment shown in Fig. 3. Reprinted from Tiller et al. (1994) Segregated mathematical model for the fed-batch cultivation of a high-producing strain of *Penicillium chrysogenum*. J Biotechnol 34: 119–131, with kind permission of Elsevier Science, Sara Burgerhartstraat 25, 1055 KV Amsterdam, The Netherlands

variables and the curve for dissolved oxygen that was regulated to a value above 60%. The growth phase during this experiment lasts about 50 h. Glucose becomes already limiting at 30 h. After that point, the growth is only supported by the glucose feed. As can be seen from Fig. 4, there is no change in the glucose feed at the transition to production. At this point, Pharma medium becomes limiting in the model and production starts because the specific growth rate falls below the upper critical value $2 \mu_{P2}$ (see Fig. 2). An interesting aspect of the simulation is the striking correspondence of the growing biomass (X_1) and the pellet sieve fraction with diameter >250 µm in Fig. 1.

The model is surely not fully predictive, since it contains a number of not directly measurable parameters and some assumptions on the growth mechanisms. Nevertheless, an accurate simulation can reveal phenomena during the cultivations that are overseen or not valued otherwise. The main conclusion drawn from the simulation studies was that the process cannot be described without consideration of complex medium and of structural changes within the population. These may be related to morphological changes and the inhomogeneities introduced by the pellets, which are both clearly not independent of each other. There is a clear variation in yield and activity of the cell from growth to production phase that cannot simply be explained by growth limitation. The presence of pellets in the early stages and their break-up in the later stages of the cultivation must have a great influence on the kinetics. The very low maintenance requirements in the early process phases is explained by the high percentage of younger cells. Cells within pellets may also be protected from shear forces that would otherwise induce repair activity. On the other hand it is very clear from the simulations that pellet breakage in the later phases correlates with a strong decrease in yield – or vice versa increase of maintenance requirements – and reduction in growth activity. The results show that further progress in understanding of the complex kinetics can only be expected by more detailed models for pellet populations and morphological differentiation. Those models should also be able to explain the dependency of the specific product formation and growth rate, as shown in Fig. 2.

2.3
A Detailed Model for Growth of *Penicillium chrysogenum* Pellets

In order to proceed towards a better understanding of influencing factors for growth and product formation of *Penicillium chrysogenum* in the presence of pellets, Meyerhoff et al. [63] developed a detailed model for the growth of single hyphae up to mycelia flocs and pellets. The model adopted the basic concepts of Yang et al. [64, 65] for the growth of *Streptomyces tendae*. This is a combined mechanistic-stochastic approach for the description of the three-dimensional development of mycelia by linear extension of hyphae through apical growth, and by septation and branching. The model was aimed at the early phases of growth beginning with a single spore and does not consider effects of transport limitation. For the simulation of pellets under more realistic conditions, terms for growth limitation by space, substrate and oxygen, cell inactivation and lysis under starvation condition (e. g. in the center of pellets) and cut-off of hyphae at the surface of pellets were introduced by Meyerhoff et al. [63]. In the following the model is briefly outlined.

For the mathematical description, Yang et al. [64, 65] assumed hyphae to consist of compartments with up to three branches and one or two growing tips, which are divided into small sectors, the atom for the simulation, as shown in Fig. 5. The linear extension rates α_i of the tips are determined by diffusion of a so-called key-compound along the branches to the tips, where it is used to build new cell material. There is a homogenous flow of substrate and formation of the key-compound assumed along the branches. For this case, the diffusion equation can be solved analytically and gives for the quasi-stationary case the maximum linear extension rate of a compartment with one tip:

$$\frac{dL_1^{max}}{dt} = \alpha_{1,m} = \alpha_m \frac{\beta e^{2\kappa L_1} + \gamma}{\beta e^{2\kappa L_1} - \gamma} \tag{13}$$

with the auxiliary variables

$$\beta = 3e^{2\kappa(L_2 + L_2)} + e^{2\kappa L_2} + e^{2\kappa L_3} - 1 \tag{14}$$

$$\gamma = e^{2\kappa(L_2 + L_3)} - e^{2\kappa L_2} - e^{2\kappa L_3} - 3 \tag{15}$$

$$\kappa = \frac{\mu_m}{\alpha_m} \tag{16}$$

Similar expressions for the other types of compartments can be derived [63], but are not repeated here. After being grown to a sufficient size, the events of septation and branching can occur. The position for these events along the hyphae and the direction for growth of new tip-sectors is given by probability functions. Transport processes of substrate and oxygen from the bulk medium

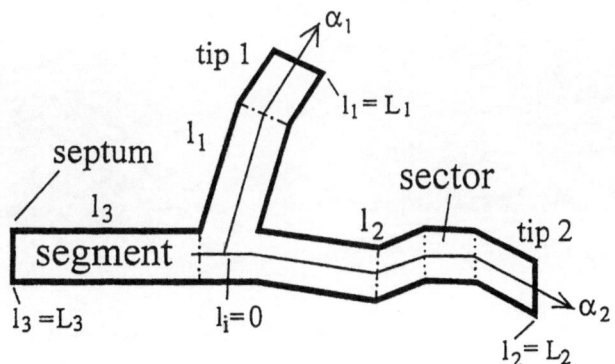

Fig. 5. Schematic representation of a mycelia compartment in the model of Yang et al. [64]. Reprinted from Meyerhoff et al. (1995), Two mathematical models for the development of a single microbial pellet. Bioproc Eng 12, Part I. Springer, Berlin Heidelberg New York, p 305–313

to the center of the pellet are described by the diffusion equation, Eq. (17), for carbon source and oxygen ($k = S$, O) with spherical symmetry:

$$\frac{\partial C_k}{\partial r}\frac{dD_{eff,k}}{dr} + \frac{D_{eff,k}}{r^2}\frac{d}{dr}\left(r^2\frac{dC_k}{dr}\right) = \left(\frac{\mu_R}{Y_{X,k}} + m_k\right)C_{X,R} \qquad (17)$$

To consider the right-hand reaction terms for substrate uptake, the pellet is divided into spherical layers with equal thickness along the radius R (see Fig. 6) and the average cell mass $C_{X,R}$ is calculated by summation over all hyphal tip segments within the layer. The average growth rate is given by the total length extension within the layer,

$$\mu_R C_{X,R} = \frac{\pi d_h^2}{4V_R}\rho_X \sum_{i=1}^{\text{No of tips}} \alpha_i \qquad (18)$$

where d_h is the mean diameter of the hyphae, V_R the volume of the layer, and the hyphae density. The growth limitation mechanism for the linear extension rates of individual tips

$$\alpha_i = \alpha_{i,m}\varphi_R(C_O)\varphi_R(C_S) \qquad (19)$$

is assumed to follow Monod-kinetics for each limiting compound,

$$\varphi_R(C_k) = \frac{C_{k,R}}{K_k + C_{k,R}} \quad \text{where} \quad k = O,S \qquad (20)$$

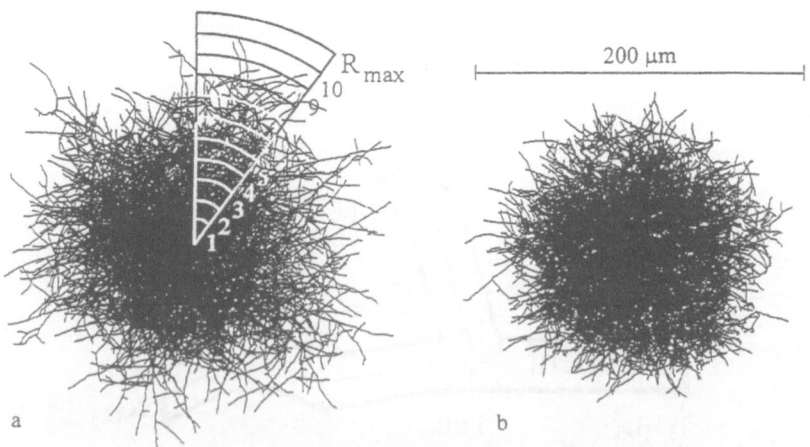

Fig. 6 a,b. Two dimensional projection of simulated pellets: a without; b with application of shear forces. The arcs represent the imaginary layers for solving the transport equations for substrate and oxygen. a: $\rho_{thr} = 0, \lambda_{Shear} = 0$; b: $\rho_{thr} = 0.02, \lambda_{Shear} = 0.5$. Reprinted from Meyerhoff et al. (1995), Two mathematical models for the development of a single microbial pellet. Bioproc Eng 12, Part I. Springer, Berlin Heidelberg New York, p 305–313

The effective diffusion coefficient is supposed to be cell density dependent as follows:

$$D_{eff,k}(\rho) = \begin{cases} 5D_{max,k} \\ D_{max,k} \dfrac{2 - 2\rho}{2 + 2\rho} \end{cases} \text{ where } \begin{array}{l} \rho < \rho_{Crit} \ \wedge \ r > r_{Crit} \\ \rho > \rho_{Crit} \ \vee \ r < r_{Crit} \end{array} \tag{21}$$

where D_{max} is the diffusion coefficient of the medium. This considers very roughly the enhanced transport by convection through the spare outer regions of the pellet (upper condition) and transport hindrance in the more dense regions (lower condition). Cell inactivation is assumed to take place when the substrate concentration falls below a critical value.

Shear forces are an important factor for the development of the pellets. The detailed model would allow for a mechanistic description of these effects for individual hyphae. In deficiency of such model conceptions a formal description was used for the simulation. It was assumed that only tips outgrowing from the denser parts of the pellet are subject to breakage according to the following probability function:

$$P_{Br} = 100 \left(1 - e^{-\lambda S_{hear} \frac{r_{tip} - r_{thr}}{r_{thr}}} \right) \tag{22}$$

where r_{thr} (ρ_{thr}) is the radius at a critical hyphae density within the pellet above which breakage can occur, r_{tip} is the radial position of the tip, and λ_{Shear} is the characteristic parameter of the probability distribution. The free parameters can be varied to adjust the model to experimental conditions.

Simulations have been carried out with the model to study its general behavior. The model parameters are given in Table 2. Figure 6 shows the comparison of two simulated pellets being grown under conditions with and without shear stress. The general appearance is quite realistic and resembles the experimental findings that pellets grown under mild shear conditions have a

Table 2. Additional model parameters used for the simulation of a single pellet by the detailed model

Parameter	Unit	Value
K_O	$g \, l^{-1}$	0.001
ϱ_X	$g \, ml^{-1}$	0.31
d_h	μm	4
α_m	$\mu m \, h^{-1}$	20
α_{ly}	$\mu m \, h^{-1}$	20
μ_m	h^{-1}	0.4
$D_{max,S}$	$cm^2 \, s^{-1}$	$0.5 \, 10^{-5}$
$D_{max,O}$	$cm^2 \, s^{-1}$	$2.0 \, 10^{-5}$
$C_{O,crit}$	$g \, l^{-1}$	0.01
$C_{S,crit}$	$g \, l^{-1}$	0.01

broader and more hairy outer region than the compact and dense pellets developing under shear stress. These can have a very smooth surface. The development of the cell volume density and the dissolved oxygen concentration over the radius in time is shown in Fig. 7 for a typical pellet without shear stress. The simulation curves are not entirely smooth because of the stochastic nature of the model. The vertical lines mark the positions along the radius, along which the growth limitation caused inactivation of the cells with a possible subsequent lysis. After the cell density has reached the maximum value of about 0.4 between 61 and 67 h, the active zone keeps a thickness of about 200 μm, the density profile moves more or less only outside without altering the shape or maximum

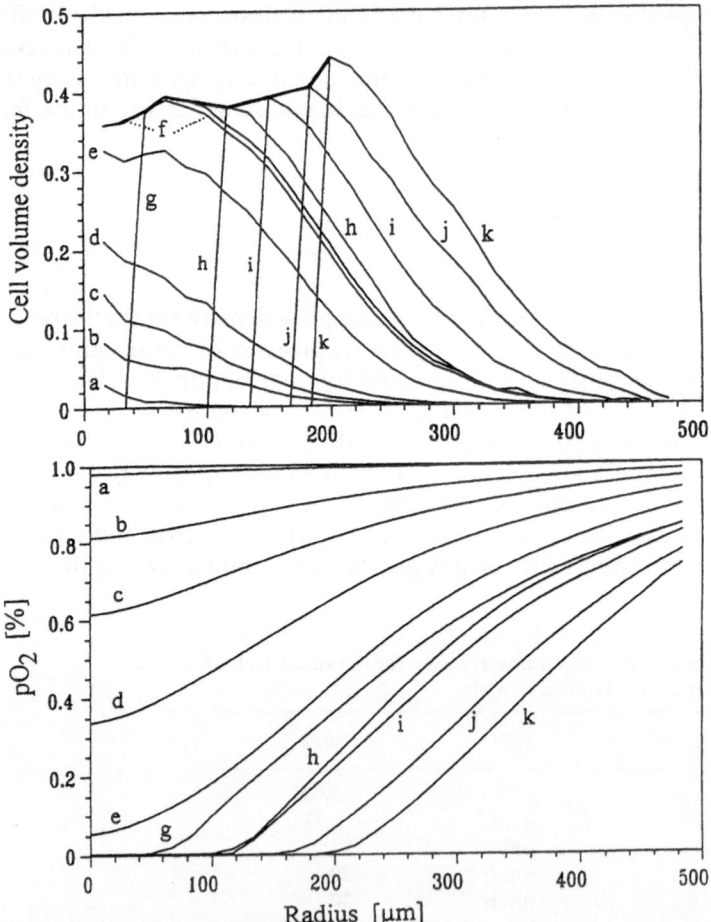

Fig. 7. Volume density profiles (*top*) and dissolved oxygen concentration (*bottom*) during the development of a typical pellet after 26 h (a), 41 h (b), 46 h (c), 51 h (d), 61 h (e), 67 h (f), 68 h (g), 70 h (h), 79 h (i), 88 h (j), and 96 h (k). Reprinted from Meyerhoff et al. (1995), Two mathematical models for the development of a single microbial pellet. Bioproc Eng 12, Part I. Springer, Berlin Heidelberg New York, p 305–313

value, since no oxygen reaches the inner parts due to transport limitation. An interesting outcome of the simulations was that it is not possible to achieve a steady state for the total biomass in the pellet for constant external conditions. Growth at the surface always over-compensates lysis in the center. Since in fed-batch cultivations the biomass becomes rather steady, other mechanism for growth inactivation have to be assumed to explain this effect.

The above model is not well suited for further simulation studies of larger pellets or pellet ensembles, which would come closer to the real process, because the demand on computer resources is tremendous. Therefore, Meyerhoff and Bellgardt [66] tried to simplify the calculations by neglecting individual hyphae and looking at average properties while keeping the morphological basis of the model and information on the microscopic structure of the pellets. In the modified model, averaging is done again in radial layers as shown in Fig. 6. The simulation proceeds in discrete time steps. While in the detailed model the primary variables were the number of tips and the length and position of each individual hyphal branch, the simplified model considers only the total length of hyphae within a layer as an integral value. From this variable, the hyphal growth unit and the number of tips are reconstructed using a correlation derived from the detailed model. Similarly, the linear extension is calculated as an average and not individually for each tip.

For the actual state of the simulation the total length of hyphae in a layer, L_R, is known. The number of tips is then

$$n_R = \frac{L_R}{hgu_R} \tag{23}$$

where the hyphal growth unit is obtained from the above-mentioned correlation as a function of the radius by

$$hgu_R = hgu^* + \frac{360\ \mu m - 2r_R}{\sqrt[3]{\sum_{R=1}^{R_{max}} n_R^*}} \tag{24}$$

The total length is updated according to

$$L_R(t) = L_R(t - \Delta t) + \Delta L_R^+(t) + \Delta L_R^-(t) \tag{25}$$

where growth (the two upper lines in Eq. (26)) and inactivation (third line in Eq. (26)) are given by

$$\Delta L_R^+ = \begin{cases} \alpha n_R \varphi_R(C_O)\ \varphi_R(C_S)\Delta t & \rho < 0.6,\ C_S > C_{S,Crit} \wedge C_O > C_{O,Crit} \\ \alpha n_R \varphi_R(C_O)\ \varphi_R(C_S)\ (0.79 - \rho)\ \Delta t & \text{where } \rho \geq 0.6 \\ -\alpha_{ly} n_R \Delta t & C_S \leq C_{S,Crit} \vee C_O \leq C_{O,Crit} \end{cases} \tag{26}$$

The first condition is for unlimited growth of the n_R tips while the second considers spatial limitation. The densest packing of cylinders of equal diameter is 0.79. The third condition describes reduction of the active hyphal length by inactivation at growth limitation. The inactive parts could then be subject to lysis. Length reduction by breakage is given by

$$\Delta L_R^- = -l_{Shear,\,av} n_{Shear,\,R} \Delta t \tag{27}$$

where the number of broken tips, $n_{Shear,\,R}$ is calculated as in the detailed model. The average mycelia length loss per break event, $l_{Shear,\,av}$, is a new model parameter. Finally, spatial growth of the pellet is considered by a diffusion-like tip flow and the concentration profiles for oxygen and substrates are calculated by the diffusion equation.

A further simplifying model assumption is employed for starting the simulation from single spores. For this case, any limitation is neglected, the hyphal growth unit is assumed to be constant at hgu*, and the tips to be distributed along the radius according to a Gauss-function. In analogy to a chain of stiff interconnected polymer molecule segments with rotating bounds, the actual radius of the filament is calculated from the Stiff-Chain-Model [67] by

$$r_{max}^2 = LL_m \frac{1 + \cos\gamma}{1 - \cos\gamma} \tag{28}$$

where L is the total length of the hyphe, L_m the length of a stiff straight segment, and $\gamma = 29°$ the average absolute angle between subsequent segments. The resulting profile is put into the above model after 150–330 tips have developed.

The modified model as outlined by Eqs. (23)–(28) reduces the simulation time by a factor of about 100 which makes it possible to carry out systematic parameter variations. For a far reaching verification of the model one would ideally need profile measurements over time along the pellet radius for the number of tips, mycelia length or cell density, and for the substrate and oxygen concentration. This is absolutely impossible at present. What can be achieved are microprobe measurements at a single pellet in diluted medium and cell density estimation after preparation of ultra-thin cuttings by digital image analysis. The time course during pellet development cannot be directly followed since both methods destroy the pellet.

Figures 8 and 9 give simulation results for glucose and oxygen profiles in comparison to microprobe measurements. The cell density profile was obtained by image analysis of an evidently similar pellet. The simulation can reproduce the measured profiles quite well although the model parameters, as given in Table 3, had to be adapted, since the growth activity in the diluted medium for the measurements is obviously lowered compared to cultivation medium. In a second step it was tested whether the model yields realistic cell density profiles. Shake flasks were inoculated with spores and cultivated up to pellets. Pellets were taken out at different times and the cell density determined. Simulations over an identical time interval yielded the very similar cell density profiles (Figs. 10 and 11) after slight variations of the parameters. While the small pellet (Fig. 11) is just becoming oxygen limited in the center, the large pellet in Fig. 10 already showed lysis effects in the center. Since lysis was not included in the

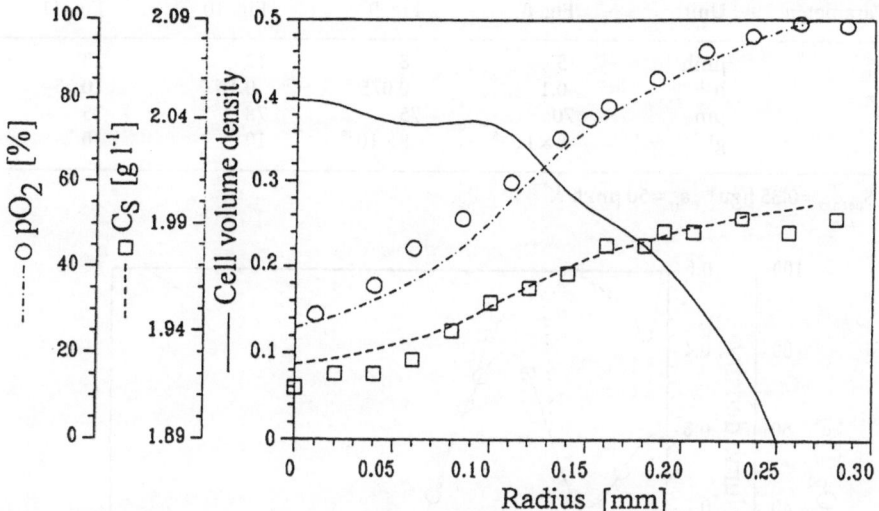

Fig. 8. Comparison between oxygen and glucose profiles as obtained by microprobe measurements (*symbols*) within a small pellet to simulation results by using the estimated cell density; 216 h after inoculation. Reprinted from Meyerhoff and Bellgardt (1995) Two mathematical models for the development of a single microbial pellet. Bioproc Eng 12, Part II. Springer, Berlin Heidelberg New York, p 315–322

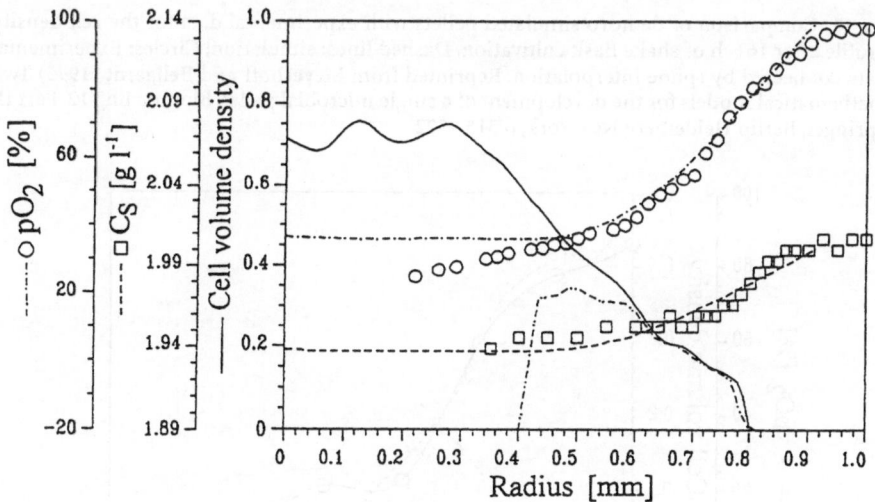

Fig. 9. Comparison between oxygen and glucose profiles as obtained by microprobe measurements (*symbols*) within a large pellet to simulation results by using the estimated cell density; 216 h after inoculation. Reprinted from Meyerhoff and Bellgardt (1995) Two mathematical models for the development of a single microbial pellet. Bioproc Eng 12, Part II. Springer, Berlin Heidelberg New York, p 315–322

Table 3. Parameters for pellet simulations with the simplified model

Parameter	Unit	Fig. 8	Fig. 9	Fig. 10	Fig. 11
α_m	μmh^{-1}	5	8	12	18
μ_m	h^{-1}	0.1	0.075	0.16	0.35
h_{gu*}	μm	70	75	78	75
$C_{O,Crit}$	gl^{-1}	8×10^{-5}	8×10^{-5}	10^{-5}	10^{-5}

[1] $S_{hear,av} = 0.35\ hgu*$, $a_{ly} = 50\ \mu m\ h^{-1}$.

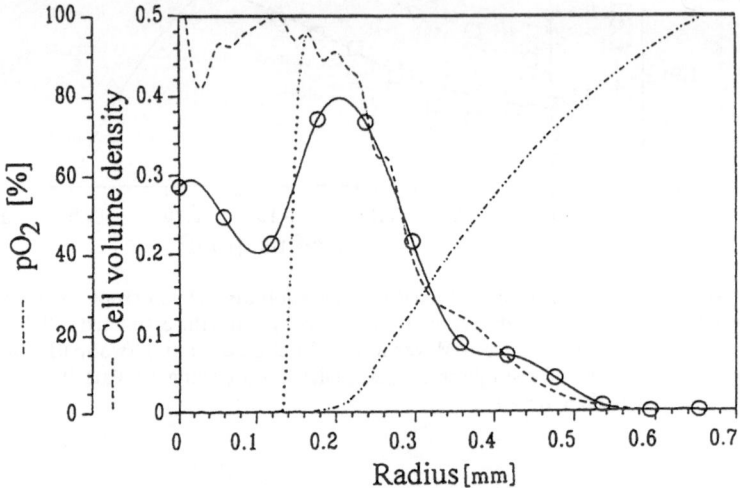

Fig. 10. Comparison of de novo simulated pellets with experimental data for the cell density profile after 161 h of shake flask cultivation. Dashed lines: simulation; Circles: Experimental data connected by spline interpolation. Reprinted from Meyerhoff and Bellgardt (1995) Two mathematical models for the development of a single microbial pellet. Bioproc Eng 12, Part II. Springer, Berlin Heidelberg New York, p 315–322

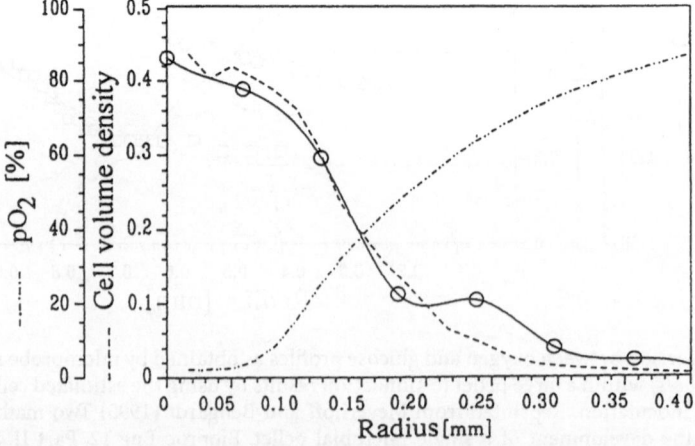

Fig. 11. Comparison of de novo simulated pellets with experimental data for the cell density profile after 115 h of shake flask cultivation. Symbols as in fig. 10. Reprinted from Meyerhoff and Bellgardt (1995) Two mathematical models for the development of a single microbial pellet. Bioproc Eng 12, Part II. Springer, Berlin Heidelberg New York, p 315–322

model, the simulated cell density in the center is higher than the measured one. The size of the simulated active outer region, denoted by the vertical line at 0.14 mm, is in a realistic range.

2.4
Integrated Process Model

The simplified pellet model presented in the previous section provides a tool for an integrated, morphology-based simulation of fed-batch cultivations in the presence of pellets. The remaining gap to the process was closed by Meyerhoff and Bellgardt [68]. The development of pellets during the cultivation, and even the preculture, is never uniform. There is always a certain variability in size and density. A simple method to model this distribution function is the simulation of a great number of pellets with slightly different properties. In the work of [68] up to 100 pellets were chosen to represent all of the pellets in the cultivation. Since it is at present not possible to get a complete picture of a pellet population used for inoculation of a cultivation by measurements, the simulation starts from scratch, i.e., inoculation of spores in the preculture. This opens the additional chance of studying the influence of variations in the preculture conditions on the main cultivation. For this task, the model of [66] had to be extended by a description of the germination time distribution of spores. A modified truncated Gauss-distribution was found to be in agreement with experimental data, as shown in Fig. 12. For the simulation, the actual germination time t_G is the given by

$$t_G = (\beta \ln x + \gamma) \, t_{G,max} \quad x > 0 \quad t_G \geq 0 \tag{29}$$

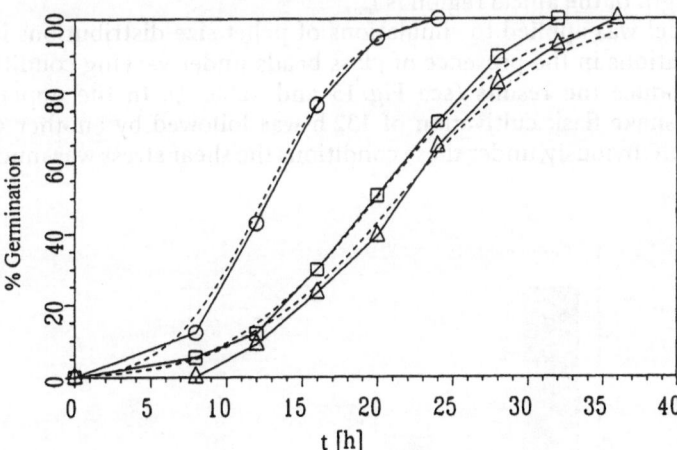

Fig. 12. Experimental data (*symbols*) for germination time distributions [102] compared to Eqs. (29) and (30) (*dashed lines*). Different treating methods and media: (o) spores S1, medium M4; (□) spores S2, medium M4; (△) spores S1, medium M2. Model parameters: $t_{G,max}$ = 25, 40, 43 h; β= 0.19 γ = 0.5. Reprinted from Meyerhoff and Bellgardt (1995) A morphology-based model for fed-batch cultivations of *Penicillium chrysogenum* growing in pellet form. J Biotechnol 38:201, with kind permission of Elsevier Science, Sara Burgerhartstraat 25, 1055 KV Amsterdam, The Netherlands

where the random number is obtained from the distribution function

$$\Phi(\ln x) = \frac{1}{\sqrt{2\pi}} \int_0^{\ln x} e^{\frac{-\ln x^2}{2}} \, d(\ln x) \tag{30}$$

For simulation of a production process the breakage of pellets by shear forces must be considered in the model, otherwise it would not be possible to explain the observed pellet size distributions. Breakage events are basically of a stochastic nature although the actual state of the pellet with respect to the cell density profile, the size and age will influence the breakage probability. The model provides this information, but as a first step, a simple linear correlation of the breakage probability in a given time interval, p_{Br}, with respect to the intensity of the shear forces and the entire biomass, C_X, was assumed to be valid for all pellets:

$$p_{Br} = K_{Br} \lambda_{Shear} C_X \frac{\Delta t}{0.2h} \tag{31}$$

The microkinetics of hyphal growth were adopted from Tiller et al. [63] with slight modifications. The production of penicillin in this model was related to the morphology by

$$\frac{dP}{dt} = q_P\left(C_X - \pi d_h^2 \rho_X \frac{n_{tip}l_{ap}}{V_L}\right) - k_h P - DP \quad \text{where} \quad q_P = \begin{cases} 0 \\ k_p \end{cases} \text{when } C_S \begin{matrix} > \\ \leq \end{matrix} 0.4 \text{ gl}^{-1} \tag{32}$$

which considers product synthesis only in sub-apical regions of the hyphae. The average length of the apical region is l_{ap}.

The model was applied to simulations of pellet size distributions in shake-flask cultivations in the presence of glass beads under varying conditions and could reproduce the results (see Fig. 13 and Table 4). In the upper part of Fig. 13a, a shake flask cultivation of 132 h was followed by another 48 h in a stirred tank. Obviously, under these conditions the shear stress was much higher

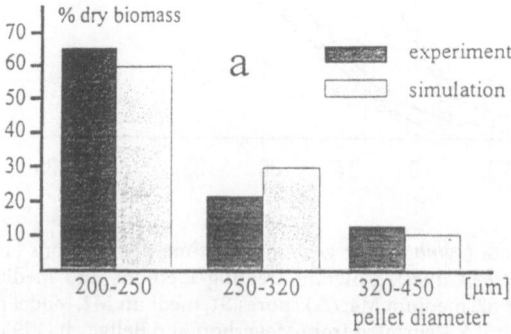

Fig. 13. a Comparison of experimental data for pellet size distributions from shake flask cultivations and simulation results: a after 190 h [27]

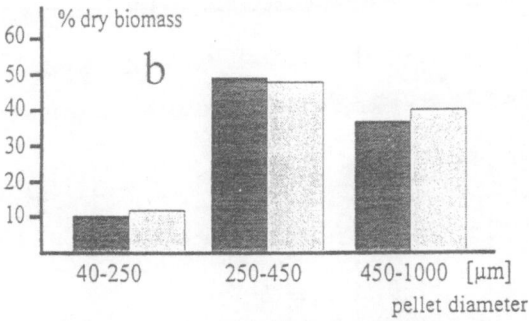

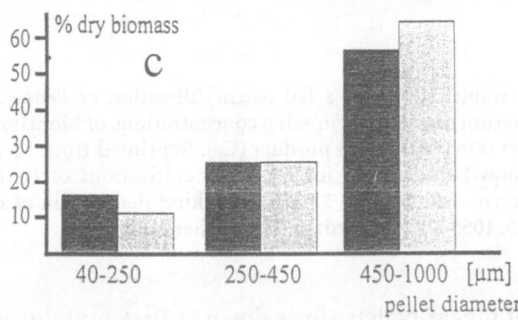

Fig. 13. b, c after 126 h [103]. Reprinted from Meyerhoff and Bellgardt (1995) A morphology-based model for fed-batch cultivations of *Penicillium chrysogenum* growing in pellet form. J Biotechnol 38: 201, with kind permission of Elsevier Science, Sara Burgerhartstraat 25, 1055 KV Amsterdam, The Netherlands

Table 4. Morphological parameters for simulation of pellet size distributions and fed-batch cultivation

Parameter	Unit	Fig. 13a	Fig. 13b	Fig. 13c	Figs. 14,15
λ_{Shear}	–	0.7	0.1	0.1	0.1
$t_{G,max}$	h	55	70	109	50
μ_m	h^{-1}	0.1	0.075	0.16	0.135, 0.154
hgu*	μm	70	70	70	70

$a_m = 20\ \mu m\ h^{-1}, a_{ly} = 20\ \mu_m\ h^{-1}, \mu_{S,max} = 0.078\ h^{-1}, \mu_{PM,max} = 0.057\ h^{-1}, K_{age} = 55\ h, K_{Br} = 0.003\ gl^{-1}.$

and breakage events have moved the maximum of the size distribution to low diameters. In Fig. 13b and c, the pellet size distribution can be explained smoothly without breakage. Simulation results of a fed-batch process are given in Figs. 14 and 15. After the precultivation (simulation not shown) there is an exponential growth phase with a significant formation of pellets with diameter > 250 μm. During the production phase with slightly increasing total biomass,

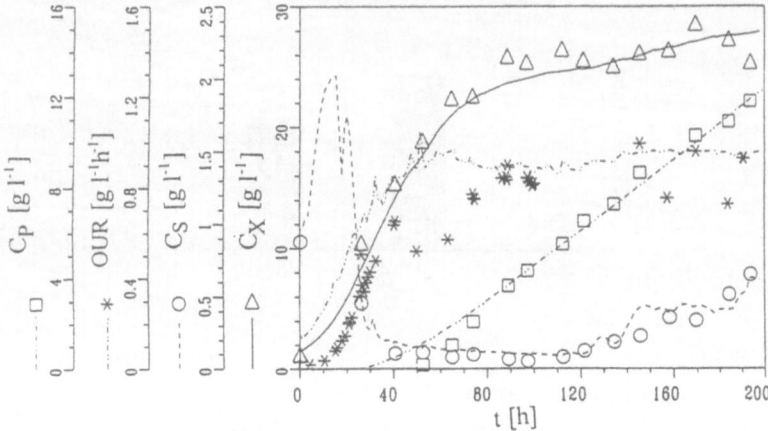

Fig. 14. Simulation results (lines) of a fed-batch cultivation of *Penicillium chrysogenum* in comparison to experimental data (symbols); concentrations of biomass (C_X), sugar substrate (C_S), oxygen uptake rate (OUR), and product (C_P). Reprinted from Meyerhoff and Bellgardt (1995) A morphology-based model for fed-batch cultivations of *Penicillium chrysogenum* growing in pellet form. J Biotechnol 38:201, with kind permission of Elsevier Science, Sara Burgerhartstraat 25, 1055 KV Amsterdam, The Netherlands

the formation of bigger pellets slows down at first, and during the later phases of the cultivation (after about 100 h) the pellets disappear due to breakage and autolysis. After 140 h, there remain only filaments and pellet fragments with small diameter. During this cultivation, the pellets are most susceptible to oxygen limitation in the early phases of the cultivation although the dissolved oxygen concentration remains high. The reasons for that effect are the high

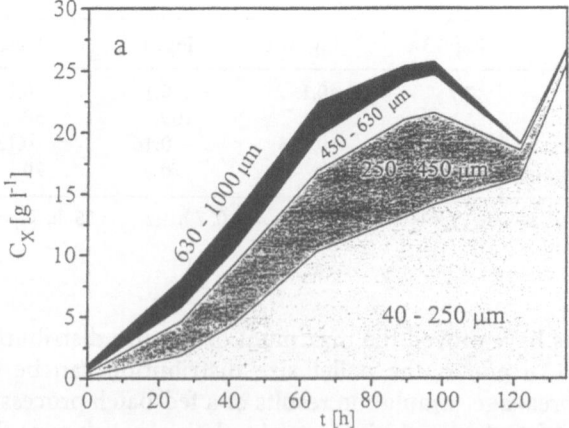

Fig. 15. a Development of the pellet size distribution for the cultivation in Fig. 14: a experimental data

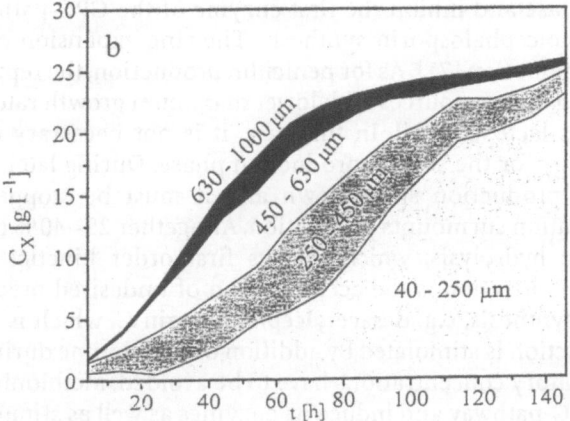

Fig. 15. b simulations results. Reprinted from Meyerhoff and Bellgardt (1995) A morphology-based model for fed-batch cultivations of *Penicillium chrysogenum* growing in pellet form. J Biotechnol 38:201, with kind permission of Elsevier Science, Sara Burgerhartstraat 25, 1055 KV Amsterdam, The Netherlands

growth rates and the high mean pellet diameter at the beginning of the cultivation. Later on, oxygen transport limitation in the pellets plays no role because of the reduced growth rates and decreasing mean diameter. The simulation shows that there is a potential for further process improvement by controlling more carefully the development of the pellets during the cultivation. Here, the model can help to find proper conditions for preculture and main cultivation, and to guide further experiments.

3
Cephalosporin Process

3.1
Overview

The antibiotic Cephalosporin C (CPC) is produced by the fungus *Acremonium chrysogenum*. The metabolism and biosynthesis are in many respects similar to those of *P. chrysogenum*. The metabolism and its regulation are reviewed in Queener et al. [3] and Demain et al. [69]. The synthesis steps of penicillin and cephalosporin up to isopenicillin N are identical. This is then converted to penicillin N without precursor and in a further oxygen-dependent reaction sequence to cephalosporin. Under oxygen limitation, these further steps are inhibited and penicillin N is accumulated.

The antibiotics synthesis is subject to catabolite repression during growth phases with high rates of carbon source uptake [6, 70, 109], which leads to the known control strategy with a first phase at high growth rates and a second phase with reduced growth but maximum production. Glucose and other rapidly utilized carbon sources repress the ring-expansion enzyme desacetoxycephalosporin-C-synthetase and inhibit the first enzyme of the CPC-pathway, the ACV synthetase in the cephalosporin synthesis. The ring expansion enzyme is not inhibited after induction [71]. As for penicillin production, the repression can be avoided by using carbon sources with lower maximum growth rate compared to glucose, such as lactose or oil. In this case, it is not necessary to establish a fed-batch strategy for the second production phase. During later phases of the cultivation the production slows down and it must be stopped before the product degradation surmounts production. Altogether 25–40% of the product undergoes such hydrolysis, which follows first order kinetics [2]. Another problem of the cultivation is the accumulation of undesired precursors of the cephalosporin synthesis, e.g. desacetylcephalosporin C, which is rather stable. The CPC-production is stimulated by addition of methionine during the growth phase, but inhibitory concentrations have to be avoided. Methionine can supply sulfur to the CPC-pathway and induce its enzymes as well as stimulate morphological differentiation from thin hyphae to producing swollen hyphae.

In submerged cultures *A. chrysogenum* can be found in three main morphological types: slim thin-walled hyphae (mycelia), swollen hyphae or hyphal fragments and arthrospores. The organism has no clear tendency to pellet formation. At high growth rate mainly slim hyphae and a low percentage of swollen hyphae are formed. Gradual limitation of the carbon source accelerate the transition from the slim filamentous to the second thicker form, which can further differentiate to spores under growth limitation. Cephalosporin production is usually maximum in process phases with a high rate of differentiation from the swollen hyphae form to arthrospores [2, 72, 73, 104–106]. Thus it correlates with a high number of swollen hyphae as well. The arthrospores themselves seem to be non-producing [74].

Matsumara et al. [72] presented a model describing the morphological differentiation and regulation of the cephalosporin-synthesis by the carbon source and methionine. They discriminate the three cell types slim hyphae, swollen hyphae and arthrospores. A fictive enzyme in the swollen hyphae that is repressed by glucose is assumed to be responsible for the CPC synthesis. The model emphasizes the importance of an intracellular methionine pool. The model was used to simulate qualitatively a process with linear-increasing substrate feedrate over time. By the feeding of glucose and methionine the process duration could be extended and the product concentration was increased.

A model considering only thin hyphae and swollen hyphae was developed by Chu and Constantinides [71] for batch cultivations on glucose and sucrose. In the model, production of CPC is repressed by glucose. The diauxic lag-phase for the switch to growing on sucrose is modeled only by a formal delay function in a non-mechanistic way. Therefore, the model is not really predictive. Also here, a fictive rate-limiting enzyme is assumed to be responsible for an additional lag-phase before start of the CPC-production after depletion of glucose. The model is extended by introducing a dependency of kinetic parameters on temperature and pH. The extended model was then used to generate optimal control profiles for these control variables. Although the model did not fit the experimental data very well, the model based control strategy could increase the productivity by more than 50%.

Malmberg and Hu [75] presented a detailed kinetic model for analysis of rate limiting steps in the CPC biosynthetic pathway. The model includes the six reaction steps to cephalosporin, starting with the amino acids L-α-aminoadipic acid, L-cysteine and L-valine. Cell growth is simulated according to a formal balance with fixed specific growth rate. In vivo data of enzyme activities were converted to the intracellular conditions required for the model simulations. From the simulation results the authors conclude that the ACV-tripeptide formation catalyzed by the ACV-synthetase is the rate-limiting step of CPC-synthesis. Increase of the intracellular level by genetic engineering methods should enhance the production.

3.2
A Structured Segregated Model for Cephalosporin Production on Complex Substrates

None of the above models can be directly applied to industrial cultivations on complex substrate mixtures. A segregated-structured model aimed for this application was developed for cephalosporin C production by Meyerhoff [76]. In the experiments, dry glucose syrup, which contains mainly maltose and oligo-meres of this sugar, and soy oil were used as carbon source [77]. There is no clear separation of the growth phases on sugars and oil, both substrates are used in parallel. Nevertheless, after limitation of the sugars a lag-phase can be observed, where the cells adapt further to growth on oil as sole carbon source. The high-producing strain used in the experiments also showed no clear separation of growth and production phase. Production starts very early when still significant amounts of sugars are present in the medium.

The model considers initially inactive, growing (but non-producing) and swollen (producing) hyphae, as well as inactive arthrospores. The inactive form is used to model an initial lag phase after inoculation of the reactor. These are transformed into the active form, which can then differentiate to swollen hyphae, and, after limitation of the sugars, to arthrospores. The mass balance equations are as follows. For the inactive cells,

$$\frac{dm_i}{dt} = - k_{ih}m_i - F_{Offline} X_i \tag{33}$$

and for the growing hyphae that use the sugar and oil substrates with specific growth rates μS and μ_{Oil},

$$\frac{dm_h}{dt} = (\mu_S + \mu_{Oil}) m_h - k_{hs}m_h + k_{ih}m_i - F_{Offline} X_h \tag{34}$$

The swollen hyphae are formed from growing hyphae at low sugar concentrations,

$$\frac{dm_s}{dt} = - k_{sa} \frac{K_I}{K_I + M} m_s + k_{hs}m_h - F_{Offline} X_h \tag{35}$$

which then differentiate further to arthrospores,

$$\frac{dm_a}{dt} = k_{sa} \frac{K_I}{K_I + M} m_s - F_{Offline} X_a \tag{36}$$

The total biomass concentration is given by

$$X = \frac{m_i}{V_L} + \frac{m_h}{V_L} + \frac{m_s}{V_L} + \frac{m_a}{V_L} = X_i + X_h + X_s + X_a \tag{37}$$

The volume balance for fed-batch is determined by the flows of ammonium, soy oil and the sampling streams for on-line and off-line analyses.

$$\frac{dV_L}{dt} = F_M + F_{Amm} + \frac{K_{Oil}}{\rho_{Oil}} F_{online} - F_{offline} \tag{38}$$

where F_{Amm}, F_{Online} and $F_{Offline}$ are in lh^{-1} and F_{Oil} in gh^{-1}. The dilution rate is:

$$D = \frac{1}{V_L} \left(F_M + F_{Amm} + \frac{F_{Oil}}{V_L} \right) \tag{39}$$

The production is coupled to the concentration of the swollen hyphae and assumed to follow a first order kinetics. The product is also subject to hydrolysis via first order kinetics

$$\frac{dp}{dt} = k_p \frac{Oil}{K_{Oil}^+ + Oil} X_s - (k_{hp} + D) P \tag{40}$$

The sugar uptake, mainly maltose, is determined by growth and product formation in the respective morphological states,

$$\frac{dM}{dt} = -\frac{\mu_S}{Y_{XS}} X_h - \frac{k_P}{Y_{XP}} X_s \frac{M}{K_M + M} + \frac{F_M}{V_L} M_R - D M \tag{41}$$

Since in the measured sugar concentration some sucrose is contained, which cannot be metabolized by the fungus, the total sugar concentration is modeled as sum of maltose and a constant value of sucrose. The balance equation for soy oil is

$$\frac{dOil}{dt} = -\frac{\mu_{Oil}}{Y_{XOil}} X_h - \left(m_{Oil} (X_h + X_s) - \frac{k_p X_s}{Y_{POil}} \right) \frac{Oil}{K_{Oil}^* + Oil} - D\, Oil + \frac{F_{Oil}}{V_L} \tag{42}$$

which considers the effects of growth, maintenance, production and feeding. The CO_2 concentration in the exhaust gas is calculated by a simplified scheme under the assumption of RQ=1 by using the following expression for the CO_2 production rate,

$$CPR = \left(\frac{\mu_{Oil}}{Y_{COil}} + \frac{\mu_S}{Y_{CS}} \right) X_h + m_c(X_h + X_s) + \frac{k_P}{Y_{PC}} X_s \qquad (43)$$

with the aeration rate F_G as follows:

$$xCO_2^A = 0.0003 + \frac{CPR V_L}{F_G \ [l/h] \ \rho_{CO_2} \ [g/l]} \qquad (44)$$

In reality, RQ may deviate somewhat from unity, especially for growth on oil. To consider the induction of oil catabolizing enzymes, the growing hyphae X_h are structured in a simple two-compartment model. The first normalized compartment E_1 (not shown) represents metabolism of sugar, the second (E_2) of oil, which is repressed as long as sugar is present in high concentrations in the medium, as follows

$$\frac{dE_2}{dt} = \mu_{Oil,max} \frac{Oil}{K_{Oil} + Oil} E_2 + u_{max} \frac{K_I}{K_I + M} (1 - E_2) - (\mu_{Oil} + \mu_M) E_2 \qquad (45)$$

The experimental results suggest that the oil uptake in parallel to sugar follows a different mechanism than for oil as a sole carbon source. The first one is assumed to be controlled by the sugar uptake rate while for the second a Monod-kinetic is chosen. The total specific growth rate on soy oil is then

$$\mu_{Oel} = \mu_{Oel,max} \frac{Oil}{K_{Oil} + Oil} E_2 + \mu_{Oil,max M} \frac{M}{K_M + M} \qquad (46)$$

The growth on sugar is also assumed to follow a Monod-kinetics:

$$\mu_S = \mu_{S,max} \frac{M}{K_M + M} \qquad (47)$$

The model is able to describe experimental data for cell mass, sugar substrate, product, oil substrate and CO_2 in the exhaust gas (X, S, P, Oil, xCO_2^A) for several cultivations with only small variations in the parameters. As an example, the simulation results of one cultivation are given in Figs. 16 and 17. The model parameters can be found in Table 5.

The pattern of the CO_2-curve shows clearly the switch from growth on sugar to oil: between 65 and 80 h the value of xCO_2^A falls since sugar becomes limiting. After 80 h there is again an increase as a consequence of the adaptation to the new substrate. The temporary reduction of the growth rate can also be seen on the biomass. Maltose is the preferred substrate for this organism. Therefore, the oil uptake rate in the first growth phase is much lower than after the sugar limitation and adaptation of the oil catabolizing system.

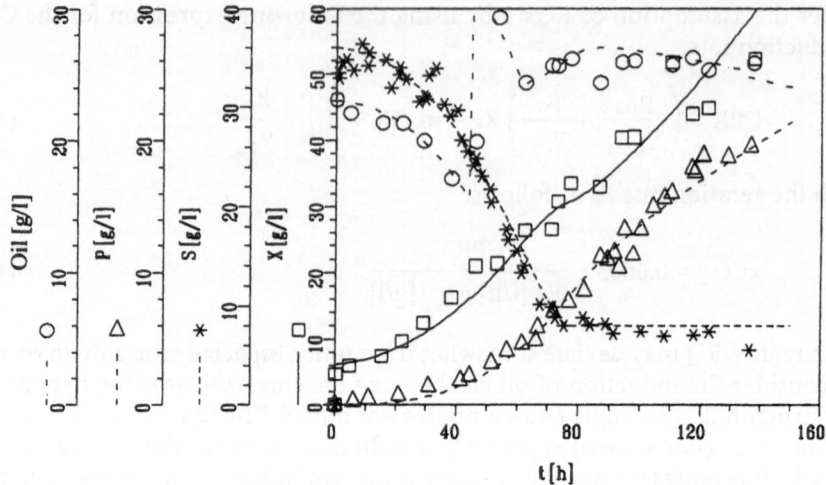

Fig 16. Experimental data (*symbols*) and simulation results (*lines*) for biomass (X), total sugar (S), Cephalosporin (P) and soy oil (Oil) during a fed-batch cultivation of *Acremonium chrysogenum*

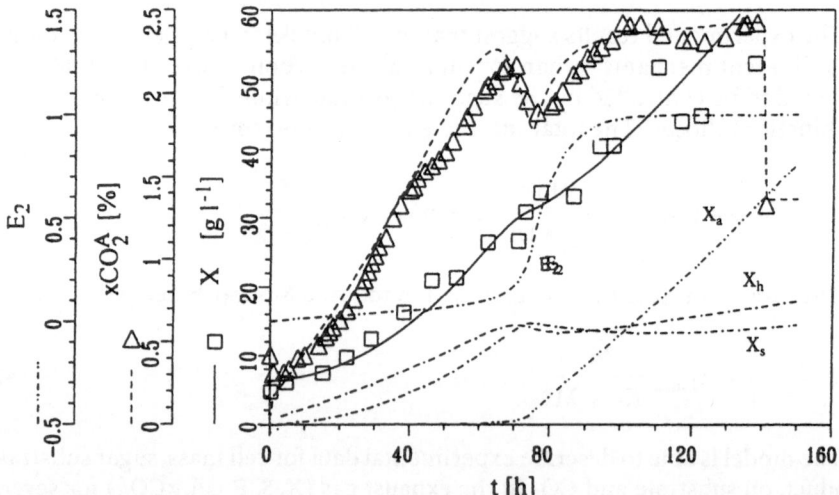

Fig. 17. Experimental data (symbols) and simulation results for biomass (X) and CO_2 in the exhaust gas (xCO_2^A) during a fed-batch cultivation of *Acremonium chrysogenum*. Only as simulation: X_h, X_s and E_2

For this process, the unique discrimination in growth and production phase is difficult. The presence of sugar (here mainly maltose) has no strong repression effect on the product synthesis. Therefore, cephalosporin is formed just from the beginning. After sugar becomes limiting, arthrospores are formed with a higher rate, which is similar to the model of Matsumara et al. [72]. In this model,

Table 5. Model parameters for a fed-batch cultivation
of *Acremonium chrysogenum*

Parameter	Unit	Value
$\mu_{M,max}$	h^{-1}	0.028
$\mu_{Oil,max}$	h^{-1}	0.032
u_{max}	h^{-1}	0.115
Y_{XM}	–	0.53
Y_{XOil}	–	0.8
Y_{COil}	–	0.48
Y_{CS}	–	1.6
Y_{PC}	–	1
Y_{PM}	–	1.5
Y_{POil}	–	1.2
K_M	$g \, l^{-1}$	3.3
K_I	$g \, l^{-1}$	0.3
k_{ih}, k_{hs}	h^{-1}	0.027
k_{sa}	h^{-1}	0.032
k_p	h^{-1}	0.023
k_{hp}	h^{-1}	0.0075
m_C	h^{-1}	0.032
m_{Oil}	h^{-1}	0.0027
E_{20}	–	0.72

no precursors were considered, since during the experiments limitation of
methionine, ammonia or phosphate were avoided.

3.3
Application of the Model to Process Optimization

The model presented above was used for investigations to improve the control
strategy for the fed-batch cultivation of *Acremonium chrysogenum* [76]. The
general strategy for optimal control of the antibiotics production is determined
by the properties of the biological system, i.e. a maximum productivity below
the maximum growth rate. The control in roughly two phases, an initial growth
phase and a second production phase with reduced growth rate to avoid
catabolite repression – which also helps in avoiding oxygen limitation – is considered as optimal and used in practice [1, 78, 79]. The biomass produced during
the growth phase should be as high as possible to obtain afterwards a high productivity. The optimal initial substrate supply and the maximum reachable cell
mass is to a great extent determined by the maximum oxygen transfer rate of the
reactor [11, 32, 61, 80, 81]. Depending on the growth kinetics, the boundary conditions and performance criterion for the optimization, the ideal optimal control can require additional phases. The general properties of optimal control
strategies for fed-batch processes were already subject to extensive studies
[82 – 84, 107].

The task of optimal control is to find suitable conditions for inoculation and
dynamic profiles for the control variables that maximize productivity, e.g. sub-

strate or precursor feeding, supply of oxygen and nitrogen source, or other operating parameters such as pH and temperature. Several authors have dealt with optimization of fed-batch antibiotics production, mostly based on simple unstructured models [31, 61, 85–91]. In most cases the substrate feeding or substrate concentration is subject to optimization, sometimes pH or temperature [30, 71, 92]. In the work of Meyerhoff [76] the presented model was applied for dynamic optimization of the sugar and oil feeding by using the Iterative Dynamic Programming (IDP) algorithm [93–96]. Compared to methods based on Pontryagin's Maximum Principle, the IDP-algorithm is easier to use, especially for models of higher order, because it does not require the calculation of the adjoint system and additional information about bang-bang phases. Furthermore, it has good convergence properties. A disadvantage is that it only generates discrete open-loop control profiles. The somewhat irregular control profiles in the following figures can be referred to the optimization method, which uses a grid of discrete intervals in the state space, since for a given accuracy limit the optimization results for different grid points may vary slightly, especially in regions where the sensitivity of the control variable is low.

Several performance criteria were tested: total product mass, product yield per supplied substrate, and economic profit. During the simulation studies, a maximum filling volume of the reactor of $V_F < 30$ l and a maximum oil concentration of 8 g l^{-1} were introduced as additional restrictions. Since the oil promotes coalescence of the air bubbles, the sense of the latter restriction is to maintain a sufficient oxygen supply to the reactor. For the optimization, the total process duration ($t_f = 150$ h) and maximum feeding rates were fixed ($F_M < 0.05$ lh^{-1}, $F_{Oil} < 12$ gh^{-1}). In a first step, it was tried to improve the process mainly by variation of the soy oil feed, the entire sugar substrate was added at the beginning. For the economic type performance index that also considers the cost of the oil substrate,

$$I(t_f) = PV_L + 50P - \int_0^{t_f} F_{Oil}\, dt \qquad (48)$$

the results are given in Fig. 18. The control keeps the oil level much lower than in the following simulation run (Fig. 19). During almost the entire growth phase, the oil feed is very moderate. At about 80 h, it jumps up to almost the maximum value to support the adaptation to the new substrate when sugar becomes limiting. At the last 40 h of the cultivation the oil feed decreases again, to slow down growth and support formation of swollen hyphae.

Figure 19 shows the optimization results for both of the control variables, sugar (F_M) and soy oil feed (F_{Oil}) for the performance index considering the total mass of the produced antibiotics:

$$I(t_f) = PV_L \qquad (49)$$

Compared to the original process, the growth phase is prolonged to about 105 h by a higher sugar feed under near-batch conditions. The sugar flow rate ensures sufficient growth during the entire process. Soy oil is used up to the

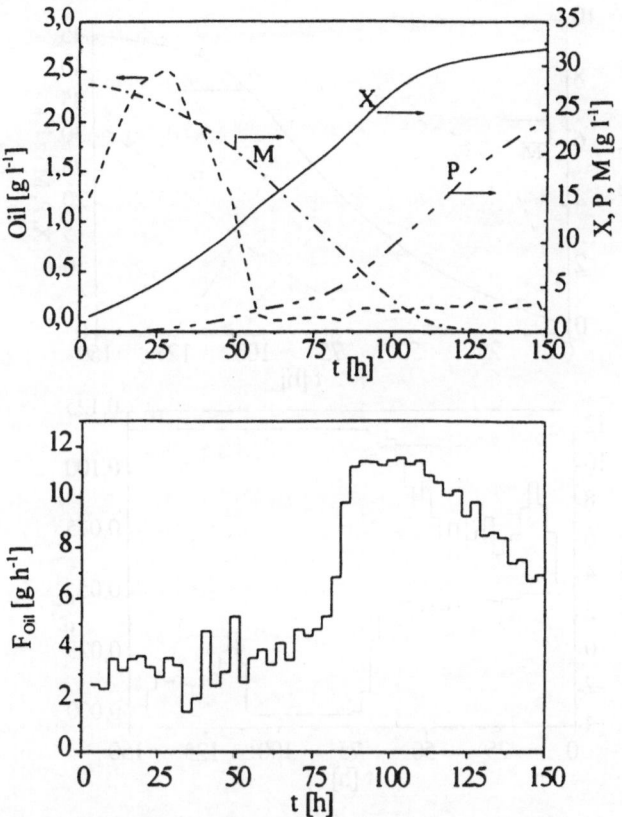

Fig. 18. Optimization results of a fed-batch cultivation of *Acremonium chrysogenum* with soy-oil feeding for the performance criterion at Eq. (48)

maximum amount that can be afforded under the given restrictions on states and feeding rate. This can be explained by the fact that the oil with its high energy content and biomass yield allows for higher biomass in the reactor. The restricted oil feed cannot avoid carbon source limitation at the end of the cultivation. From the simulation, the optimized process shows a clearly increased product concentration and productivity. In comparison to Fig. 18 one should note that the initial filling volume is much lower to give room for the sugar feed.

The profit-optimization of industrial multi-reactor plants for mycelia fed-batch cultivations is further discussed by Yuan et al. [97], considering different kinds of directly production related costs (e.g. substrate or energy supply) and indirect costs (e.g. investment, labor cost). To deal with the natural variability of the cultivations, they suggest a quasi-on-line estimation of the profit function, either by using a mathematical model of the process or measurements of the product concentration. Generally, after inoculation the profit function is negative. During the process, with increasing product concentration,

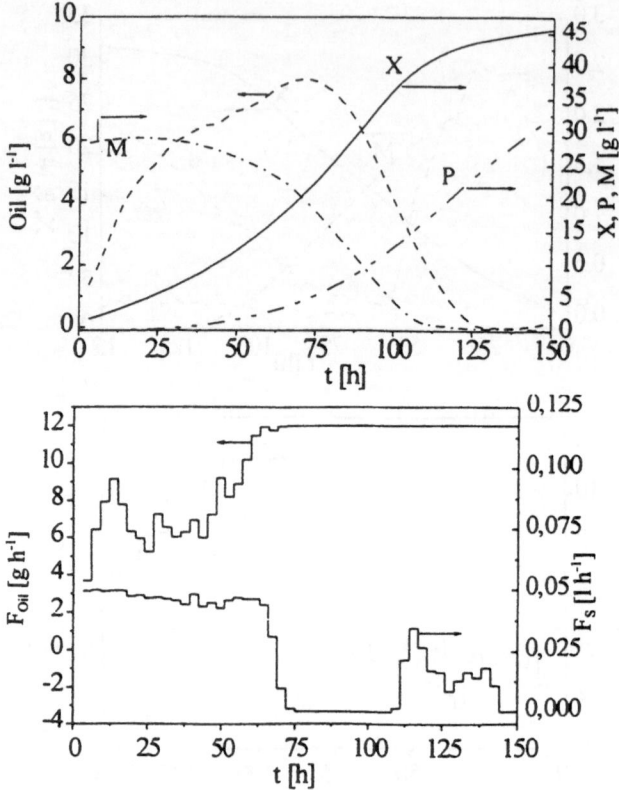

Fig. 19. Optimization results of a fed-batch cultivation of *Acremonium chrysogenum* with sugar and soy oil feeding for the performance criterion Eq. (49)

it normally passes through zero and reaches a maximum near the end of the cultivation when product synthesis slows down. This is the optimum point for starting the down-stream processing. The optimization scheme can follow two directions: on the cultivation level, optimal conditions for inoculation and dynamic profiles for control variables can be determined – as shown in the above example – to obtain the maximum profit for each individual run. Model inaccuracies and process variability can be accounted for by adapting the model to the running process using a moving-window parameter identification and subsequent recalculation of the control profiles. On the plant level, the optimization can be done by dynamic scheduling and determination of suitable harvest times. Using the on-line estimated profit, Yuan et al. [97] suggest a strategy for deciding when a cultivation should be stopped. In comparison to a set of historic data, the actual run is classified as bad, normal or good. By stopping bad runs earlier and prolonging the cultivations classified as good ones, it is estimated that the profit can be increased up to 5 % without changing the control scheme for the cultivations. Further improvements can be expected by combining the dynamic scheduling with optimal dynamic control of each run.

3.4
Artificial Neural Network for Process Modelling

The previous examples have shown that the effort for mathematical modelling of the antibiotics production with its complex kinetics is very high, even though the models cover only some of the important aspects of the process. Many phenomena that can be observed during the cultivations remain unexplained. At least for industrial cultivations, where generally substrates from natural sources with many components are used, and where the reactors may not be completely mixed, it is impossible to consider all influencing factors in a formal or true mechanistic way for several reasons. Not all of those factors are known, but even if so, a lack of analytical techniques may hinder their quantification and the clarification of their interaction with the cellular metabolism. If no reasonable assumptions are available to replace the lack of knowledge, the conventional modeling attempt must fail. But for many applications it is not necessary to establish a model that is able to explain process phenomena, e.g. for process control and optimization purposes it might be sufficient to have some kind of a formal, but accurate black-box description of the process.

In such cases, purely data driven modeling methods that can in principle use all information from available measurements – without using a priori knowledge of the process – can simplify the modelling procedure, and furthermore, have advantages with respect to the accuracy of the process description. The application of the well known concept of feed-forward artificial neural networks for biotechnical processes has been investigated in a great number of papers (e.g. [98 – 100, 110]) and shown to be suited for this field. This very general type of non-linear model is fitted to the data by adjusting the weighting factors for signal transmission between the basic model elements, the neurons. To be robust, the missing built-in knowledge about the process behavior and structure must be substituted in this system by a great number of data sets for training of the network in the whole range of operating conditions which should be covered by the model.

Such an Artificial Neural Networks (ANN) was successfully used by Käse [101] for the modelling of cephalosporin C production. The feed-forward network was constructed of an input layer, four hidden neuron layers and an output layer, which generates the prediction for the process state, as shown in Fig. 20. The outputs of the network are the concentrations of biomass, product, sugar substrate and CO_2 in the exhaust gas. The initial conditions of the same variables are used as input, together with the running time. The network was trained with data of eight cultivations with varying initial conditions. Up to now the application is restricted to cases with invariant feeding strategy. During the training phase, the network learns how the initial conditions influence the course of the cultivation. For simulation purposes and process prediction, the network is supplied with the actual initial conditions from a running process and the desired process time of prediction. The simplicity of this approach has to be weighed against its sensitivity with respect to errors on the initial conditions. Further improvements by self-adaptation mechanisms are discussed by Käse [101].

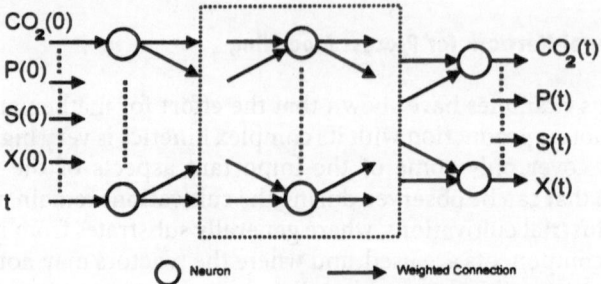

Fig. 20. Structure of the feed-forward artificial neural network for prediction of cultivations of *Acremonium chrysogenum*

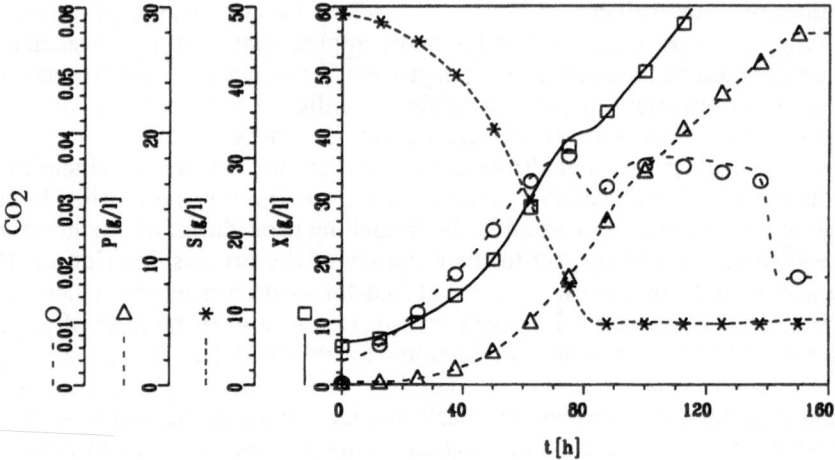

Fig. 21. Simulation results (*lines*) with a feed-forward neural network as shown in Fig. 20 compared to data points (*symbols*) of a *Acremonium chrysogenum* cultivation. Time course of Product (P), Cell mass (X), Substrate (S) and CO_2 in the exhaust gas

An example for simulation results of a cultivation of *Acremonium chrysogenum* is given in Fig. 21. The quality of the process description is rather high and very difficult to achieve with a conventional mathematical model, while the effort for the neural network is much lower. Once the method is established, the training of the network requires only several hours on a workstation compared to several months for the development of the mathematical model by an experienced scientist.

4
Conclusions and Further Prospects

Many aspects of the growth of filamentous microorganisms beginning from single hyphae, filaments, pellets up to pellet populations can be covered by the available mathematical models. The integration of the modelling of microscopic phenomena and mechanisms on the cellular level into the macroscopic description on the population and bioreactor level opens the chance for a better understanding of the key factors that control the complicated kinetics of such cultivations. The examples presented have shown that the tools for a very detailed but efficient simulation of the processes are available. In particular, the methods of digital image analysis have provided a lot of new quantitative information about growth of hyphae, but the crucial problem for further improvement and validation of the models is still the availability of quantifiable experimental data on all levels, which are covered by the models. Due to the nature of the measuring principles, the data may represent deterministic, e.g. concentration profiles in pellets, or statistical information, e.g. distributions of pellet size or locales of branching in growing hyphae. Deterministic models combining a mechanistic or formal deterministic description with stochastic elements can make direct use of both types of information. Another important aspect of modelling and simulation is the unified interpretation of unrelated or not directly related measurements, e.g. cell density profiles from image processing and concentration profiles measurements by micro-probes under conditions that deviate from the cultivation.

Although the models can help to make best use of the available experimental data, it is absolutely necessary to develop improved and new experimental techniques and measuring methods for the still unrevealed phenomena:

- mechanisms of breakage of hyphae and its probability as a function of the shear stress;
- kinetics of growth and lysis of single hyphae under steady limiting conditions;
- development – and finally disruption – of single pellets under defined shear stress;
- truly non-destructing measurement of concentration profiles in pellets, including substrates, products and cell density.

This would allow further progress towards a full understanding of processes with filamentous microorganisms and establishment of process control schemes on a true physiological basis.

5
References

1. Hersbach GJM, Van der Beek CP, Van Dijck PWM (1984) In: EJ Vandamme (Ed) Biotechnology of Industrial Antibiotics. Part 3: The Penicillins: Properties Biosynthesis, and Fermentation. Marcel Dekker, New York, p 46
2. Smith A (1985) Cephalosporins. In: Moo-Young M (ed) Comprehensive Biotechnology. vol. 3: The practice of biotechnology, Current Commodity Products. Pergamon, Oxford, New York
3. Queener SW, Wilkerson S, Tunin DR, et al. (1984) Cephalosporin C fermentation: Biochemical and regulatory aspects of sulfur metabolism. In: EJ Vandamme (ed) Biotechnology of Industrial Antibiotics. Marcel Dekker, New York Basel
4. O'Sullivan CY, Abraham EP (1981) Biosynthesis of beta-lactam antibiotics. In: Antibiotics, Vol. IV, Springer, Berlin Heidelberg New York
5. Queener S, Neuss N (1982) The biosynthesis of beta-lactam antibiotics. In: The Chemistry and Biology of beta-lactam Antibiotics. Vol 3. Academic Press, New York, p 141

6. Martin JF, Revilla G, Zanca DM, Lopez-Nieto MJ (1983) Carbon Catabolite Repression of Penicillin and Cephalosporin Biosynthesis. In: Trends in Antibiotics Research, Antibiotics Research Association, Japan, p 258
7. Küenzi MT (1980) Arch. Microbiol. 128:78
8. Court JR, Pirt SJ (1981) J Chem Technol Biotechnol 31:235
9. Sanchez S, Paniagna L, Nestaas RC, Lara F, Moru J (1980) Nitrogen Regulation of penicillin G biosynthesis in *Penicillium chrysogenum*. Proc. 5th Int Ferment Symp London, Canada, July 20–25
10. Niehoff J (1987) On-line Analytik bei der Produktion von Penicillin V durch *Penicillium chrysogenum*. Dissertation thesis. University of Hannover
11. Wittler R, Schügerl K (1985) Appl. Microbiol. Biotechnol 21:834
12. White RL, John EMM, Baldwin JE, Abraham EP (1982) Biochem J 203:791
13. Hilgendorf, P., Heiser, V., Diekmann, H., Thoma, M. (1987). Appl. Microbiol. Biotechnol. 27:247–251
14. Schügerl K (Ed) (1991) Measuring, Modeling and Control of Biotechnological Processes. In: Rehm HJ, Reed G (vol eds) Biotechnology, Vol. 4, VCH-Verlagsgesellschaft, Weinheim
15. Schügerl K (1991) Bioreaction Engineering Vol. 2. John Wiley, Chichester
16. Moser A (1988) Bioprocess Technology. Springer, Vienna New York
17. Bellgardt K-H (1991) Cell models. In: Rehm HJ, Reed G (eds) Biotechnology, Vol. 4, VCH Verlagsgesellschaft, Weinheim, p 267–298
18. Nielsen J (1992) Adv. Biochem Eng Biotechnol 46:187
19. van Suijdam JC, Metz B (1981) Biotechnol Bioeng 23:111
20. Metz B, de Bruiijn EW, van Suijdam JC (1981) Biotechnol Bioeng 23:149
21. van Suijdam JC, Metz B (1981) J Ferment Technol 59:329
22. Trinci APJ (1970) Arch Microbiol 73:353
23. Metz B, Kossen NWF (1977) Biotechnol Bioeng 19:781
24. van Suijdam JC, Kossen NWF, Paul PG (1980) Europ J Appl Microbiol 10:211
25. Whitaker A (1987) Int Ind Biotechnol June/July. 2:85
26. Wittler R, Baumgartl H, Lübbers DW, Schügerl K (1986) Biotechnol Bioeng 28:1024
27. Hotop S, Möller J, Niehoff J, Schügerl K (1993) Proc Biochem 28:99
28. Edwards N, Beeton S, Bull AT, Merchuck JC (1989) Appl Microbiol Biotechnol 30:190
29. Wittler R (1983) Penicillinproduktion in Mammutschlaufenreaktoren. Dissertation thesis, University of Hannover
30. Constantinides A, Spencer JL, Gaden EL jr (1970) Biotechnol Bioeng 12:803–830
31. Fishman VM, Biryukov VV (1974) Kinetic model of secondary metabolite production and its use in computation of optimal conditions. Biotech. Bioeng. Symp. 4. Interscience. Wiley, New York, p 647
32. Calam CT, Russel DW (1973) J Appl Chem Biotechnol 23:225
33. Megee RD, Kinoshita S, Fredrickson AG, Tsuchiya HM (1970) Biotechnol Bioeng 12:771
34. Trinci APJ (1974) J Gen Microbiol 81:225
35. Righelato RC (1979) The kinetics of mycelial growth. Fungal walls and hyphal growth. Cambridge University Press, London, p 385
36. Bull AT, Trinci APJ (1977) Adv Microbiol Physiol 15:1
37. Hegewald E, Wolleschensky B, Guthke R, Neubert M, Knorre WA (1981) Biotechnol Bioeng 23:1133
38. Bajpai RK, Reuss M (1980) J Chem Technol Biotechnol 30:332
39. Montague A, Morris AJ, Wright M, Aynsley M, Ward A (1986) Can J Chem Eng 64:567
40. Nicolai BM, Van Impe JF, Vanrolleghem PA, Vandewalle J (1991) Biotechnol Letters 13:489
41. Heijnen JJ, Roels JA, Stouthamer AH (1979) Biotechnol Bioeng 21:2175
42. Genon G, Saracco G (1992) Chem Biochem Eng Q 6:75
43. Calam CT, Ismail BAK (1980) J Chem Technol Biotechnol 30:249
44. Calam CT, Ismail BAK (1980) Calculation of growth from carbon dioxide output and problems with early growth stages in penicillin production. In: 7th Symposium on Continuous Cultivation of Microorganisms, Prague, 1978. Czechoslovak Academy of Science Prague, p 745

45. Cooney CL, Mou DG (1982) Application of computer monitoring and control to the penicillin fermentation. In: 3rd international conference on computer applications in fermentation technology. Manchester, England, 1981, p 217
46. Mou DG, Cooney CL (1983) Biotechnol Bioeng 25:225
47. Nelligan I, Calam CT (1983) Biotechnol Letters 5:561
48. Calam CT (1979) Folia Microbiol 24:276
49 Germain B, Mele M, Kremser M (1980) Biotechnol Bioeng 22:255
50. Calam CT (1982) Factors governing the production of Penicillin by Penicillium chrysogenum. In: Overproduction of Microbial Products. Academic Press, New York, p 89
51. Nestaas E, Wang DIC (1983) Biotechnol Bioeng 25:781
52. Guthke R, Knorre WA (1987) Bioprocess Eng 2:169
53. van Suijdam JC, Hols H, Kossen NWF (1982) Biotechnol Bioeng 24:177
54. Bhavaraju SM, Blanch HW (1976) J Ferment Technol 54:466
55. Kluge M, Siegmund D, Diekmann H, Thoma M (1992) Appl Microbiol Biotechnol 36:446
56. Ainsley M, Ward AC, Wright AR (1990) Biotechnol Bioeng 35:820
57. Prosser JI, Trinci APJ (1979) J Gen Microbiol 111:153
58. Nielsen J (1992) Biotechnol Bioeng 41:715
59. Paul GC, Thomas CR (1996) Biotechnol Bioeng 51:558
60. Tiller V, Meyerhoff J, Sziele D, Schügerl K, Bellgardt KH (1994) J Biotechnol 34:119
61. Bajpai RK, Reuss M (1981) Biotechnol Bioeng 23:717
62. Pirt SJ, Righelato RC (1967) Appl Microbiol 15:1284
63. Meyerhoff J, Tiller V, Bellgardt K-H (1995) Bioproc Eng 12:305
64. Yang H, Reichl U, King R, Gilles ED (1991) Biotechnol Bioeng 39:44
65. Yang H, King R, Reichl U, Gilles ED (1991) Biotechnol Bioeng 39:49
66. Meyerhoff J, Bellgardt K-H (1995) Bioproc Eng 12:315
67. Tanford C (1961) Physical chemistry of macromolecules. John Wiley, New York, p 156
68. Meyerhoff J, Bellgardt K-H (1995) J Biotechnol 38:201
69. Demain AL, Ahronowitz Y, Martin JF (1983) Metabolic Control of Secondary Biosynthetic Pathways. In: Vining LC (ed) Biochemistry and Genetic Regulation of Commercially Important Antibiotics. Addison-Wesley. Reading, MA
70. Matsumara M, Imanaka T, Yoshida T, Tagichi H (1978) J Ferment Technol 56:345
71. Chu WBZ, Constantinides A (1987) Biotechnol Bioeng 32:277
72. Matsumara M, Imanaka T, Yoshida T, Tagichi H (1981) J Ferment Technol 59:115
73. Matsumara M, Imanaka T, Yoshida T, Tagichi H (1980) J Ferment Technol 58:197
74. Pirt SJ (1969) Microbial growth and product formation. In: Microbial growth. 19th Symposium of the Society to General Microbiology. Cambridge University Press, London p 199
75. Malmberg LH, Hu WS (1992) Appl Microbiol Biotechnol 38:122
76. Meyerhoff J (1995) Mathematische Modelle für Bioprozesse mit filamentösen Pilzen. Dissertation thesis, University of Hannover
77. Tollnik C (1993) Durchführung und Charakterisierung von Kultivierungen mit Hochleistungsstämmen von Acremonium chrysogenum in einem 10L-Rührkesselreaktor unter Verwendung modifizierter Kulturmedien. Diploma thesis, University of Hannover
78. Swartz RW (1979) The use of economic analysis of penicillin G manufacturing costs in establishing priorities for fermentation process improvement. In: Annual Reports on Fermentation Processes, Vol. 3, Academic Press, New York, p 75
79. Lorenz T, Diekmann J, Früh K, Hiddesen R, Möller J, Niehoff J, Schügerl K (1987) J Chem Tech Biotechnol 38:41
80. Ryu DDY, Humphrey AE (1972) J Ferment Technol 50:424
81. Giona AR, De Santis R, Marelli L, Toro L (1976) Biotechnol Bioeng 18:493
82. Modak JM, Lim HC (1987) Biotechnol Bioeng 30:528
83. Modak JM, Lim HC, TayebYJ (1986) Biotechnol Bioeng 28:1396
84. Modak JM, Lim HC (1989) Biotechnol Bioeng 33:11
85. Constantinides A, Rai VR (1974) Biotechnol Bioeng Symp 4:663
86. Yamane T, Kume T, Sada E, Takamatsu T (1977) J Ferment Technol 55:587
87. Choi CY, Park SY (1981) J Ferment Technol 59:65

88. Okabe M, Aiba S (1975) J Ferment Technol 53:730
89. Hong J (1986) Biotechnol Bioeng 28:1421
90. Vicik SM, Fedor AJ, Swartz RW (1990) Biotechnol Prog 6:333
91. Guthke R, Knorre WA (1981) Biotechnol Bioeng 23:2771
92. Constantinides A, Spencer JL, Gaden EL jr (1970) Biotechnol Bioeng 12:1081
93. Luus R (1989) Hung J Ind Chem 17:523
94. Luus R (1990) Int J Control 52:239
95. Bojkov B, Luus R (1994) Chem Eng Res Des 72:72
96. Bojkov B, Luus R (1992) Ind Eng Chem Res 31:1308
97. Yuan JQ, Guo SR, Schügerl K, Bellgardt K-H (1997) J Biotechnol 54:175
98. Glassey J, Montague A, Ward AC, Kara BV (1994) Process Biochem 29:387
99. Linko P, Zhu YH (1991) J Biotechnol 21:253
100. Linko P, Zhu YH (1992) Process. Biochem. 27:275
101. Käse A (1996) Modellierung von Bioprozessen mit künstlichen Neuronalen Netzwerken. Dissertation thesis. University of Hannover
102. Paul GC, Kent CA, Thomas CR (1993) Biotechnol Bioeng 42:11
103. Hotop S (1991) Penicillin V Produktion im Rührkessel – Verfahrenstechnische Aspekte der Prozeßoptimierung. Dissertation thesis. University of Hannover
104. Nash CH, Huber FM (1971) Appl Microbiol 22:6
105. Qeener SW, Ellis LF (1975) Can J Microbiol 21:1981
106. Bartoshevic YE, van den Heuvel JC (1990) Chem Eng J 44:B1
107. Lim HC, Tayeb YJ, Modak JM, Bonte P (1986) Biotechnol Bioeng 28:1408
108. Prosser JI, Tough AJ (1991) Crit Rev Biotech 10:253
109. Keim J, Shen YQ, Wolfe S, Demain AI (1984) Europ J Appl Microbiol Biotechnol 19:232
110. Montague G, Morris J (1994) Trends Biotechnol 12:312

Received April 1997

Influence of the Process Parameters on the Morphology and Enzyme Production of *Aspergilli*

K. Schügerl · S. R. Gerlach · D. Siedenberg

Institut für Technische Chemie der Universität Hannover, Callinstr. 3, D-30167 Hannover, Germany

Several papers have been published dealing with various fungi to determine their morphology, enzyme production or process performance. However, no publication considered all of these aspects simultaneously. In the case of the production of xylanase by *Aspergillus awamori* the interrelationship of various key parameters are investigated. The influence of the reactor type (shake flasks, stirred tank and airlift tower loop reactor), the medium composition (semisynthetic and complex medium with wheat bran of different sizes, respectively as well as different concentrations of phosphate), and the specific power input (stirrer speed) on the growth, morphology, physiology, and productivity of the fungus are investigated. The results reveal a complex interrelationship which explains why the published results are contradictory. Without considering all of the relevant parameters, it is not possible to make general conclusions.

Advances in Biochemical Engineering/
Biotechnology, Vol. 60
Managing Editor: Th. Scheper
© Springer-Verlag Berlin Heidelberg 1998

List of Symbols and Abbreviations

(L = length, M = mass, T = time, 1 dimensionless)

A	empirical constant 1
a	empirical constant 1
C	empirical constant 1
CPR	CO_2 production rate $M L^{-1} T^{-1}$
D	shear rate T^{-1}
D_R	vessel diameter L
d	impeller diameter L
d_{imax}	maximal internal diameter of the flasks L
d_p	pellet diameter L
d_{VF}	shear stress factor (diameter of the flocs) L
E	eccentricity of the shaker L
E_X	pellet volume fraction 1
E_M	maximal pellet volume fraction 1
e	shaking radius L
F	degree of filling 1
Fr	Froude number 1
g	acceleration of gravity $L T^{-2}$
K	fluid consistency index $M L^{-1} T^{-n}$
$k_L a$	volumetric mass transfer coefficient T^{-1}
k	empirical constant 1
N	rotation speed T^{-1}
Ne	Newton number 1
Ne_G	Newton number in aerated medium 1
n	flow behavior index 1
P	power input $M L^2 T^{-3}$
P_G	power input in aerated medium $M L^2 T^{-3}$
Q_G	gas throughput number 1
q_G	gas throughput $L^3 T^{-1}$
R	roughness factor 1
Re	Reynolds number 1
V_S	medium volume L^3
V_R	active volume of the impeller L^3
w_{SG}	superficial gas velocity $L T^{-1}$
X	cell mass concentration $M L^{-3}$
$Y_{X/S}$	yield coefficient of growth with respect to the substrate consumption 1
$Y_{P/S}$	yield coefficient of product formation with respect to the substrate consumption 1
$Y_{P/X}$	yield coefficient of product formation with respect to the cell mass formed 1
η_a	apparent dynamic viscosity $M L^{-1} T^{-1}$
μ	specific growth rate T^{-1}
ν_L	kinematic viscosity of the medium $L^2 T^{-1}$
ρ_L	density of the liquid medium $M L^{-3}$

1
Introduction

In the industry *Aspergilli* are used for production of enzymes and organic acids [1, 2]. A large number of publications have dealt with the production of citric acid (e.g. [3]). Only few of them have considered the morphology of the fungus (e.g. [4, 5]). In the case of citric acid production, the influence of manganese on the mycelial morphology and the citric acid production has been discussed [3].

Several publications have reported on enzyme production by *Aspergilli*, e.g., on the production of amylase [6], pectinase [7], glucoamylase [8], xylanase [9–12], ribonuclease [13], etc. On the other hand, several research groups have been engaged in the determination of morphology of fungi and streptomyces, especially *Penicillium chrysogenum* *(for review see Paul and Thomas in this volume) and *Streptomyces tendae* (for review see King in this volume), but no paper has been published, which treats the relationship between fungal morphology and enzyme production.

The aim of the present review is to report on the recent results in this field, which were obtained with *Aspergillus awamori*.

2
Methods

2.1
Determination of Fungal Morphology

Metz et al. [14] used manual digitizing to evaluate the morphology of molds. Adams and Thomas [15] developed an interactive digital image analysis method. Reichl [16] and Reichl et al. [17] also investigated the development of mycelia by using interactive digital image analysis. They developed an Acridine Orange (AO)-staining to determine the septation of the hyphae [18]. Yang et al. [19] applied a mathematical model based on these measurements to describe the spore germination and the growth of hyphae. Packer and Thomas [20] developed a fully automatic digital image analysis method.

Reichl and Gilles [21], Reichl et al. [22], Tucker et al. [23] and Durant et al. [24] used explicit techniques to characterize mycelium clumps and pellets and distinguish between filamentous and pellet shaped mycelia. Olsvik et al. [25] and Tucker and Thomas [26] investigated the interrelationship between rheology and fungal morphology. Cox and Thomas [27] differentiated between smooth and hairy pellets. Treskatis [28] developed a fully automatic sampling and image analysis. He carried out an object characterization based on a Bayes-classification as well. Nielsen et al. [29] investigated the influence of the fragmentation of *Penicillium chrysogenum* on penicillin production.* (For more details see Capter 1.

In the example presented a standard digital image analysis was used, which was combined with an artificial neural net (ANN) processing [30]. Figure 1 shows the schematic digital image analysis procedure. By this technique object

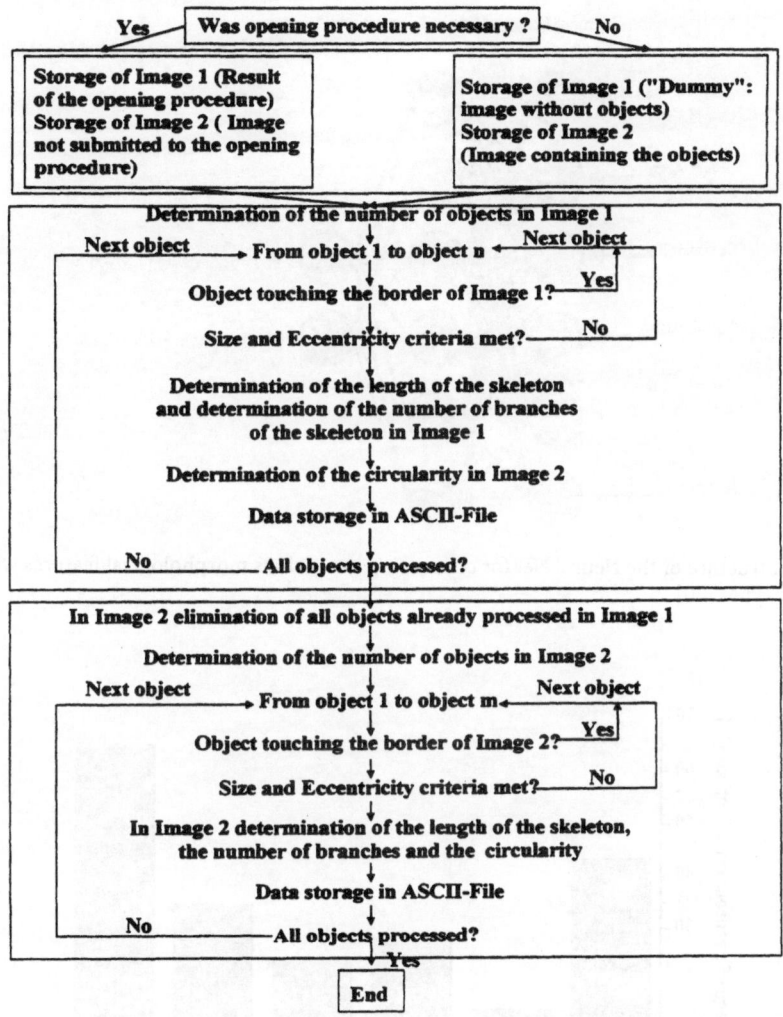

Fig. 1. Digital image analysis procedure [30]

area, eccentricity, circularity (norm. form factor), length of the skeleton and the number of branches of the skeleton were determined. In order to classify the mycelial morphology, four object types were defined by the combination of pellet (1), filamentous (-1), globular (1) and elongated (-1) objects: globular pellets (1/1), elongated pellets (1/-1), clumps (-1/1) and filamentous mycelia (-1/-1). In Fig. 2 the combination of the morphological features and the cultivation time with these four object types by means of an ANN is shown. The training of the ANN was performed by back-propagation, which was recommended for feed-forward nets by Rumerhart et al. [31]. A combined gradient method was used, which minimizes the mean quadratic error of the net [30].

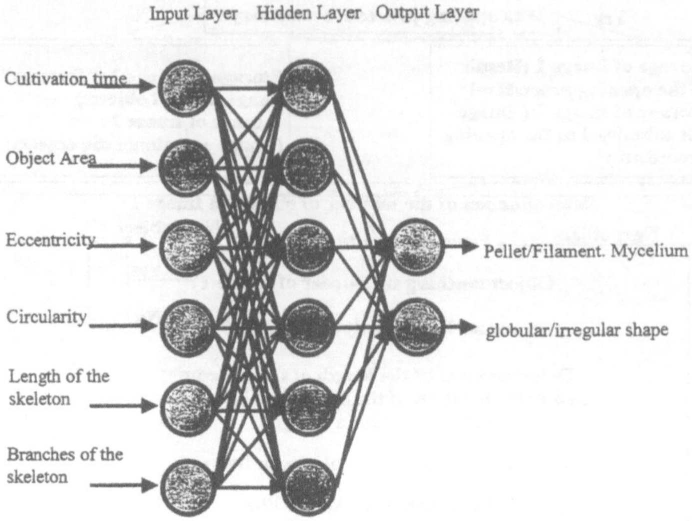

Input Layer   Hidden Layer   Output Layer

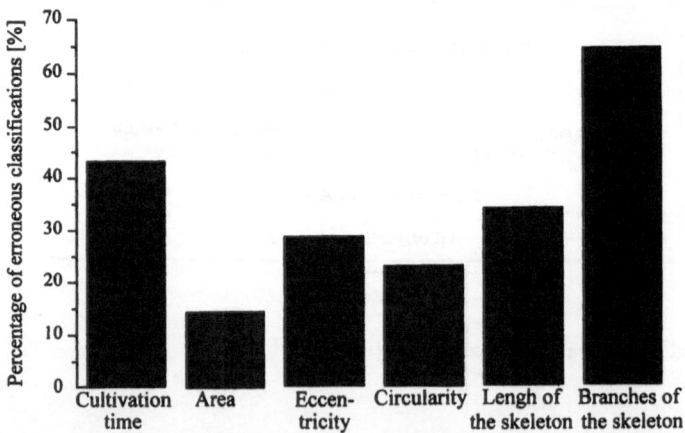

Fig. 2. Structure of the Neural Net for connecting the various morphological features with the object types [30]

Fig. 3. Significance (probability of erroneous classification) of the various morphological features [30]

To evaluate the significance (probability for erroneous classification) of the various morphological parameters, cluster analysis was applied [30, 32, 33]. The application of this technique to fungal morphology is shown on the example *Aspergillus awamori* (Fig. 3). According to this analysis, the area has the highest significance and the branches of the skeleton the lowest. Investigation of the significance of the combination of these morphological measurements reveals

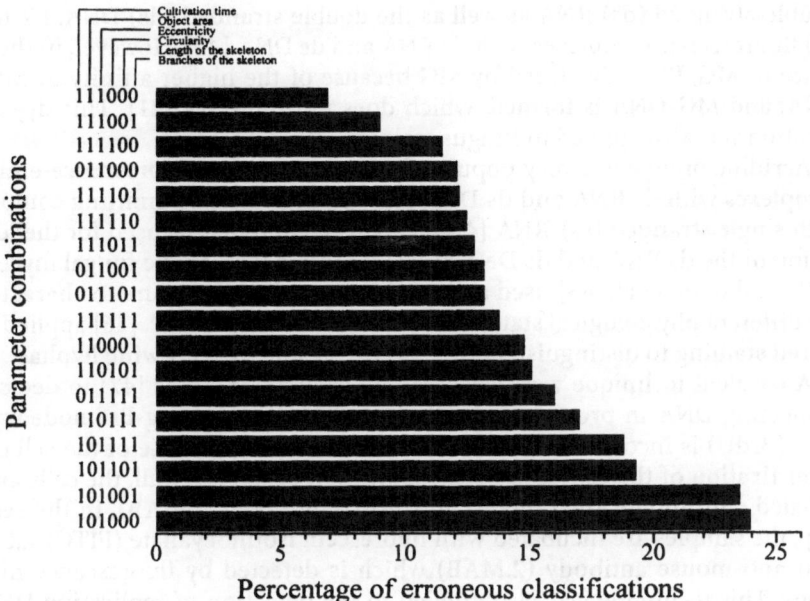

Fig. 4. Significance (probability of erroneous classification) of the combination of the various morphological features and time [30]

that the classification, including time, object surface area and eccentricity, yields the highest significance (lowest erroneous classification) (Fig. 4).

2.2
Determination of Fungal Physiology

The physiological state of the spores and the mycelium was investigated by a few authors. Paul et al. [34] discriminated between germinated and non-germinated spores. Tucker and Thomas [35] investigated the influence of the inoculum concentration on the morphological development of the mycelium. Packer et al. [36] estimated the cell volume and cell mass by taking the degeneration of the old hyphae into account. The development of vacuoles in the fungal hyphae were considered by Paul et al. [34, 37]. Olsvik et al. [25] investigated the relationship between rheology and morphology of *Aspergillus niger*. Mohseni and Allen [38] evaluated the influence of the rotation speed in shake flask culture and in a stirred tank reactor on the morphology of *Aspergillus niger* determined by digital image analysis.

Staining techniques were developed in histology and applied in flow cytometry.

The combination of fluorescence diacetate (FDA) and ethidiumbromide (EB) or propidium iodide (PI) is used for vitality investigations: vital cells exhibit green and dead cells red fluorescence. This technique was applied to fungus as well [39]. Pyronin Y (PY) and methylgreen (MG) are used for the detection of double-stranded (ds) RNA as well as the double stranded (ds) DNA. PY forms red fluorescence complexes with ds RNA and ds DNA [40]. However, in the presence of MG, PY is displaced by MG because of the higher affinity of MG for DNA, and MG-DNA is formed, which does not fluoresce [41]. This dye combination was also applied to fungus staining [39].

Acridine orange is a very popular dye. It forms green fluorescence-emitting complexes with ds RNA and ds DNA and red fluorescence-emitting complexes with single stranded (ss)-RNA [42–44]. This technique was used for the localization of the ds RNA and ds DNA as well as the ss RNA in the fungal mycelium [39]. Vanhoutte et al. [45] used a differential staining procedure to characterize the different physiological states of the fungus. Matsuoka et al. [46] applied congo red staining to distinguish between growing and non-growing hyphae.

A covalent technique was developed by Dien and Scrienc [47] to detect the replicating DNA in propagating cells. The thymidine analog Bromodeoxyuridine (BrdU) is incorporated into the DNA during the S-phase of the cell cycle. After fixation of the cells and permeabilization of the cell wall, the cells are incubated with monoclonal mouse anti-BrdU-antibody (1. MAB). In the second step, the samples are incubated with fluorescein isothiocyanate (FITC)-labelled goat anti-mouse antibody (2.MAB) which is detected by fluorescence microscopy. This technique was also applied to the detection of replicating DNA in fungal mycelium [39].

2.3
Determination of Fungal Interactions with Solid Substrates

When solid substrate is involved, the process performance depends on the structure and composition of the substrate and the interaction between the substrate and the fungus.

The ultrastructure of *Aspergillus niger* has been investigated by only a few authors [48–50]. Several industrial production processes involving fungi use complex media, consisting of agricultural byproducts (corn steep flour, peanut flour, wheat bran, sugar beet molasses, pharma medium, etc.). Some of these complex medium components are solid plant fragments and are used in solid state and suspension cultures, respectively. The structures of various plant fragments have been determined by botanists (e.g. [51, 52]).

In spite of the common use of these solid substrates, no electron microscopic investigation has been carried out on the interaction between the fungus and the solid substrate, except those of Adolph et al. [53].

The semithin sectioning for light microscopy was performed by Pyramitom 11 800 using glass knives, and the ultrathin sectioning was carried out with Ultracut E and glass knives.

The samples were prepared for the electron microscopic measurements by four different techniques (Table 1). Preparation 1 yielded shrinkage artifacts due to the hypertonic fixing solution used. Preparation 2 led to good preservation of the ultrastructure. Preparation 4 caused strong plasmolysis which partly destroyed the ultrastructure. Preparation 3 preserved the utrastructure and yielded excellent immunocytochemical reactivity [53]. The preparations for

Table 1. Applied preparation techniques

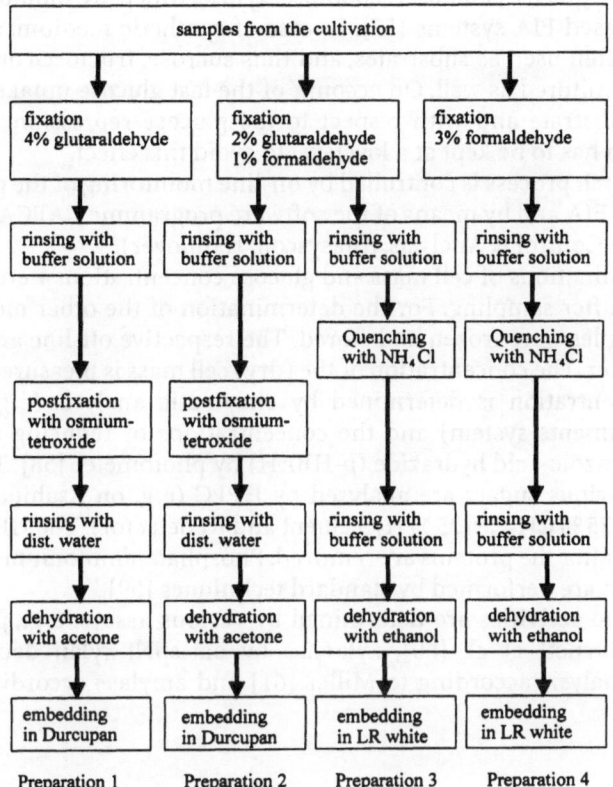

samples from the cultivation			
fixation 4% glutaraldehyde	fixation 2% glutaraldehyde 1% formaldehyde	fixation 3% formaldehyde	
rinsing with buffer solution	rinsing with buffer solution	rinsing with buffer solution	rinsing with buffer solution
		Quenching with NH₄Cl	Quenching with NH₄Cl
postfixation with osmium-tetroxide	postfixation with osmium-tetroxide		
rinsing with dist. water	rinsing with dist. water	rinsing with buffer solution	rinsing with buffer solution
dehydration with acetone	dehydration with acetone	dehydration with ethanol	dehydration with ethanol
embedding in Durcupan	embedding in Durcupan	embedding in LR white	embedding in LR white
Preparation 1	Preparation 2	Preparation 3	Preparation 4

electron spectroscopy imaging (ESI) and electron energy loss spectroscopy (EELS) [54] were cut with diamond knives.

For the detection of the fungal enzyme, xylanase, fresh ultrathin sections of Preparations 3 and 4 (Table 1) were labelled with colloidal gold according to the immunological technique described by Roth [55] and modified by Jäger and Bergman [56] (for more details, see Adolph et al. [53]).

2.4
Process Monitoring Methods

Complex media with agricultural byproducts contain cellulose, hemicellulose, starch, pectin, lignin, and several other compounds in lower concentrations. *Aspergilli* can produce all of the enzymes which are necessary to utilize these substrates, except lignin. In the case of solid substrates, they have to be dissolved prior to their utilization. It is therefore expected that the fungus produces enzymes which are necessary for the dissolution and utilization of the complex substrate.

The enzymatic decomposition of starch, cellulose, and hemicellulose yields various sugars: glucose, maltose, cellobiose, xylose, which are monitored on-line by enzyme based FIA systems [57]. In case of synthetic medium, sucrose and glucose are often used as substrates, and thus sucrose, fructose concentrations have to be monitored as well. On account of the fast glucose uptake in the case of glucose substrate and with respect to the glucose repression, the glucose concentration has to be kept at a low level to avoid this effect.

The fed-batch process is controlled by on-line monitoring of the glucose concentration by FIA and by means of the software programme CAFCA (Computer Assisted Flow Control & Analysis, Anasyscon, Hannover).

The determinations of cell mass and glucose concentrations were performed immediately after sampling. For the determination of the other medium components, samples were frozen and stored. The respective off-line analyses were performed later. The concentration of the (dry) cell mass is measured by weight, glucose concentration is determined by enzymatic analysis (e.g. by Yellow Spring Instruments system) and the concentrations of reducing sugars with p-hydroxy-benzoic-acid hydrazide (p-HBAH) by photometer [58]. The concentrations of various sugars are analysed by HPLC (e.g. on Asahipak-NH2P-50 column, with 75% CH_3CN/25% H_2O eluent and RI detector). The HPLC analysis is carried out after the proteins are removed. Phosphate, ammonium and protein determination are performed by standard techniques [59].

The enzyme activities are determined by various assays: e.g., protease according to Burnett et al. [60], xylanase by oat-spelt-xylan decomposition and xylose analysis according to Miller [61] and amylase according to Bergmeyer [62].

2.5
Product Identification Methods

During the process several proteins are secreted into the medium. The activities of the key enzymes, amylase, hemicellulase (xylanase), and pectinase, are monitored during the production process by enzyme assays. The composition of the proteins are analysed by SDS electrophoresis and (reversed phase, ion exchange, and size exclusion) chromatography. After chromatographic separation of the components, the activities of the fractions corresponding to the different peaks are determined by enzymatic assays. The non-active proteins are indentified by their molecular weights.

2.6
Determination of the Power Input in Shake Flasks

The torque imposed on the shaker with a large number of shake flasks was determined with liquid and with the same volume of solid agar in the shake flasks. The difference between the torques of the liquid and the solid agar in the flasks yields the power input into the liquid by means of the friction between the liquid and the wall of the flask. The power input was calculated by a suitable programme [63, 64]. Based on measurements in shake flasks of various sizes, with and without baffles with water and aqueous solutions of different viscosity, the following relationships were developed by Milbradt [64] and Zoels [63]:

$$\frac{P}{V_S} = 1.0025 \cdot 10^{-10} \frac{N^3 \, d^4_{i,max}}{V_S^{2/3}} \tag{1}$$

and

$$\frac{P}{V_S} = 1.553 \cdot 10^{-10} \frac{N^{3.12} \, d^{3.455}_{i,max}}{V_S^{0.636} \cdot e^{0.012}} \tag{2}$$

respectively, where P/V_S = specific power input [kW m^{-3}], N=rotation speed of the shaker [min^{-1}], V_S=working volume of the flasks [cm^3], $d_{i,max}$=maximal internal diameter of the flasks [cm] and e=shaking radius [cm].

2.7
Determination of the Volumetric Oxygen Transfer Coefficient in Shake Flasks

The following relationship was recommended for the volumetric mass transfer coefficient in shake flasks [65]:

$$k_L \, a = A \left(\frac{d^3_{i,max}}{V_S} \right)^{8/9} \left(\frac{N^2 \sqrt{(d_{i,max} \cdot E)}}{g} \right)^a \tag{3}$$

where $k_L a$=volumetric mass transfer coefficient [h^{-1}], A = constant [h^{-1}], a = constant [–], N = rotation speed of shaker tablar [s^{-1}], g=acceleration due to gravity [cm s^{-2}], E=eccentricity [cm]. For cultivation media A = 26 h^{-1} and a = 0.666.

According to the investigations of Henzler and Schedel [65] and Schultz [66] the diffusion through the cotton plug of the shake flasks is about 20% lower than in the free gas volume. The measurements reveal that the metabolic activity of the microorganisms is limited by the transport process in the shaking flasks. The major transport resistance is in the gas/liquid interface.

The volumetric mass transfer coefficient is enhanced by a factor of four with increase of the shaker speed from 100 to 400 rpm [65]. The ratio of the medium volume to the shaking flask volume also influences the $k_L a$ values. At low volume ratio the $k_L a$ values are in the range of 100–200 h^{-1}, whereas at high volume ratio, they are much lower: 20–60 h^{-1}. These measurements were carried out in shake flasks with low viscosity media at 100–200 rpm. Therefore, in this review, both the shaker speed and the medium volume to flask volume ratio were given.

2.8
Determination of Power Input in Stirred Tank Reactors

The specific power input is controlled by the stirrer speed. The power input due to the aeration has a relatively low share in the total power input. However, it reduces the power input by the stirrer. The specific power input can be calculated

by empirical relationships based on the dimensions of the agitator and vessel as
well as the medium viscosity. The Newton number (Ne) is given as a function of
the Reynolds number (Re): $Ne = f(Re)$ [67]:

$$Ne = \frac{P}{N^3 d^5 \rho_L} \quad \text{and} \quad Re = \frac{N d^2}{\nu_L}$$

where P=power input, N = stirrer speed, d = diameter of the impeller, ρ_L = density
of the liquid, and ν_L = kinematic viscosity of the liquid

For Reynolds numbers larger than 100, the Newton number is nearly independent of Re. For flat bladed disc turbines with four baffles, for example,
Ne=3.0 for Re=333 (200 rpm), Ne=3.5 for Re=666 (400 rpm) and Ne=4.0 for
Re=1000 (600 rpm) for low viscosity non aerated media. By aeration of the
cultivation medium, the power input diminishes. The influence of the aeration
rate on the Newton number is given by Judat [68]:

$$Ne_G = f (Q_G, Fr, D_R/d)$$

$$Ne_G = \frac{P_G}{\rho_L N^3 d^5} \quad \text{Newton number for aerated medium,}$$

P_G = power input in aerated medium,

$$Q_G = \frac{q_G}{N d^3} \text{ the gas throughput number, } Fr = \frac{N^2 d}{g} \text{ Froude number}$$

q_G = gas throughput and D_R = vessel diameter d = impeller diameter.

2.9
Determination of Shear Stress in Stirred Tank Reactors

To determine the shear stress imposed on cells, a test system consisting of the
aqueous suspension of "Blue clay" (Witterschlicker Blauton HFF from Co. H.J.
Braun Tonbergbau, particle size 98 wt% < 2 μm) was used, which was co-
agulated with the flocculating agent Praestol 650 BC (a copolymer of acryla-
mide and a cationic comonomer) (Co. Stockhausen). The size of the flocs in
the range 2 to 1000 μm depends on the mechanical stress imposed on them. In
the applied 1 g l^{-1} NaCl solution, the variation of ion strength influences the
floc size only slightly. With a constant mechanical stress imposed on the flocs,
a constant mean floc size is obtained within 60 min, the size of which is measured with laser optical system of BASF [69–71]. The floc size d_{VF} (μm) is
called the shear stress factor. The shear stress caused by the stirrer and imposed on the cells depends on the specific power input (P/V_S) according to
Biedermann [70]:

$$d_{VF} = K \left(\frac{P}{V_S} \right)^a \left(\frac{V_S}{V_R} \right)^b \tag{4}$$

where d_{VF} is the diameter of the applied clay flocs, which were used for the determination of the shear stress, V_R is the active volume of the stirrer [71], K, a and b are constants, which depend on the applied stirrer, vessel and medium.

3
Characterization of Reactor Performances

3.1
Stirred Tank Reactor

In stirred tank reactors various agitators are used depending on the vessel size, the medium viscosity and the shear sensitivity of the cells. At low viscosity media, tangential and radial acting flat bladed disc turbines (Rushton turbine) are applied. The usual diameter of the agitator is 1/3 of the vessel diameter. In slender reactors, the axial mixing is enhanced by pitched blade impellers. In cultivation media with intermediate viscosity, agitators are used with a diameter which is 2/3 of the vessel diameter.

In the investigations reviewed here, two-stage flat bladed disc turbine agitators with an agitator/vessel diameter ratio 0.33 were applied. The specific power input and the shear stress factor were evaluated to 0.05 kW m^{-3} and $d_{VF}=200$ μm at 200 rpm, to 0.50 kW m^{-3} and $d_{VF}=150$ μm at 400 rpm and to 2.0 kW m^{-3} and $d_{VF}=120$ μm at 600 rpm for the nonaerated bioreactor in low viscosity medium.

With increasing viscosity, the specific power input increases and with aeration it decreases. In presence of low concentration of solid particles (bran) the power input is further reduced [72]. During the cultivation, the viscosity of the cultivation medium increases. This increase is generally lower if pellet suspension prevails. For suspensions of fungal pellets a relationship was recommended by Wittler et al. [73], which was confirmed by Mitard and Riba [74] for *Aspergillus niger* pellet suspensions. The apparent viscosity η_a is given by

$$\eta_a = \left(\frac{E_X/E_M}{1 - E_X/E_M} \right)^2 C^2 D \exp \left[-b(E_X/E_M) \right] \tag{5}$$

where E_X is the pellet volume fraction, E_M the maximum pellet volume fraction, D the shear rate, and C and b are constants.

Mitard and Riba [74] identified these constants for *Aspergillus niger* pellets of various mean diameters (d_P):

$d_P=0.84$ mm, 　b$=2.88$, 　$E_M=1.34$, 　C$=5.53$ with a relative deviation (RD)$=27\%$

$d_P=1.0$ mm, 　　b$=2.35$, 　$E_M=1.015$, C$=3.83$ with RD$=20\%$

$d_P=2.25$ mm, 　b$=1.56$, 　$E_M=0.61$, 　C$=1.74$ with RD$=15\%$

$d_P=2.45$ mm, 　b$=1.15$, 　$E_M=0.46$, 　C$=0.8$ with RD$=14\%$.

Olsvik and Kristiansen [75] described the viscosity of *Aspergillus niger* cultivation medium with the Oswald-de Waele (power law) model and related it to the fungal morphology [25]:

$$\eta_a = K\ D^{n-1} \tag{6}$$

where K is the consistency index, D the shear rate, and the n the flow behavior index.

According to their investigations more than 87% of the hyphae were clumps. The consistency index K and the flow behavior index n of the power law were correlated to the cell mass concentration X, the roughness factor R [circularity/(4 times area within the clump diameter] and the compactness factor F [area/area of clumps with voids filled] for each individual clump.

K = -0.94 + 0.06 R, n = -0.51 - 0.005 R; K = -0.35 + 35 × 10^{-5} clump area.

K = -0.56 + 0.0018 R X$^{1.7}$.

In stirred tank reactors, the shear rates D can be calculated according to Metzner et al. [76]: D = k N, where N is the stirrer speed and k an empirical constant which is only slightly altered by different types of stirrer type and stirrer/vessel diameter ratio. For two-stage flat bladed disc turbines of 10 cm diameter k = 11.5 ± 1.4 holds true [67]. The shear rates were determined to:

D = 36.3 s^{-1} (200 rpm), D = 63.2 s^{-1} (400 rpm) and D = 110 s^{-1} (600 rpm).

3.2
Airlift Tower Loop Reactor

Bubble column and airlift tower loop reactors are characterized by low specific power input and high efficiency of the oxygen transfer with respect to the specific power input [77, 78]. They are used in industry when low specific power input is sufficient for the operation [67]. The specific power input is given by $P_G/V_S = \rho_L\ g\ w_{SG}$. At the specific aeration rate 0.5 vvm (w_{SG} = 3.25 cm s^{-1}) the specific power input was determined to be P/V_S = 0.0162 kW m^{-3} and the shear stress factor was determined to be d_{VF} = 250 μm for low viscosity media without solid content [70]. At high viscosity and in solid suspensions the d_{VF} value is considerably higher. According to Nishikawa the average shear rate in bubble colum reactors is given by D = k w_{SG}, where w_{SG} is the superficial gas velocity in the riser [cm s^{-1}]. The original value for k, recommended by Nishikawa was gradually reduced to k = 15 (in the relationship recommended by Henzler, see [79].

At the typical gas velocity w_{SG} = 3.25 cm s^{-1}, the shear rate was evaluated to be D = 48.75 s^{-1}.

The apparent viscosity of the cultivation medium calculated according to the model of Wittler et al. [73] and model parameters recommended by Mitard and Riba [74] was evaluated as η_a = 164 mPa s at w_{SG} = 3.25 cm s^{-1}.

The volumetric mass transfer coefficient depends on the aerator and medium properties. In the low viscosity cultivation medium the volumetric mass transfer coefficient was determined as k_L a = 0.03 s^{-1} at w_{SG} = 3.25 cm s^{-1}. At intermediate and high viscosities the following relationship was recommended by

Deckwer et al. [80]: $k_L a = 2.08 \times 10^{-4} w_{SG}^{0.59} \eta_a^{-0.84}$. The volumetric mass transfer coefficient was determined by this relationship to be $k_L a = 0.0132 \text{ s}^{-1}$ at $w_{SG} = 3.25 \text{ cm s}^{-1}$ and $\eta_a = 164 \text{ mPa s}$ in the airlift tower loop reactor.

4
Fungal Cultivations

The cultivations in different media and under various fluid dynamic conditions indicate a strong interrelation between them. Therefore, at first, the cultivations in solid state cultures, shake flasks, stirred tank and airlift tower loop reactors will be considered separately. The comparison of the results is undertaken in Chap. 5.

4.1
Solid State Cultivations

Solid state cultivation (SSC) is popular in the food industry (e.g., koji fermentation) and sometimes practiced in biotechnological processes (cellulase/hemicellulase and citric acid production), but only a few investigations have been carried out with monitoring of only one or two process parameters [81–85].On account of the lack of power input and because of the geometric limitation of the growth of mycelia in the solid substrates, only low packing density could be obtained with *Trichoderma reesei*, *T. lignorum* and *C. cellulolyticum* in wheat straw cultures. Scanning electron microscopic images indicate the growth of the hyphae on the loose layer of parallel oriented milled wheat straw [86]. The enzyme production by *Aspergillus oryzae* or *soyae* were investigated in solid state culture, but without the determination of the morphology [87]. A comparison of the performances of solid state cultivations of *Aspergillus niger van Tiegham* on mixtures of wheat straw and bran (1:1) and submerged cultivations (SC) with xylan indicated that the pH optimum of the xylanase produced by SSC is lower (3.8) than that of SC (5.0). The temperature optimum of the xylanase produced by SSC is higher (50 °C) than that of SC (45 °C). The yield of xylanase production in SSC is lower than in SC. However, in SSC higher enzyme activities were obtained: 2500 U g^{-1} (dry) cell mass with SSC and only 145 U g^{-1} (dry) cell mass with SC [11]. Because of the strong mass transfer limitation at high cell density, the growth and product formation rates are considerable lower than in submerged cultivations. During the batch cultivation the thickness of the biofilm increases and the cells in the deeper regions die away and lyse because of the lack of oxygen and substrate.

4.2
Cultivations in Shake Flasks in Semisynthetic Medium (Table 2)

Cultivations in shake flasks are usually performed under loosely defined conditions. Neither the pH value and dissolved oxygen concentration nor the concentrations of substrates and products are monitored during the cultivation. There-

Table 2. Composition of the semisynthetic medium

Sucrose (Glucose)	10 g l^{-1}
NH_4Cl	9 g l^{-1}
KH_2PO_4	1.5 g l^{-1}
$NaNO_3$	1 g l^{-1}
$MgSO_4 \, 7H_2O$	1 g l^{-1}
$CaCl_2 \, 2H_2O$	0.3 g l^{-1}
yeast extract	1 g l^{-1}

fore, these cultivations are suitable only for preliminary investigation of the influence of various parameters on the process performance. The effect of temperature, pH value, various nitrogen sources (yeast extract, protease peptone, urea, ammonium chloride, sodium nitrate, cornsteep liquor, etc.) as well as the efficiency of different inducers (ball-milled oat straw, ball milled barley straw, delignified ball milled oat straw, oat spelt xylan, larch wood xylan, avicel, cornsteep liquor and different sugars) and their concentration on the activitie s of xylanase, β-xylosidase, protease and extracellulary protein concentration produced by *Aspergillus awamori* have been determined in shake flask cultures [10].

4.2.1
Induction of the Xylanase Production

With a semisynthetic medium, the induction of enzyme production is decisive for the process performance. Investigations reveal that oat spelt xylan is an excellent inducer. It is better than xylose [88]. Cultivations were performed with 10^2 spores ml^{-1} (filamentous mycelium) and 10^5 spores ml^{-1} (pellet suspension) in 5-1 shake flasks at 25 °C on an orbital shaker at 150 rpm. Various concentrations of xylan were added to the medium, after the reducing

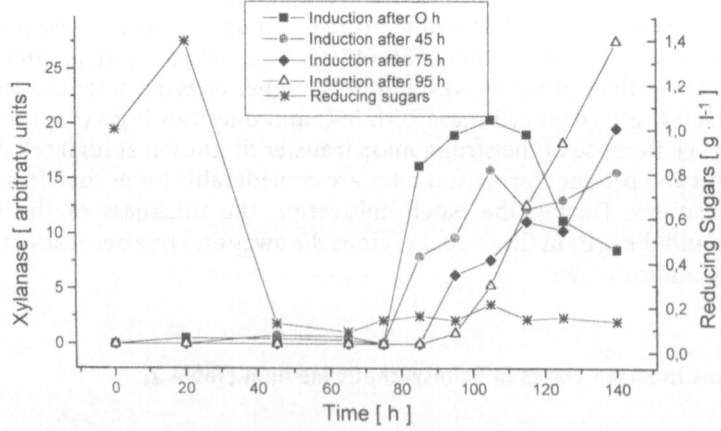

Fig. 5. Concentration of reducing sugars and xylanase activity as a function of time with various induction times [88]

sugars were consumed by the fungus. With increasing xylan concentration the enzyme activity increased in the concentration range 1 to 3 g l⁻¹.

In order to evaluate the optimal induction time, 3 g l⁻¹ xylan was added to the cultivation medium at different times between 0 and 115 h. In Fig. 5, the xylanase activity courses are shown during the cultivation, which were obtained with different induction times. On the same diagram, the concentration courses of the reducing sugars are shown as well. The xylanase production started 30 h after the sugar was consumed and the highest activities were obtained with early (t=0 h) and late (t=75 and 95 h) xylan additions. However, with early xylan addition the enzyme activity decreased after 110 h. The later the xylane was added to the medium, the shorter was the time lag between xylan addition and start of the xylanase production. This revealed that cells under substrate limitation reacted faster to the inductor than those without substrate limitation [88].

4.2.2
Influence of the Morphology on Enzyme Production

By the use of various initial spore concentrations in the shake flask cultures, the fungal morphology can be influenced. At low spore concentrations large pellets were formed and with increasing spore concentration, the pellet size diminished. However, the pellet size distribution was rather broad. Therefore, *Aspergillus awamori* was cultivated with 10^2, 10^3, and 10^5 spores ml⁻¹, respectively, in a stirred tank reactor with 1.5–l working volume at 25 °C, 400 rpm and 1 vvm. After the end of the growth phase and consumption of the reducing sugars, the pellets, clumps and filamentous mycelium were separated by a set of test sieves with different mesh sizes, rinsed with 0.9% NaCl solution under aseptic conditions and the different fractions were distributed into shake flasks, in which the fractions were further cultivated at 25 °C on orbital shaker at 150 rpm (Fig. 6). The xylanase production was induced at various times with xylan. Samples were taken at different times and the xylanase activities of the samples were determined. In Fig. 7 the specific enzyme activities are shown for the six fractions of the three runs. The fractions with the filamentous mycelium and the smallest clumps and pellets exhibited the highest specific xylanase activity [88].

4.2.3
Influence of Phosphate Concentration on Morphology and Xylanase Production

The glucose uptake is closely correlated with the phosphate consumption. The orthophosphate concentration decreased to zero simultaneously with the glucose concentration during the growth phase. During the enzyme production no glucose was present and the phosphate concentration increased again. To investigate the influence of the initial phosphate concentration on the morphology and xylanase productivity, three runs were performed with 0.3, 1.0, and 2.0 g l⁻¹ initial phosphate concentrations with 10^4 spores ml⁻¹ at 25 °C and 200 rpm. With 0.3 g l⁻¹ initial phosphate concentration, the phosphate concentration in the medium decreased to zero at 20 h. With the initial phosphate concentrations of 1.0 and 2.0 g l⁻¹, a reduction to 0.6–0.7 g l⁻¹ and 1.4–1.5 g l⁻¹, respectively, were observed. The cell mass concentration and the fungal morphology were not

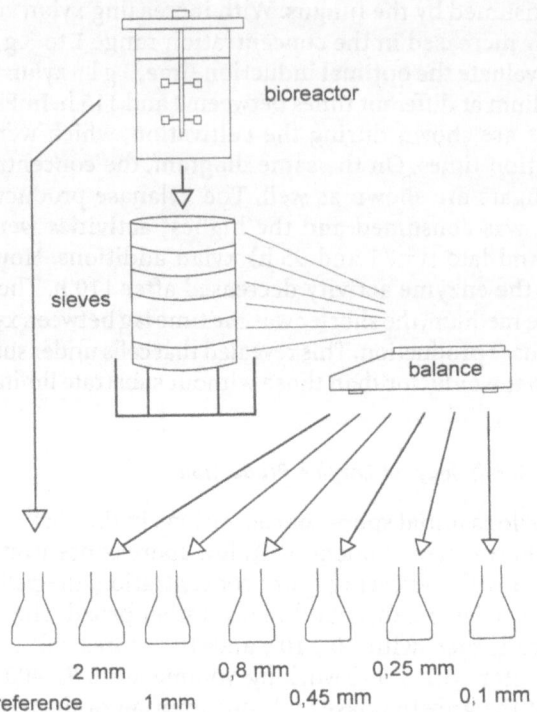

Fig. 6. Outline of the setup for the determination of enzyme activity as a function of fungal morphology [88]

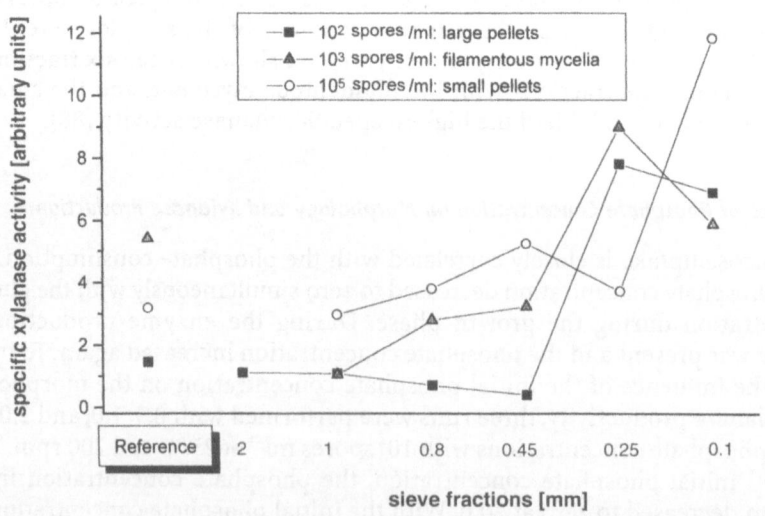

Fig. 7. Specific xylanase activities in shake flask cultures with different sieve fractions of the fungus [88]

influenced by the phosphate concentration for up to 60 h. In all of the applied phosphate concentrations only pellets were observed. After this time, the cell concentrations deviated: with the lowest initial phosphate concentration the lowest cell concentration was obtained.

After the reducing sugars were consumed, the xylanase production was induced with 2 g l⁻¹ xylan. The xylanase activity started to increase at 100 h and attained the highest level with the lowest phosphate concentration (Fig. 8).

4.2.4
Influence of Rotation Speed of the Shaker on Fungal Morpholgy

For a comparison of the results in shake flask and stirred tank cultures, it is necessary to evaluate the specific power input into the shake flasks. Figure 9

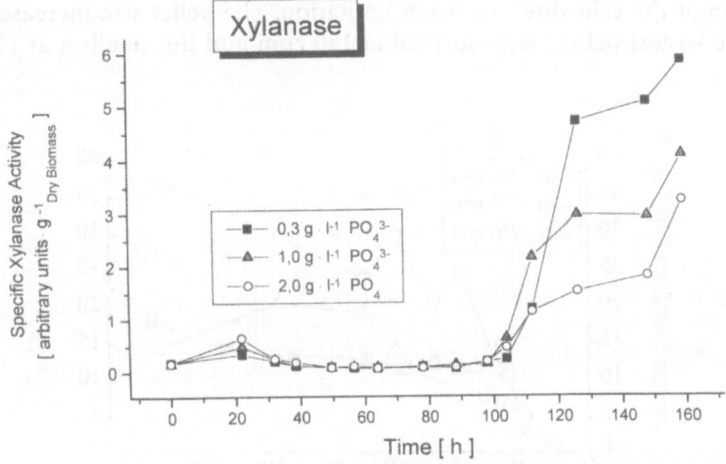

Fig. 8. Specific xylanase activities in shake flask cultures at different phosphate concentrations [88]

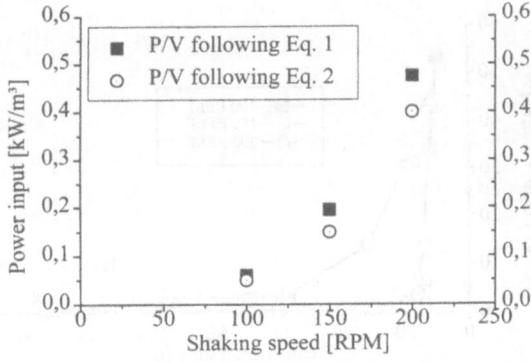

Fig. 9. Specific power input in shake flask cultures as a function of rotational speed of the shaker [30]

shows the specific power input in baffle-free shake flasks as a function of the
rotational speed of the shaker. By increasing the shaker speed from 100 to
200 rpm, the specific power input increases by a factor five [30]. At 150 rpm
$P/V = 0.2$ kW m^{-3} was evaluated. The effect of the rotation speed of the shaker
was investigated with cultivations using 10^6 spores ml^{-1} in flasks without baffles
at 100, 150, and 200 rpm. The cell volume fraction increased with the time at all
of the three rotation speeds. The highest cell volume was obtained at 150 rpm
and the lowest at 100 rpm (Fig. 10 A). At the beginning of the cultivation, a relative high volume fraction of the cells formed clumps (Fig. 10B), and only a low
fraction exhibited filamentous morphology (Fig. 10C). Both of them quickly
disappeared with increasing time and after 20 h, only pellets were present at
150 and 200 rpm (Fig. 10 D). With 100 rpm the transition from clumps and filamentous to pellet morphology was delayed at 40 h and after 100 h a fraction of
the pellets decomposed into filaments and clumps, probably because of the
starvation of the cells due to oxygen limitation. The pellet size increased with
time. The largest pellets were formed at 150 rpm and the smallest at 100 rpm
(Fig. 10 E).

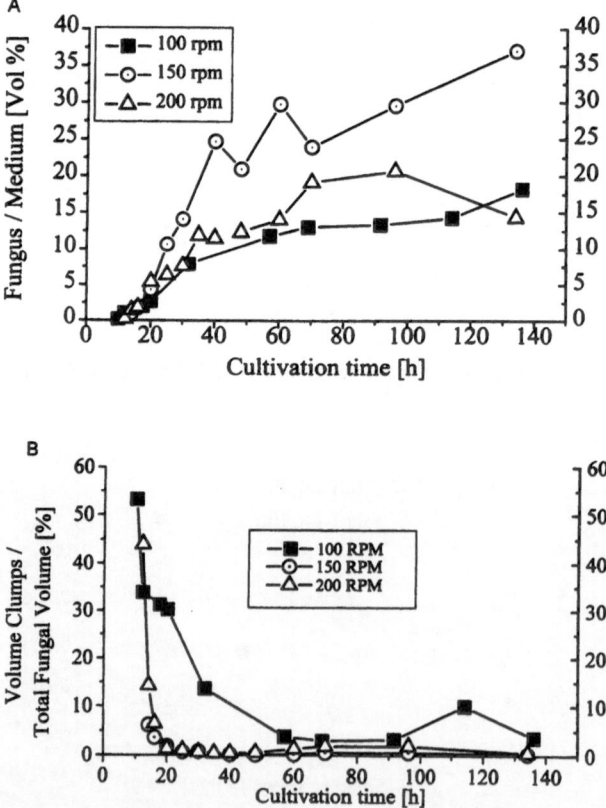

Fig. 10 A, B. Influence of rotation speed of orbital shaker on fungal morphology [30]: A relative
volumes of the fungal biomass; B volume fractions of the fungal clumps

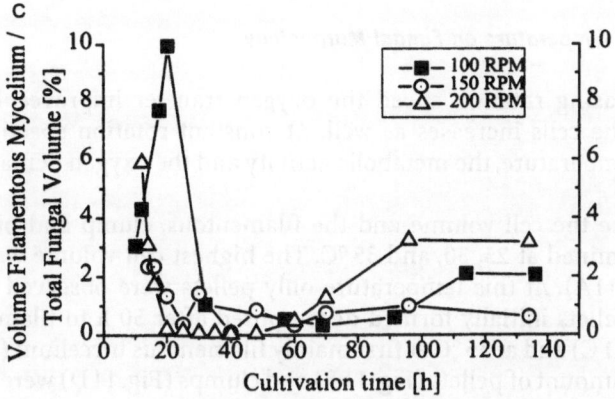

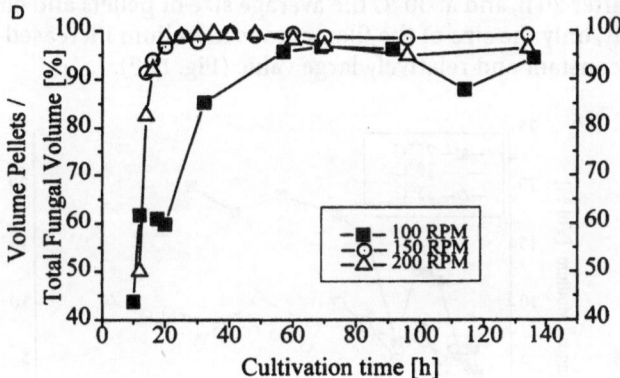

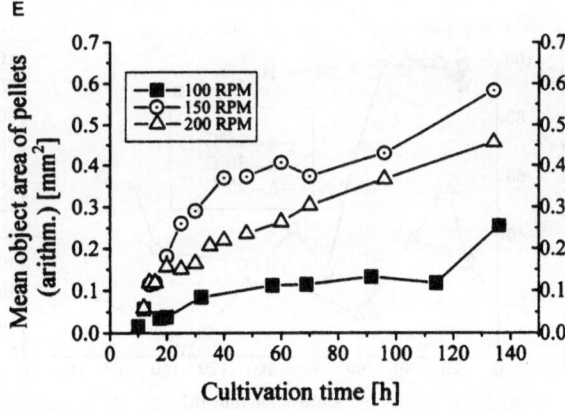

Fig. 10 C–E. C volume fractions of the filamentous mycelium; D volume fractions of the fungal pellets; E surface area of the fungal pellets

4.2.5
Influence of Temperature on Fungal Morphology

With increasing rotation speed the oxygen transfer improves, but the shear stress on the cells increases as well. At constant rotation speed, but with increasing temperature, the metabolic activity and the oxygen demand of the cells increase.

Therefore the cell volume and the filamentous, clump and pellet fractions were determined at 25, 30, and 35 °C. The highest cell volume was obtained at 25 C (Fig. 11A). At this temperature only pellets were observed (Fig. 11B). At 30 °C the pellets initially formed decomposed after 50 h to filamentous mycelium (Fig. 11 C) and at 35 °C at first mainly filamentous mycelium (Fig. 11 C), and only a low amount of pellets (Fig. 11 B) and clumps (Fig. 11 D) were formed. After 20 h, the filamentous mycelium and the pellets turned into clumps (Fig. 11 D). However, the average size of the clumps decreased with the time (Fig. 11 E). At 25 °C the average size of the pellets, clumps and filamentous mycelium increased with time after 20 h, and at 30 °C the average size of pellets and clumps changed only slightly, only the size of the filamentous mycelium increased and after 60 h attained a constant and relatively large value (Fig. 11 F).

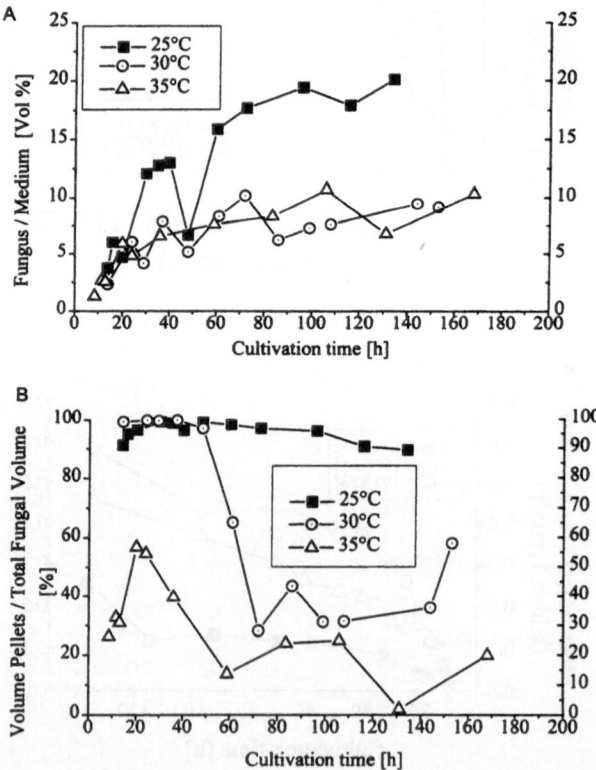

Fig. 11 A, B. Influence of temperature on fungal morphology in shake flask cultures [30]: A relative volumes of fungal biomass; B volume fraction of fungal pellets

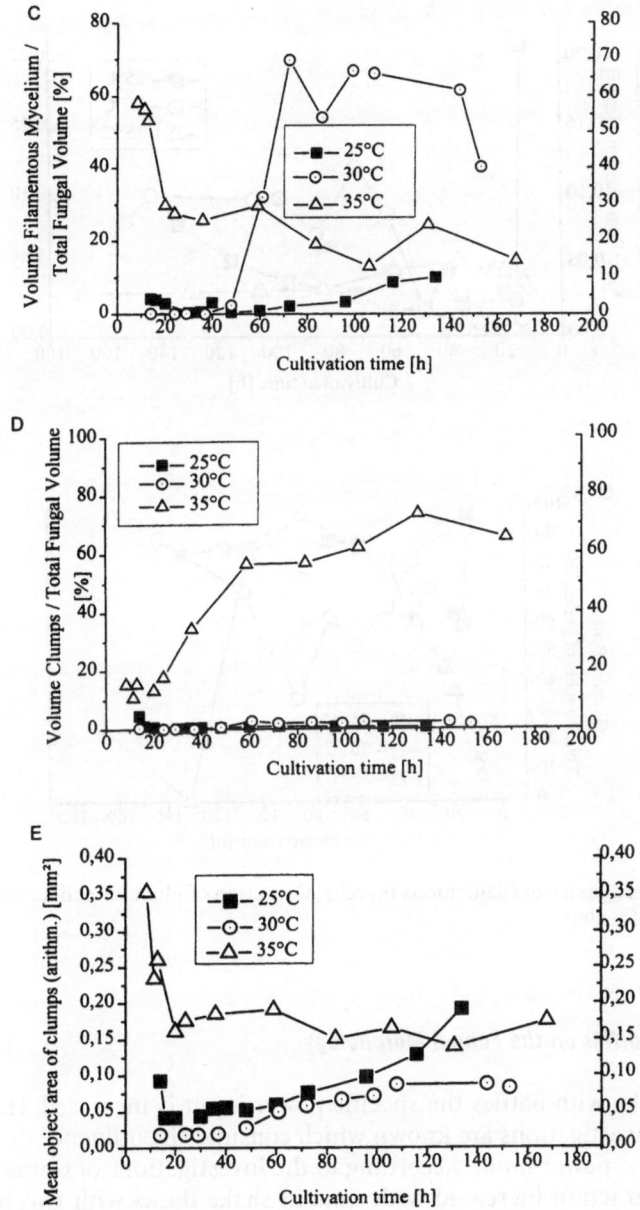

Fig. 11 C–E. C volume fractions of filamentous mycelium; D volume fractions of fungal clumps; E average size (surface areas) of fungal clumps

At 25 °C, the fraction of globular pellets was high, at 30 °C this fraction was high as well, but passed a minimum at 90 h, and at 35 °C it increased with time and attained a high value at 80 h (Fig. 11 G). These results reveal, that globular pellets were more frequent than oval ones [30].

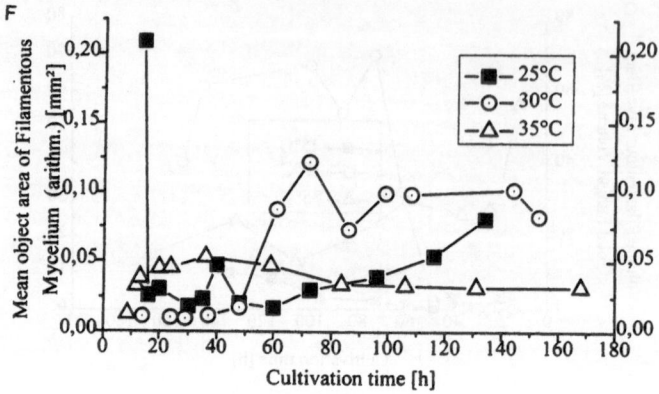

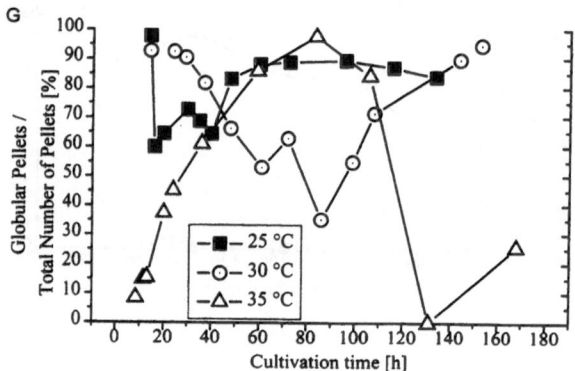

Fig. 11 F, G. F average size of filamentous mycelia; G fraction of globular pellets with respect to total number of pellets

4.2.6
Influence of Baffles on the Fungal Morphology

In shake flasks with baffles the specific power input is increased. However, no systematic investigations are known which consider the influence of the baffles on the specific power input. According to the investigations of Gerlach [30] the cell volume fraction increased with time in shake flasks with two baffles and without baffles and decreased in shake flasks with four baffles at 150 rpm (Fig. 12 A). Without baffles 90% of the cell volume were pellets, with baffles no pellets were present. No filamentous mycelium was observed in shake flasks without baffles and with two baffles, but with four baffles 60% of the cell volume had filamentous morphology. No clumps were found in shake flasks without baffles, 10% of the cell volume consisted of clumps with four baffles and in shake flaks with two baffles, the volume fraction of the clumps varied between 5 and 22% (Fig. 12 B).

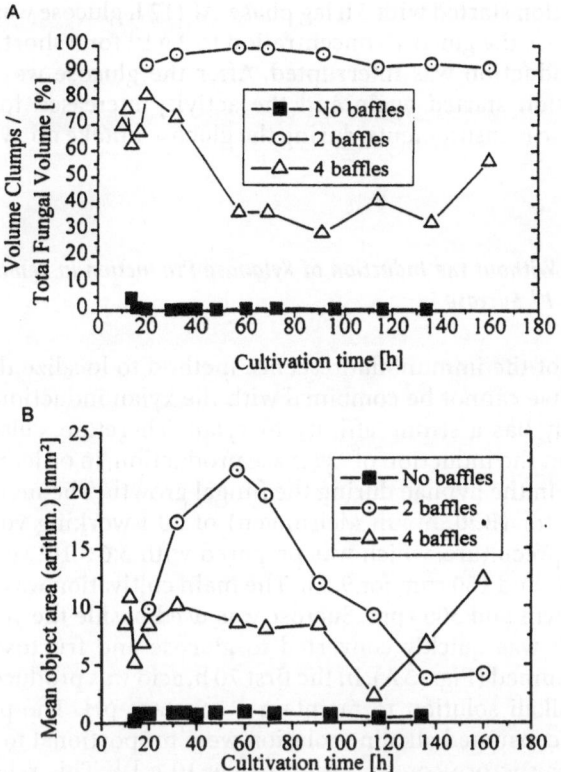

Fig. 12 A, B. Influence of baffles in shake flasks on fungal morphology [30]: A relative volumes of fungal mass; B fractions of fungal clumps with respect to total number of objects

4.3
Cultivations in Stirred Tank Reactors in Semisynthetic Medium (Table 2)

4.3.1
Repression of Xylanase Formation

The investigation on the induction of xylanase production indicated that in the presence of glucose in the cultivation medium no induction is possible. To determine a possible critical glucose concentration, at which repression occurs, a constant glucose concentration level has to be maintained. Therefore these investigations were carried out in a stirred tank reactor (Biostat M) with 1.5 l working volume.

The cultivation was performed with a preculture, which was prepared with 10^5 spores ml^{-1} at 30 °C and 150 rpm for 10 h, and the main cultivation in the stirred tank reactor was performed at 25 °C , pH 4.5, 1 vvm and 300 rpm in presence of xylan. The cultivation started with 5 g l^{-1} glucose and after the growth phase the glucose concentration was reduced to 0.14 g l^{-1}. This concentration level was maintained by fed-batch operation. During this time no enzyme was produced. The glucose feeding was stopped after 42 h and the xylanase production started with 5 h lag phase. At 112 h glucose was added to the medium to increase the glucose concentration to 2 g l^{-1} for a short time. By that, the xylanase production was interrupted. After the glucose was used up, the enzyme production started again and the activity increased to high values. According to these measurements, during the glucose uptake no xylanase induction occurs [59].

4.3.2
Growth of Fungus Without the Induction of Xylanase Production in Semisynthetic Medium with 20 g l^{-1} Sucrose

The application of the immunofluorescence method to localize the replicating DNA in the hyphae cannot be combined with the xylan induction, because the primary antibody has a strong affinity to xylan. Therefore, cultivations were perfomed without the induction of xylanase production, in order to localize the replicating DNA in the hyphae during the fungal growth. The main culture in a stirred tank reactor (B 20, Braun Melsungen) of 20 l working volume was inoculated with a preculture, which was prepared with 3.6×10^4 spores ml^{-1} and incubated at 30 °C and 150 rpm for 9.5 h. The main cultivation was performed at 25 °C, pH 4.5, 1 vvm and 300 rpm. Sucrose was used beside the yeast extract as carbon source. It was quickly converted to glucose and fructose which were successively consumed (Fig. 13 A). In the first 70 h, acid was produced, which was neutralized by alkali solution to maintain a constant pH. The produced CO_2 amount and the consumed alkaline solution were proportional to the cell mass concentration for sucrose concentration below 10 g l^{-1}. This relationship was independent of the morphology. The extracellular orthophosphate concentration decreased to zero during the glucose uptake and stayed there until the sugars were consumed. The intracellular phosphate concentration was high during this time. According to the electron energy loss spectroscopy measurements, the phosphorus (and the sulfur) were enriched in the vacuoles of the hyphae (Fig. 13 B). After 70 h the extracellular phosphate concentration increased due to cell lysis (Fig. 13 C). In spite of the high metabolic activity of the fungus (a large peak of the CO_2 concentration in the off-gas), the xylanase activity was low without induction of the enzyme production (Fig. 13 D).

The immunofluorescence measurements reveal, that the replicating DNA is mainly in the subapical and branching zones. In general, young hyphae contain much more replicating DNA zones than old ones.

4.3.3
Influence of Stirrer Speed on Morphology and Xylanase Production

According to the literature, the stirrer speed has a considerable influence on fungal morphology (e.g., [89–92]). However, the published results are contradictory, which indicates that the effect of the stirrer speed on the fungal morphology is influenced by various parameters such as preculture, medium composition and viscosity, stirrer type, etc.

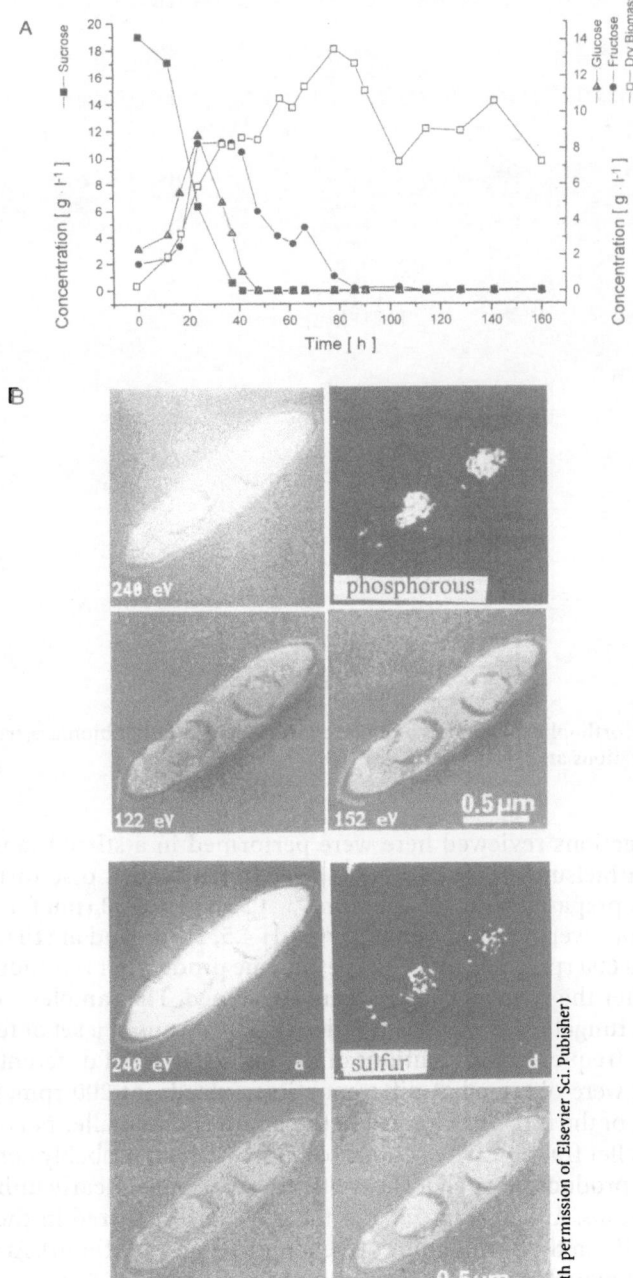

Fig. 13 A, B. Cultivation in stirred tank on synthetic medium without induction of xylanase "Standard" cultivation: **A** biomass and sugar concentrations [59]; **B** detection of phosphorus and sulfur in the vacuoles of the hyphae – determined by electron energy loss spectroscopy (EELS) technique [53]

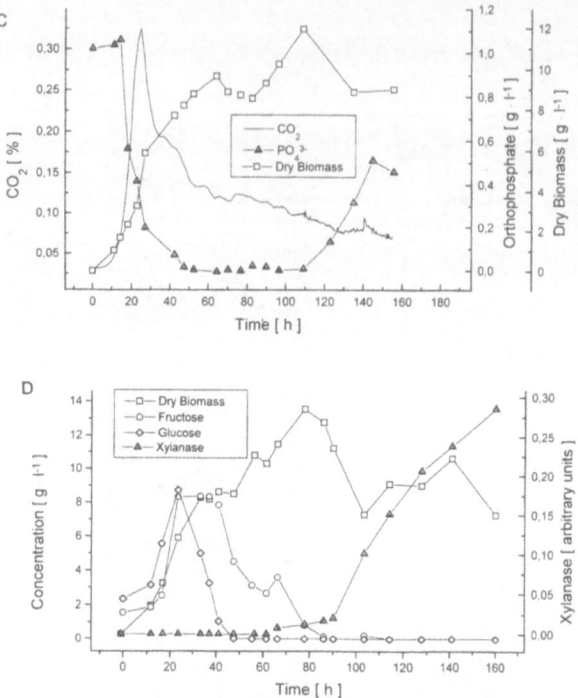

Fig. 13 C, D. C CO_2, orthophosphate, and biomass concentrations [59]; D biomass, fructose, and glucose concentrations and xylanase activity [59]

The investigations reviewed here were performed in a stirred tank reactor (B20, B. Braun Melsungen) with a two-stage six-flat-bladed disc turbine. The preculture was prepared with 10^5 spores ml^{-1} at 30 °C and 150 rpm for 10 h. The main cultivations were carried out at 25 °C, pH 4.5, 1 vvm and at stirrer speeds of 200, 400, and 600 rpm, respectively. The enzyme production was induced with 2 g l^{-1} xylan after the reducing sugars were consumed. The samples were taken at 90 h and the fungal mass was size fractionated by means of a set of test sieves. In Fig. 14 A the frequency distributions of the wet cell mass of different sizes are plotted, which were obtained at different stirrer speeds. At 200 rpm, the main portion (80%) of the fungus belonged to the 1-mm and a smaller portion to the 1.2-mm size pellet fraction. The residual cell mass, which probably consisted of cell fragments produced by pellet shear off, was distributed nearly uniformly in the other fractions. At 400 rpm, 70% of the cell mass appeared in the 0.2-mm fraction and 20% in the filamentous mycelium fraction (< 0.1 mm). At 600 rpm, 41% of the cell mass belonged to the 0.2-mm fraction and 18% to the filamentous mycelium fraction (< 0.1 mm), 30% of the cell mass formed clumps of 0.45 – 0.8-mm sizes.

During batch cultivation the apparent viscosity changed with cell mass concentration, morphology and stirrer speed. Therefore, only a few typical cases are considered.

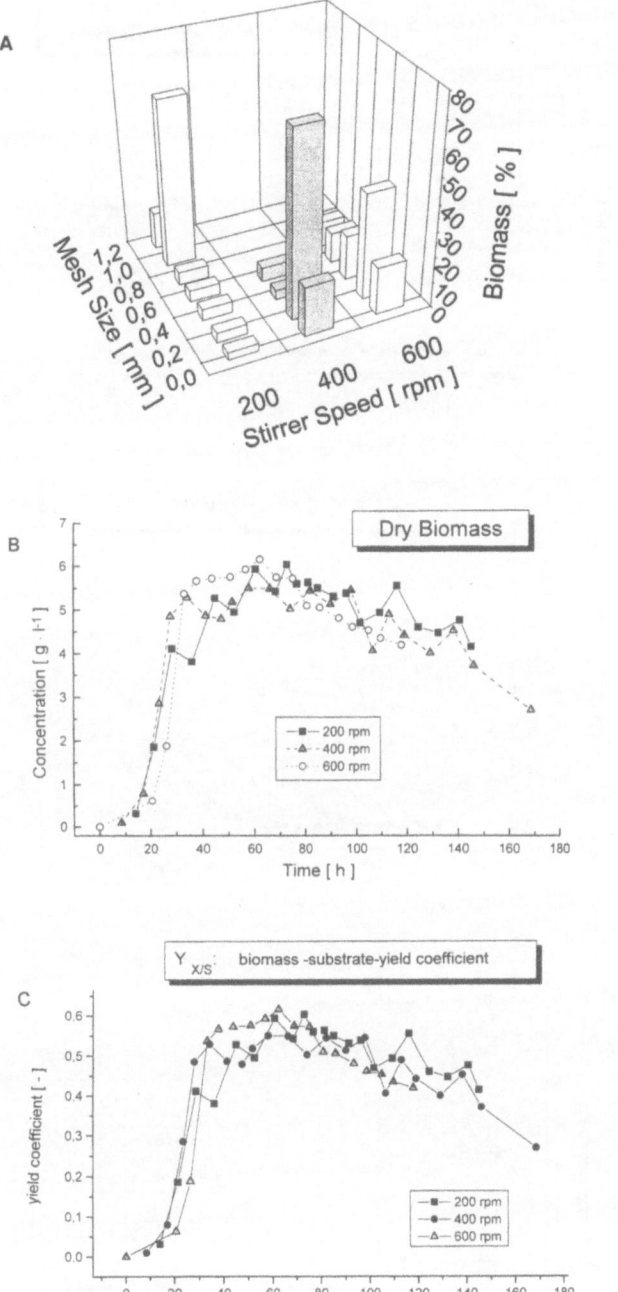

Fig. 14 A–C. Influence of stirrer speed on morphology and xylanase production in stirred tank reactors in synthetic medium [59]: **A** size distribution of fungal biomass at different stirrer speeds; **B** dry biomass concentrations; **C** yield coefficients of growth with respect to substrate $Y_{X/S}$

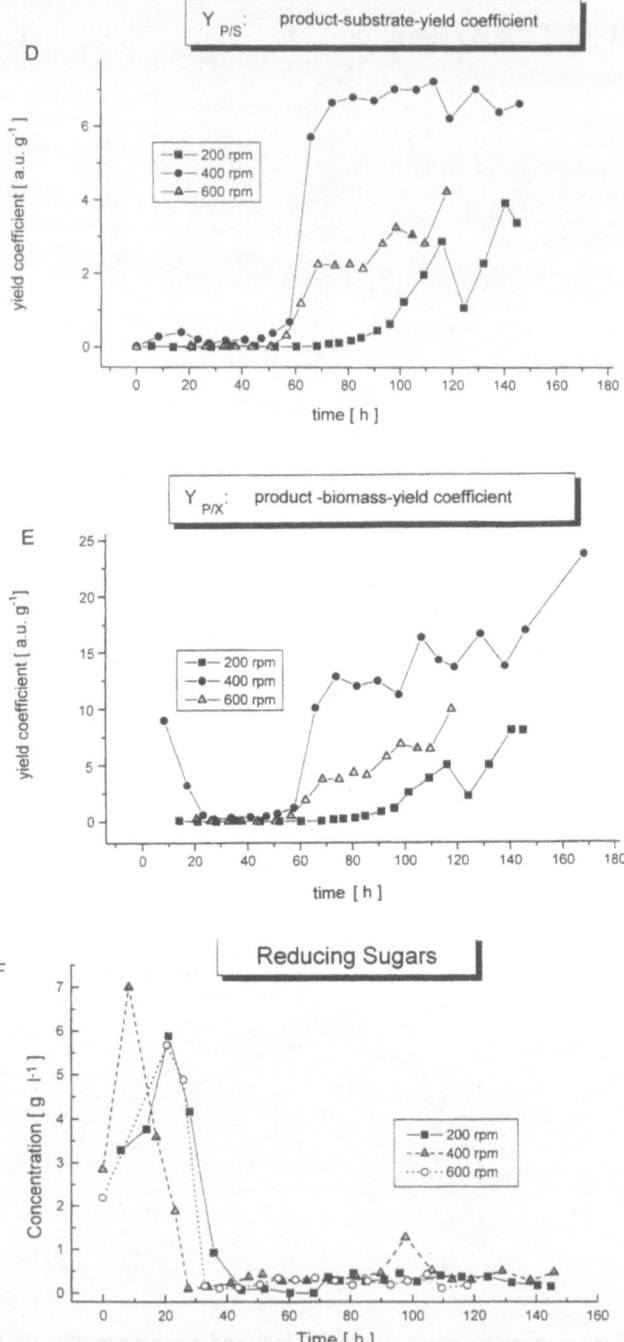

Fig. 14 D–F. D yield coefficient of product with respect to substrate $Y_{P/S}$; E yield coefficient of product with respect to biomass $Y_{P/X}$; F concentration of reducing sugars

By the relationship of Wittler et al. [73] and by the indentified model parameters of Mitard and Riba [74] for a typical run at 90 h, the apparent viscosity was evaluated at the applied shear rates to 166 mPa s (200 rpm), 110 mPa s (400 rpm) and 70 mPa s (600 rpm). The corresponding specific power inputs are: $P/V = 0.50$ kW m^{-3} at 200 rpm, $P/V = 1.42$ kW m^{-3} at 400 rpm, and $P/V = 3.0$ kW m^{-3} at 600 rpm. By using the relationship of Judat [68] for the evaluation of the specific power input with aerated medium, and the relationship of Biedermann [70] for the shear factor, the following specific power inputs and shear factors were evaluated at an aeration rate of 1 vvm: $P_G/V_S = 0.22$ kW m^{-3} and $d_{VF} = 130$ µm at 200 rpm, $P_G/V_S = 0.85$ kW m^{-3}, $d_{VF} = 100$ µm at 400 rpm and $P_G/V_S = 2.40$ kW m^{-3}, $d_{VF} = 80$ µm at 600 rpm. Obviously at specific power inputs above about 0.5 kW m^{-3} the size of the pellets is considerably reduced.

The corresponding volumetric mass transfer coefficients were calculated by the relationship recommended by Herbst et al. [94] and Schügerl [95] for high viscous cultivation media: $k_L a = 0.007$ s^{-1} (400 rpm) and $k_L a = 0.01$ s^{-1} (600 rpm). For the lowest stirrer speed (200 rpm) no relationships hold true.

The courses of the (dry) fungal cell mass concentration X (Fig. 14 B) and the yield coefficient of the growth with respect to the substrate $Y_{X/S}$ (Fig. 14 C) were influenced by the stirrer speed only slightly. Nevertheless, the yield coefficients of the product with respect to the substrate $Y_{P/S}$ (Fig. 14 D) and to the cell mass $Y_{P/X}$ (Fig. 14 E) depended considerably on the stirrer speed. This large difference in the yield coefficients of the product is caused by the high xylanase activity obtained at 400 rpm and the low activity at 200 rpm and not by the difference in the substrate consumption (Fig. 14 F). The large amount of 1-mm sized pellets formed at 200 rpm have a low enzyme productivity. The high share of filamentous mycelium at 400 rpm is responsible for the high enzyme activity. At 600 rpm a fairly large portion of the fungal cell mass appeared in form of small pellets and filamentous mycelia, but at this high shear stress the enzyme production was obviously suppressed [59]. According to results the product yield coefficients were impaired by the lack of oxygen at low stirrer speeds and by high shear stress at high stirrer speeds.

In contrast, the cell growth and the yield coefficient of the growth were obviously rather insensitive to the oxygen transfer limitation at 200 rpm and to the high stress at 600 rpm.

4.3.4
Influence of Phosphate Concentration on Fungal Morphology in Semisynthetic Medium

The phosphate concentration in shake flask cultures influenced the enzyme production, but did not influence the fungal morphology. To investigate, whether this effect depends on the reactor used, investigations were performed in the stirred tank reactor on semisynthetic medium under standard conditions with 20 g l^{-1} sucrose concentration (see Sect 4.3.2) and 0.3 g l^{-1}, 1.05 g l^{-1}, and 2.1 g l^{-1} phosphate concentrations. The volume of the fungal mass was much higher with 1.05 g l^{-1} and 2.1 g l^{-1} than with 0.3 g l^{-1} phosphate (Fig. 15). With 0.3 g l^{-1} and 2.1 g l^{-1} phosphate volume fraction of mycelial clumps and with 1.05 g l^{-1} phosphate that of pellets dominated. In contrast to the shake flask

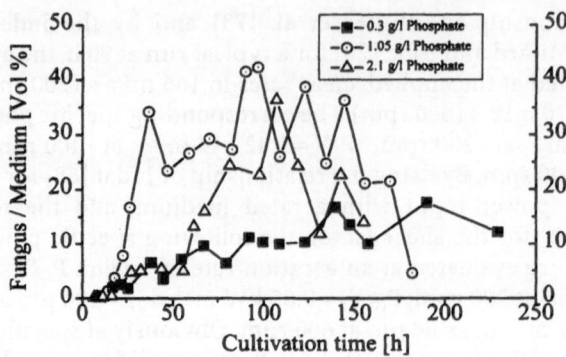

Fig. 15. Biomass volume fraction at different phosphate concentrations in stirred tank reactor on synthetic medium [30]

cultures, the initial phosphate concentration in stirred tank reactors seems to have an influence on the fungal morphology.

4.4
Cultivation in Stirred Tank Reactors in Complex Medium

4.4.1
Decomposition of Wheat Bran

Wheat bran is a byproduct of flour production and consists of the fruit shell (pericarp), a seed coat (testa) and the outer layers of the endosperm. The pericarp itself consists of the outer and inner pericarp. The former is built by epidermis and hypodermis, the latter consists of cross cells and tube cells. The testa consists of three layers, the outer cuticle, the color layers and the inner cuticle. The next layer – the so called nucellar tissue – belongs to the endosperm and consist of compressed cells. The innermost layer of the wheat bran consists of aleurone cells which are rich in fats and proteins (Fig. 16a) [51, 52].

Wheat bran consists of 28% hemicellulose, 9% starch, 8.7% cellulose, 3.2% lignin and 3% pectin [96, 97].

The wheat bran used had an average size of 5–6 mm. The decomposition of this wheat bran was investigated with precultures prepared in shake flasks at 30 °C and 150 rpm with 9.5 h incubation time and in stirred tank reactor with 7×10^4 spores ml^{-1} at 25 °C, pH 4.5, 1 vvm and 300 rpm. In the complex cultivation medium the wheat bran is the sole carbon and energy source. It is the source of growth factors and the inducer xylan as well.

At the beginning of the cultivation, most of the hyphae grew on the inner side of the wheat bran, forming dense pellets, where they were protected from the shear stress caused by the impeller (Fig. 16b). At first the aleurone layer and the nucellar tissue were degraded because the seed coat formed a barrier to fungal growth. In particular, no penetrations of the outer cuticle of the seed coat were observed. The barrier formed by the outer cuticle of the seed coat can only be

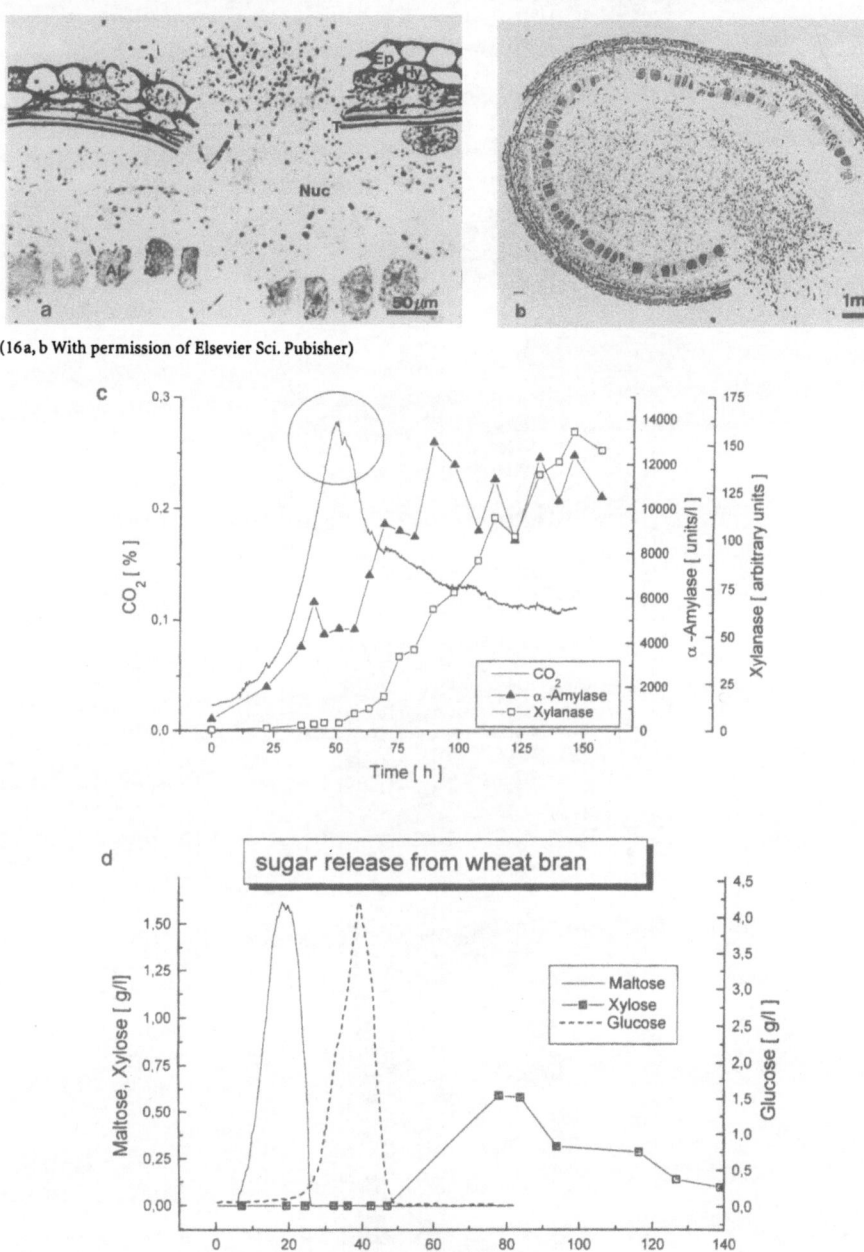

(16a, b With permission of Elsevier Sci. Pubisher)

Fig. 16 a–d. Microscopic investigations of the wheat bran decomposition and the interaction between wheat bran and *Aspergillus awamori* [53]: **a** ultrastructure of wheat bran – light microscopy [53]; **b** cross section of the wheat bran which surronds a pellet – light microscopy [53]; **c** α-amylase and xylanase activities and CO₂ concentration in the off-gas as a function of cultivation time [98]; **d** maltose, glucose, and xylose concentrations as a function of cultivation time [98]

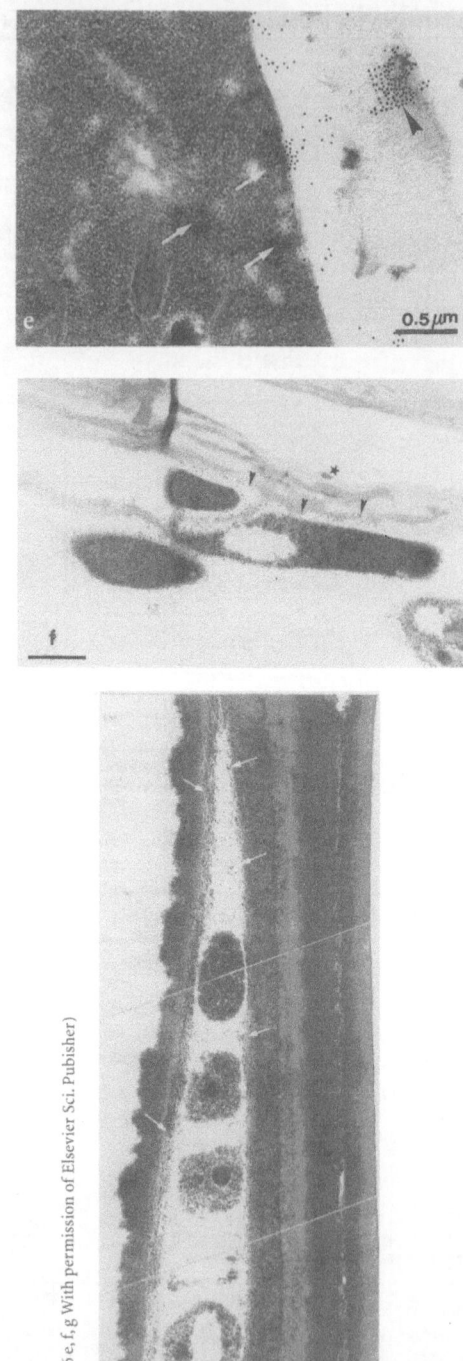

(16 e, f, g With permission of Elsevier Sci. Pubisher)

Fig. 16 e – g. e immunogold labeled xylanase in the cytoplasma, at the plasma membrane and on the surface of the wheat bran fragments [53]; **f** immunogold labeled xylanase on one side of the wheat cell wall with attached hyphae [53]; **g** four hyphae penetrating the seed coat – immunogold labeling of the xylanase (*arrows*) and xylanase halos around the hyphae detected by immunogold labeling [53]

evaded by the fungus by growing through mechanical damage or via edge of the bran. In the pericarp, the inner layers show less resistance to fungal colonization than the outer layers. In particular, the layer between the inner and outer pericarp can easily be colonized. The decomposition started with the dissolution of the pectin-containing structures and growth into this space. The complete filling of this space with dense mycelium occurred later.

In Fig. 16c the courses of α-amylase and xylanase activities as well as the CO_2 concentration in the off-gas are plotted as a function of the time. α-Amylase formation started immediately after the inoculation, xylanase appeared in the medium 50 h later. No pectinase and cellulase activities were detected.

The starch in the wheat bran was gradually decomposed to maltose and glucose. Their concentration increased, passed a sharp maximum and decreased rapidly because of consumption (Fig. 16d). The CO_2 concentration in the off-gas passed a sharp maximum during the consumption of the sugars as well.

The determination of the fungal cell mass concentration was not possible, because of the presence of the solid wheat bran and the adherence of the fungus to the wheat bran. However, because of the close relationship between the growth rate and the CO_2 production rate (CPR), the cell mass was estimated by the integration of the CPR over time [98].

Before the xylanase appeared in the medium, it was detected within the cytoplasma and at the plasma membrane by means of immunogold labeling [53]. The xylanase concentration increased in the cytoplasma with time. The enzyme migrated in direction of the plasma membrane, was enriched there and secreted into the medium. There, it was adsorbed onto the surface of the wheat bran fragments (Fig. 16e). At the same time, xylanase activity increased in the medium (Fig. 16c). At the interface of the fungus with the wheat bran, high concentrations of xylanase were detected. Figure 16f shows a layer of cell wall material of the pericarp, which was colonized by fungal hyphae at only one side. On this side only was the enzyme detection positive.

Figure 16g shows several cross sectioned fungal cells, growing inside the seed coat. The hyphae separate the seed coat by growing inside the outer layer of the color layer. The halo around the fungal cells indicate a high local concentration of xylanase (Fig. 16g). In general, enzyme activity was found only on the wheat cell walls in the immediate vicinity of the fungal hyphae. No enzyme activity was detected elsewhere.

4.4.2
Influence of the Preculture

The cultivation on complex medium was inoculated with precultures prepared with semisynthetic medium (Table 2) and alternatively with complex medium. In the preculture with complex medium, wheat bran was the sole carbon source. The spores adhered to the surface of the wheat bran and developed within 24 h to hyphae. The largest part of the fungal cell mass formed pellets and was attached to the wheat bran. The duration of the germination and the penetration of the hyphae into the wheat bran as well as the fungal morphology were independent of the spore concentration.

By using the preculture prepared with the semisynthetic medium, the lag phase in the main culture was longer, but the growth rate, the CPR peak and the substrate consumption rate were higher (Fig. 17A, B). The α-amylase activity was lower (Fig. 17C), but the xylanase activity was higher (Fig. 17D) than with the preculture with the complex medium.

4.4.3
Influence of Size of Wheat Bran Particles

The transmission electron microscopic investigations of Adolph et al. [53] indicated a direct contact between the fungus and the wheat bran particles. This was confirmed by the immunogold measurements revealing high xylanase activity at their interface. By reducing the wheat bran particle size from 5–6 mm to 1–2 mm by milling, an increase of the substrate uptake rate was expected. The

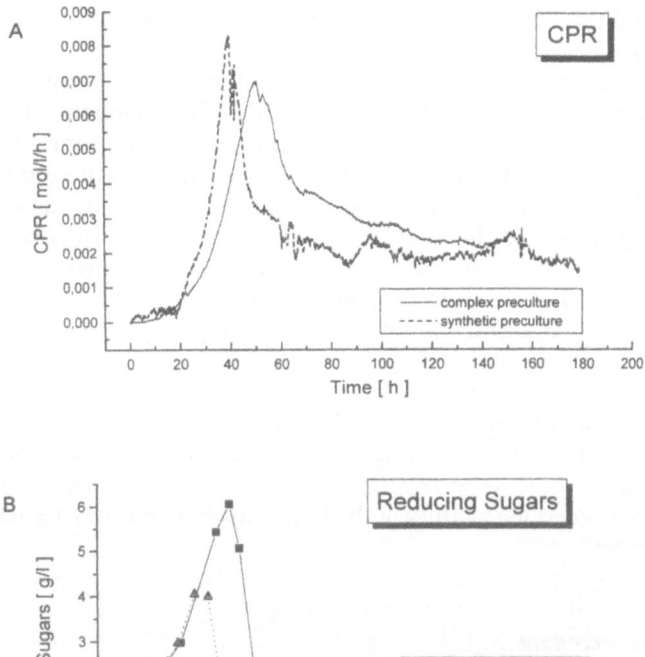

Fig. 17 A, B. Influence of precursor on cultivation in stirred tank in complex medium: A CO$_2$ production rates (CPR) [98]; B reducing sugar concentrations

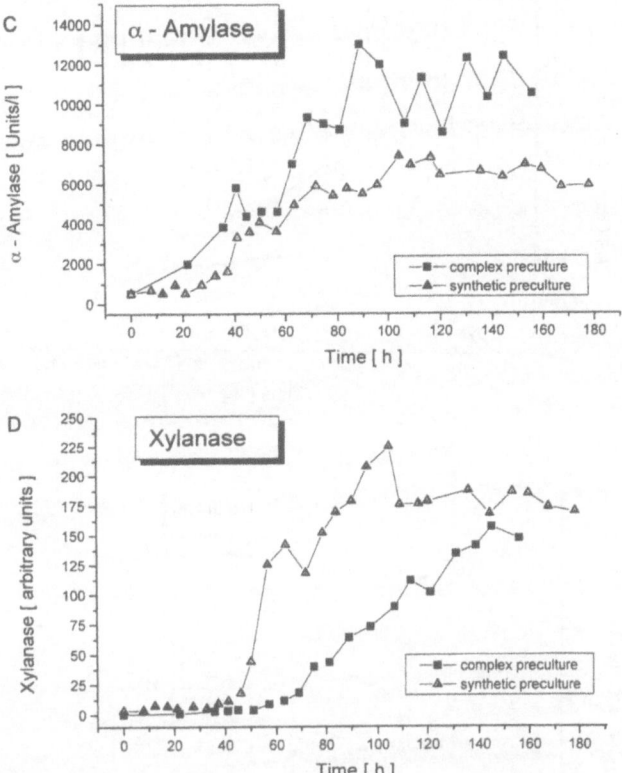

Fig. 17 C, D. C α-amylase activities; D xylanase activities

cultivations with the original wheat bran and with the milled wheat bran were carried out under standard conditions (see Sect. 4.4.1).

In the presence of the original wheat bran, fungal pellets were formed, which were attached to the wheat bran. With the milled wheat bran, no pellets were formed. The filamentous mycelium was not attached to the wheat bran, because no protection from the shear stress was possible. The larger specific surface area of the wheat bran caused an acceleration of the growth (Fig. 18 A), higher sugar and phosphate uptake rates (Fig. 18 B, C), higher ammonium uptake (Fig. 18 D), although the xylanase activity was reduced (Fig. 18 E) [98]. The low xylanase activity can be explained with an inefficient induction of xylanase production [Fig. 18 E] by the less intimate contact between the fungus and the xylan-containing wheat bran.

4.4.4
Influence of the Stirrer Speed

In a semisynthetic medium, stirrer speed had a considerable influence on fungal morphology and enzyme production. The stirrer speed improved the transport

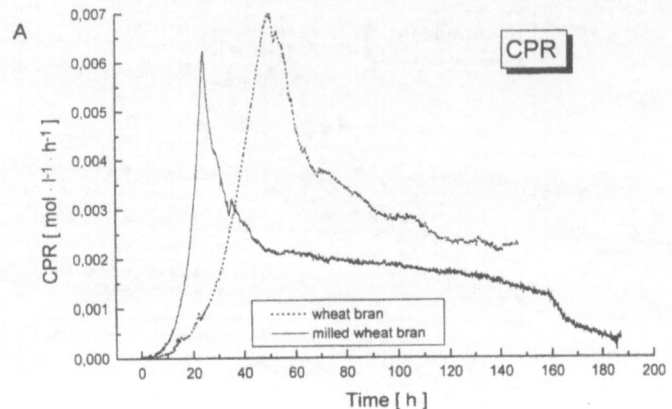

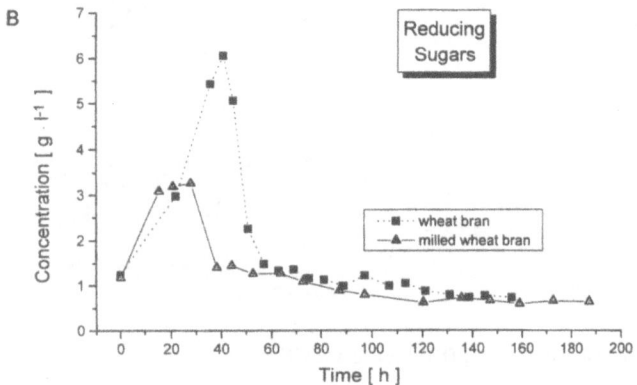

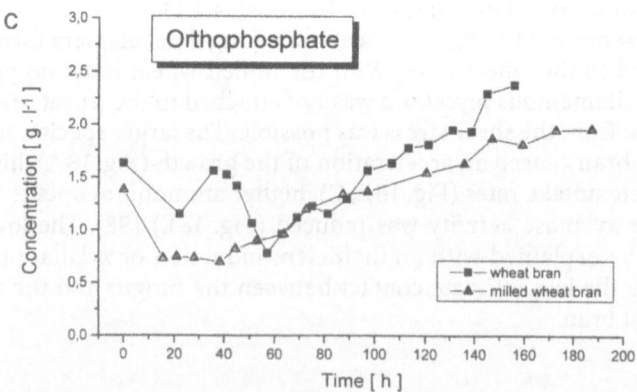

Fig. 18 A–C. Influence of wheat bran size on cultivation in stirred tank in complex medium [98]:A CO_2 production rates (CPR); B reducing sugar concentrations; C orthophosphate concentration

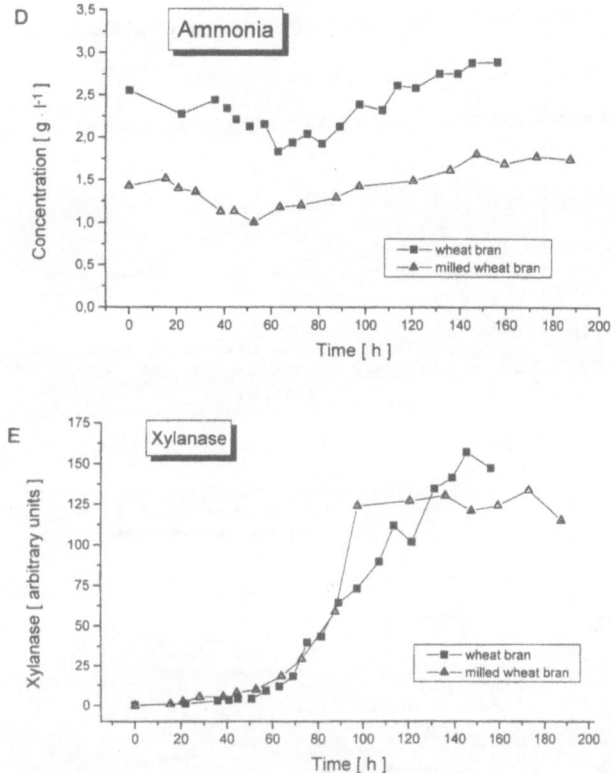

Fig. 18 D, E. D ammonia concentrations; E xylanase activities

processes and reduced the pellet size on one hand and increased the shear stress on the other. Therefore, at intermediate stirrer speed (400 rpm) the highest enzyme productivities were obtained.

In complex medium, the predominant part of the fungus was attached to the wheat bran particles, which protects it from the shear stress. To investigate the effect of the stirrer speed, the cultivations were performed under standard conditions (see Sect. 4.4.1), except for the stirrer speed, which was maintained at 300, 500, and 750 rpm, respectively. By changing the stirrer speed, the morphology did not change. Pellets were always formed which were surrounded by the wheat bran. However, at the highest stirrer speed (750 rpm) the fungus was partly separated from the wheat bran. With increasing cultivation time, the wheat bran was gradually enzymatically decomposed and mechanically disintegrated and the fungal mycelia were partly lysed, which reduced their intimate contact and protection of the hyphae from the shear effect. As a result, the solid content in the reactor diminished and the stirrer effect on the fungus increased.

The phosphate and ammonium uptake rates were not influenced by the stirrer speed. In contrast to the morphology, the fungal metabolism was strongly

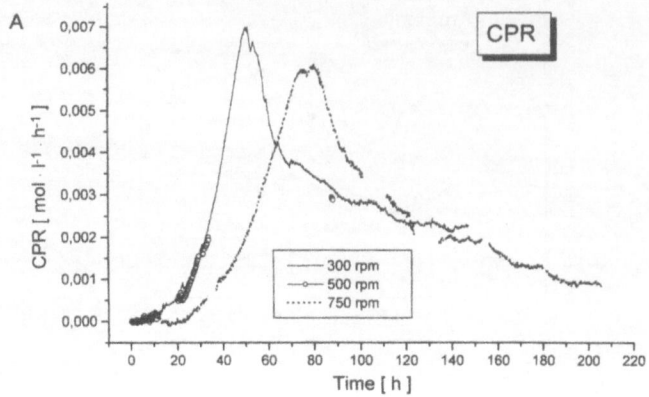

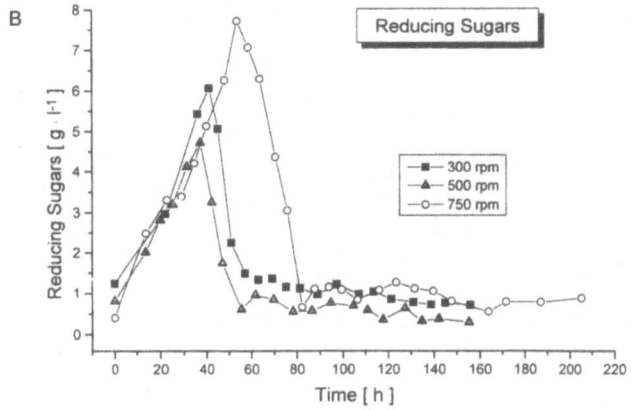

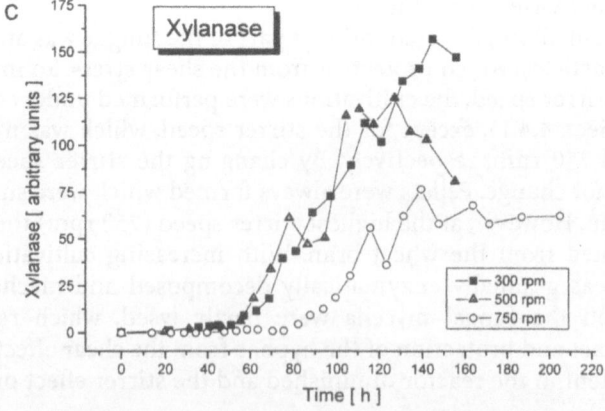

Fig. 19 A–C. Influence of stirrer speed on cultivation in stirred tank in complex medium [98]: A CO_2 production rates (CPR); B reducing sugar concentrations; C xylanase activities

influenced by the stirrer speed. At lower stirrer speeds (300 and 500 rpm) the time lag was shorter (Fig. 19 A), the growth rate (not shown), the sugar uptake rate (Fig. 19 B) and the xylanase acitivity (Fig. 19 C) were higher than at 750 rpm. The low xylanase activity at high stirrer speed can be explained by the partial separation of the fungal mycelium and the wheat bran [98].

4.5
Cultivation in an Airlift Tower Loop Reactor in Semisynthetic (Table 2) and Complex Media

In the airlift tower loop reactor, the specific power input under the applied cultivation conditions in the semisynthetic medium was lower than those in the stirred tank at low stirrer speed (200 rpm). Furthermore, in the bubble column and airlift tower loop reactors the energy dissipation rate is uniform, in contrast to the stirred tank reactors, in which the energy dissipation rate changes from the agitator edge to the vessel wall by a factor of about 100 [67]. The energy dissipation rate controls the maximum possible size of bubbles, droplets, flocs, pellets, etc. Particles, which are larger than their dynamic equilibrium size, are destroyed by the turbulence energy carried by the eddies. Therefore, in airlift reactors dynamic equilibrium pellet/clump size is constant in the reactor, in contrast to stirred tank reactors, in which the dynamic equilibrium pellet/clump size varies: at the edge of the impeller it is about one order of magnitude smaller than at the vessel wall. Therefore, large pellets, which are stable at the wall, are destroyed at the impeller edge. Low density, loose pellets/clumps are destroyed under these conditions. Only dense pellets endure this variable stress. This is the reason why in stirred tank reactors mainly dense pellets with smooth surface are observed and in airlift reactors pellets with low density and hairy surface are present as well. This difference in the pellet morphology of *Aspergillus niger* cultivated in stirred tank and bubble column reactors was confirmed by Berovic et al. [99] as well.

Pellets with hairy surface have higher viscosity than smooth pellets at the same fungal biomass concentration, which was not taken into account in the relationship of Wittler et al. [73] and Mitard and Riba [74], because they used dense and smooth pellets for their investigations. The difference between the rheology of the cultivation media with dense and smooth pellets of *Aspergillus niger* a stirred tank reactor on the one hand and low density and hairy pellets in a bubble column reactor on the other was also investigated by Berovic et al. [99]. They determined the rheology at different biomass concentrations and evaluated the consistency index K as a function of the cell mass concentration X:

$$K = A\,X^b$$

In Table 3 the constants of A and b are compiled for stirred tank reactors at different stirrer speeds and for bubble column reactors of different height. The flow behavior index n gradually decreased from $n=1$ at $X=2$ to $n=0.5$ at $X=18$ g l^{-1} in both of the reactors without large difference. This relationship, which was used in the present review for the calculation of the medium viscosity with hairy pellets, yields higher apparent viscosity than the relationship of Wittler et al. [73].

Table 3. Numerical constants of the relationship between the consistency index K and the fungal biomass concentration X : K = A X^b [99]

	A	b
Stirred tank N = 300 rpm	6.0×10^{-3}	2.47
Stirred tank N = 400 rpm	1.2×10^{-2}	2.00
Stirred tank N = 600 rpm	8.1×10^{-5}	3.25
Bubble column H = 1.5 m	4.7×10^{-4}	2.80
Bubble column H = 3.0 m	1.9×10^{-8}	6.40

4.5.1
Influence of Phosphate Concentration on Morphology and Xylanase Production in Semisynthetic Medium

The effect of phosphate concentration on fungal morphology was investigated in shake flasks and stirred tank cultures. In shake flask cultures no constant pH could be maintained and it was reduced from 5.5 to 2.5 during 80 h batch culture. In the stirred tank reactor the shear stresses were different. The use of the airlift tower loop reactor allows the maintenance of a definite pH and a low and uniform shear stress.

The precultures were prepared with 10^5 spores ml^{-1} at 30 °C and 150 rpm for 10 h. The main cultivations were performed at 25 °C, pH 4.5 and 0.2 vvm and initial phosphate concentrations of 0.3 g l^{-1}, 1.05 g l^{-1} and 2.01 g l^{-1}, respectively. In all of the three runs pellets were formed. However, the pellets were dissimilar. At 0.3 g l^{-1} initial phosphate concentration loose hairy pellets 6 mm in diameter were formed with low cell density in their center. At 1.05 g l^{-1} phosphate concentration dense pellets 4 mm in diameter and with uniform pellet density were produced. At 2.01 g l^{-1} phosphate concentration hollow pellets 7 mm in diameter and with a dense surface were observed. Only a few hairy hyphae extended from their surface.

At 1.05 g l^{-1} initial phosphate concentration, the fungal biomass concentration increased with time and followed the course similar to those observed in the stirred tank reactor (Fig. 20 A). The same holds true for the yield coefficient of the growth $Y_{X/S}$ (not shown). The fungal biomass concentrations and the yield coefficients of growth $Y_{X/S}$ with 0.3 g l^{-1} and 2.01 g l^{-1} initial phosphate concentrations followed different courses. The increase of the cell concentration and the yield coefficient were considerably delayed, thus the maximum cell concentrations were attained later, between 80 and 100 h, and thereafter, between 110 and 120 h, the cell concentration and the yield coefficient rapidly decreasing. This quick decrease of the cell mass concentration can only be caused by cell lysis. However, the courses of the protease activities in these three runs were very similar (Fig. 20 B). They started to increase between 60 and 100 h in the same way, and the protease activity was the lowest at 0.3 g l^{-1} initial phosphate concentration.

In the center of the large pellets, the cells died away and lysed, because of their insufficient oxygen and substrate supply. The courses of the CPR (Fig 20 C) and glucose concentration (Fig. 20D) reveal that the metabolic activities of the cells were lower at low and high phosphate concentrations than at intermediate ones.

At 0.3 g l^{-1} phosphate concentration the concentration decreased to zero after 60 h. At 1.05 g l^{-1} the lowest phosphate concentration (0.25 g l^{-1}) was obtained at 80 h and at 2.01 g l^{-1} the phosphate concentration dropped from 2.0 to 1.12 g l^{-1} at 70 h. After these minima, the phosphate concentrations increased again, probably due to the cell lysis. The increase was the largest with the highest, and the weakest with the lowest phosphate concentrations (not shown). The highest specific xylanase activity and product yield coefficients $Y_{P/S}$ and $Y_{P/X}$ were obtained with the lowest initial phosphate concentration (0.3 g l^{-1}) (Fig. 20E, F) and the lowest with the highest initial phosphate concentration. These results were similar to the shake flask cultivations. This high enzyme activity at 0.3 g l^{-1} phosphate concentration is probably due to the increased production and secretion of the xylanase by the hairy surface of the pellets [98].

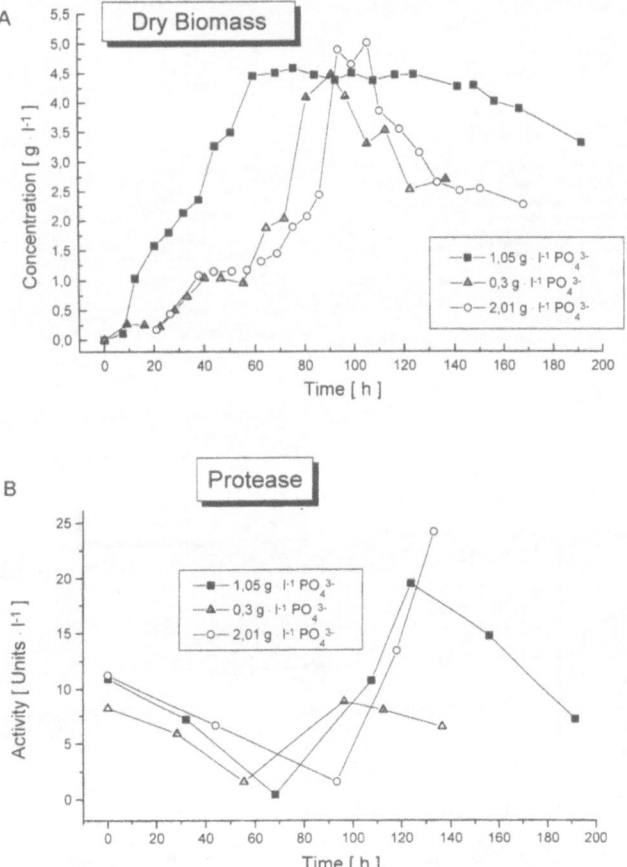

Fig. 20 A, B. Influence of phosphate concentration on cultivation in airlift tower loop reactor in synthetic medium [59]; A biomass concentrations; B protease activities

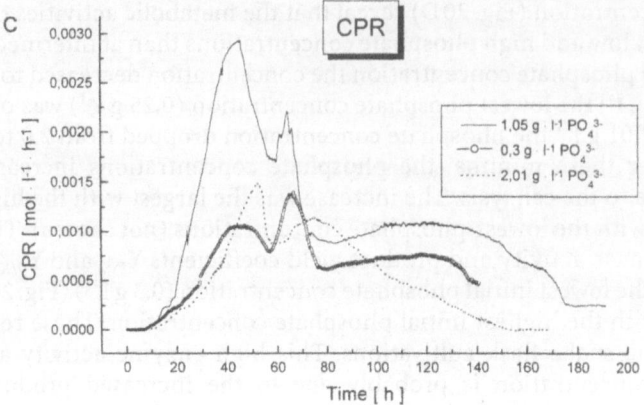

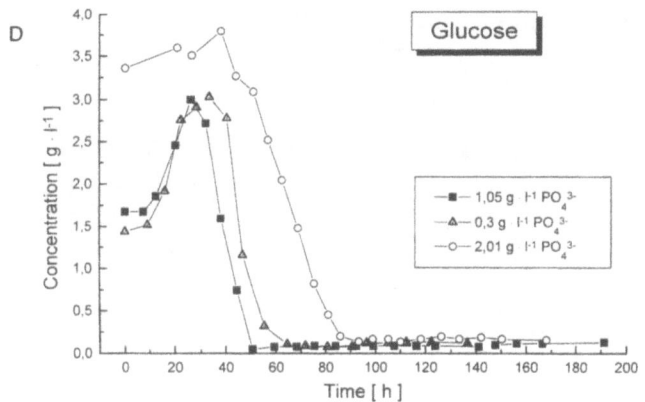

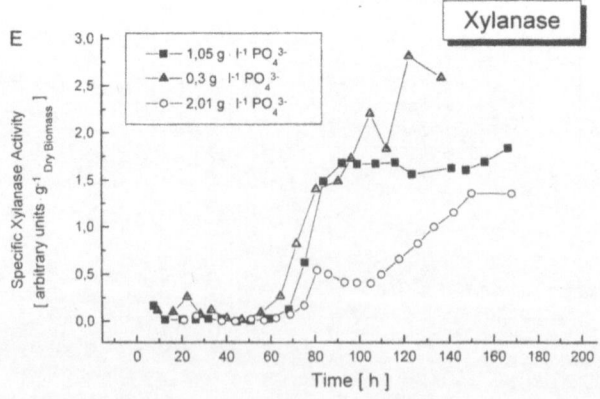

Fig. 20 C–E. C CO_2 production rates (CPR); D glucose concentrations; E yield coefficient $Y_{P/X}$

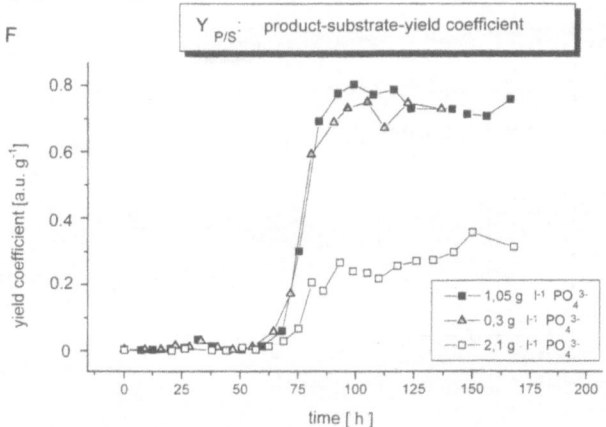

Fig. 20 F. F yield coefficient $Y_{P/S}$

4.5.2
Cultivation on Complex Medium with Original and Milled Wheat Bran, Respectively

The investigations in the stirred tank reactor with original and milled wheat bran revealed that with milled wheat bran higher metabolic activity, but lower xylanase production was obtained than with the original wheat bran. To investigate the influence of the reactor type on the process performance with 5 – 6 and 1 – 2 mm-sized wheat brans, cultivations were performed with these solid substrates in the airlift tower loop reactor as well.

The investigations were performed with a preculture, which was prepared at 30 °C and 150 rpm for 10 h in complex medium. They were inoculated with 5×10^4 spores ml^{-1} and cultivated at 25 °C, pH 4.5 and 0.5 vvm.

With the 5 – 6-mm size wheat bran filamentous mycelium was formed, which was partly suspended in the medium and partly attached to the wheat bran particles. Due to the filamentous mycelium, the medium viscosity increased considerably, but the oxygen supply of the cells was not endangered. After the glucose was consumed, the xylanase activity increased to very high values by 100 h.

The investigations with the milled wheat bran were performed under the same cultivation conditions. Again, filamentous mycelia was formed. But in this case the wheat bran fragments were completely covered by mycelia and they were integrated into clumps and pellets. This filamentous network caused an increase of the medium viscosity and a reduction of the medium circulation in the loop as well as the reduction of the dissolved oxygen concentration. This caused a decrease of the growth rate and the CO_2 production rate (Fig. 21 A). However, the release of the reducing sugars and glucose and their consumption rate differed only slightly (Fig. 21 B, C). Similar to the stirred tank investigations, the xylanase activity was higher with non-milled than with milled wheat bran (Fig. 21 D) [98].

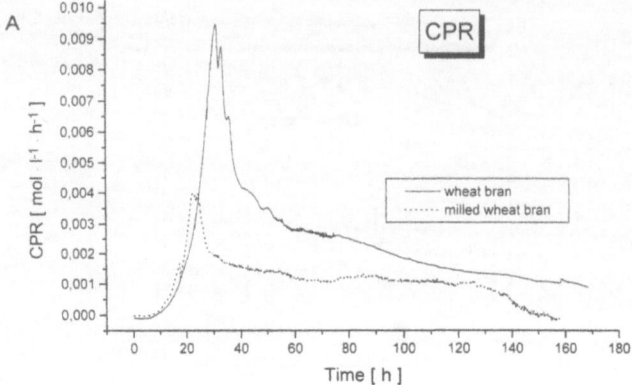

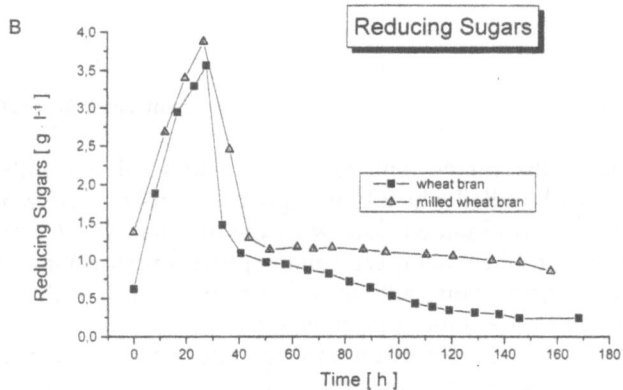

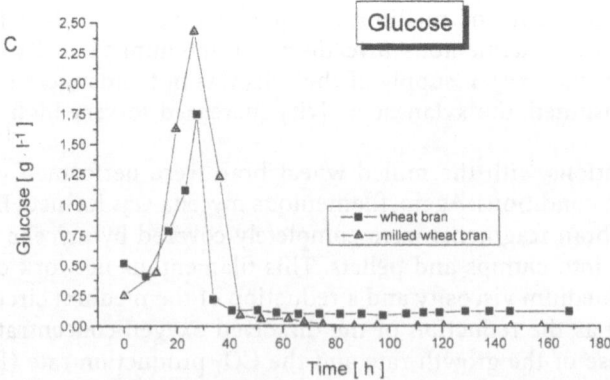

Fig. 21 A–C. Cultivation in airlift tower loop reactor on complex medium with non-milled and milled wheat bran – a comparison [98]: **A** CO_2 production rates (CPR); **B** reducing sugars; **C** glucose

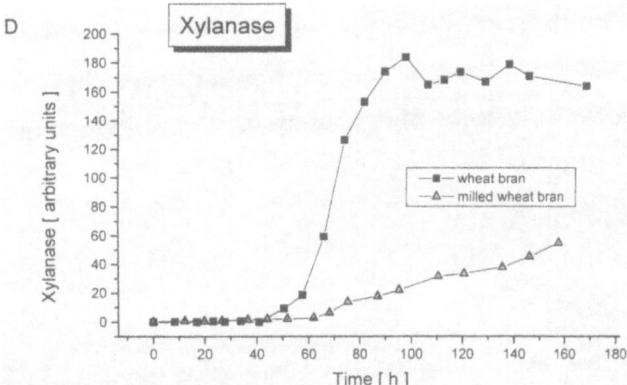

Fig. 21 D. D xylanase activity

5
Influence of Process Parameters on Morphology and Enzyme Production

In Sect. 4, the cultivations in different reactors were investigated separately and especially the influence of various parameters on the process performance was considered in them. In this section these results are compared and the differences worked out.

5.1
Spore Germination and Development to Hyphae, Clumps, and Pellets

It is well known that the concentration of the spores and the size of spore conglomerates influence the morphology of fungi. The germination of fungal spores was investigated by Smith [2] and mathematically modeled by Nielsen [92], Tucker and Thomas [35], Bosch et al. [100] and by several other authors.

The investigations with *Aspergillus awamori* revealed, that the fungal morphology can be influenced by the spore concentration (see Sect. 4). Also the germination of the spores influences the morphology. Therefore, as an example, the germination of the spores will be considered in shake flask cultures with 10 g l^{-1} glucose at 25 °C and 150 rpm and with a final spore concentration of 10^6 ml^{-1} [30].

The inoculation was performed with non-agglomerated spores. During the first 4 h no changes could be detected, after 6.5 h few swollen spores were observed. With time the fraction of the swollen spores and the number of aggregates increased. After 11.5 h, a large fraction of the spores germinated and non-branched hyphae were formed. The branching started at 12.5 h and at 14.5 h aggregates could be observed with the naked eye.

In dormant spores, no protein synthesis (absence of mRNA) could be observed by means of AO-staining indicated by red fluorescence. The protein synthesis started in spore aggregates earlier than in single spores. Figure 22a shows (yellow) spore aggregates, in which the protein synthesis just started.

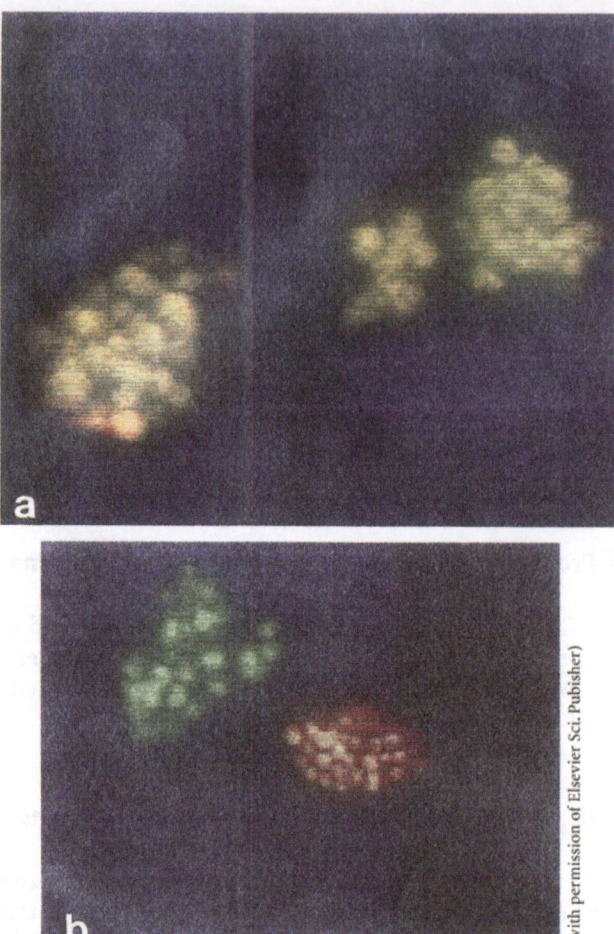

(with permission of Elsevier Sci. Pubisher)

Fig. 22 a, b. AO-staining of spores after 6 h cultivation [101]; **a** spores which just start to germinate *(yellow)*; **b** non germinating spores *(green)* and germinating spores *(red)*

Figure 22 b shows dormant (green) and germinating (red) spore aggregates. In Fig. 23 A the fraction of the spores and germinated hyphae are shown, and in Fig. 23 B the average object surface area of the hyphae and spores are are illustrated. With increasing cultivation time the frequency distribution of the eccentricity became more and more symmetric (not shown).

The AO-staining revealed that during germination and the exponential growth phase, the mRNA concentration increased within the entire length of the young hyphae (not shown). Later on, the protein synthesis was restricted to local regions of the hyphae (Fig. 24). It is remarkable that the protein synthesis occurs in subapical regions and not in the tips of the hyphae [39]. The vacuoles are easy to recognize, because they exhibit no autofluorescence. The first vacuoles had appeared after 7.5 h cultivation time.

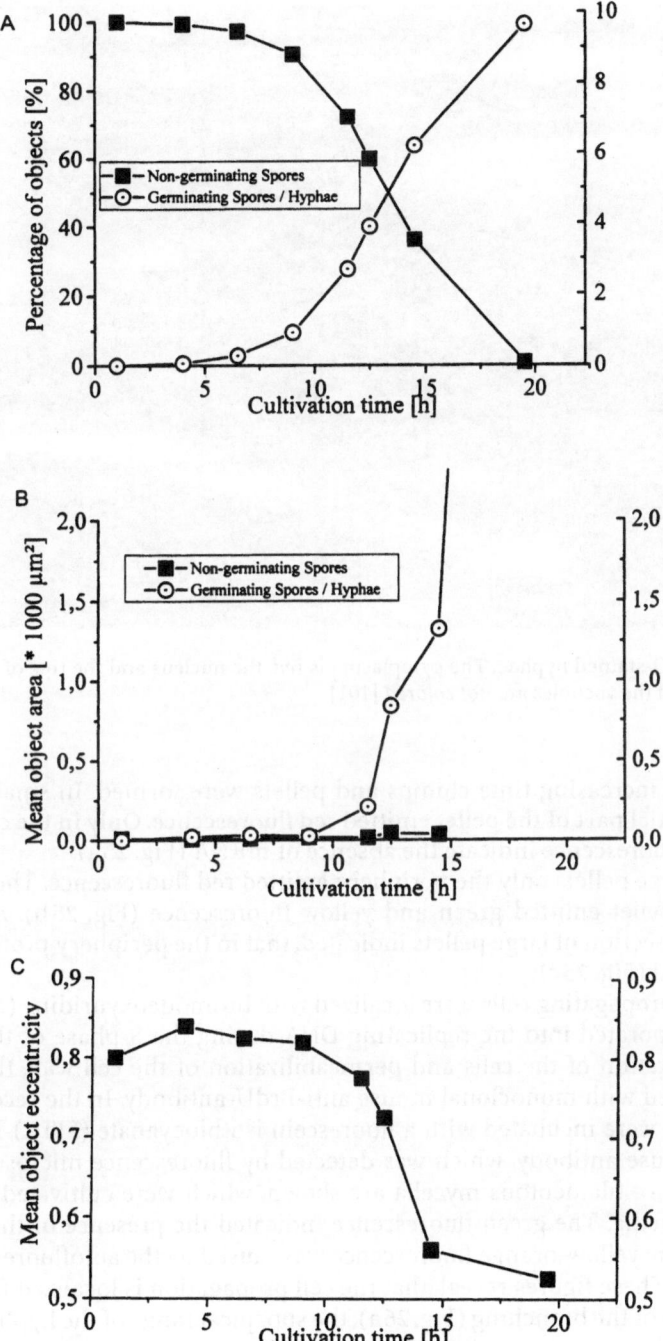

Fig. 23 A–C. Spore germination [30]; A fraction of non-germinated and germinated spores as a function of cultivation time; B mean object area of hyphae and spores as a function of cultivation time; C mean eccentricity of all objects as a function of cultivation time

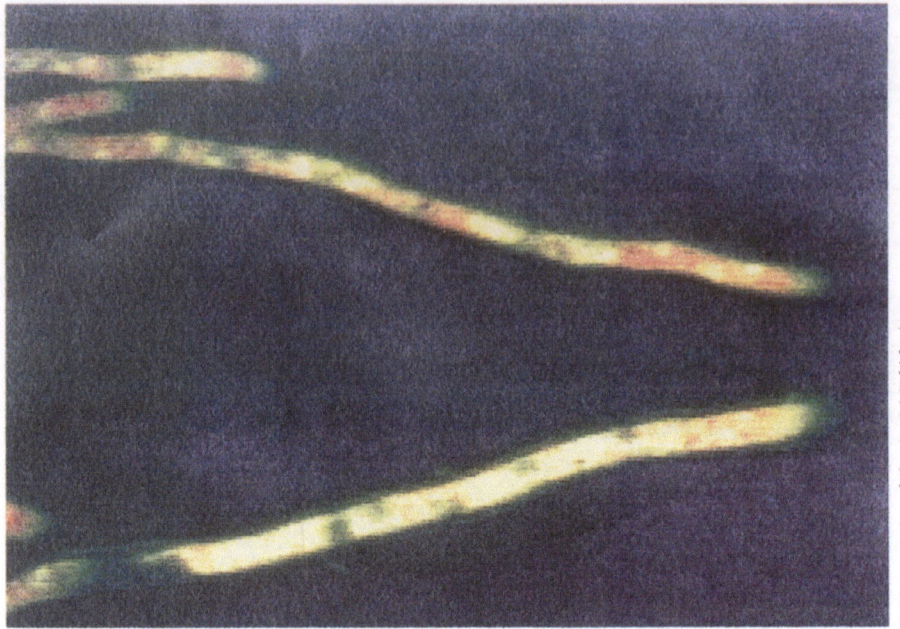

(with permission of Elsevier Sci. Pubisher)

Fig. 24. AO-stained hyphae. The cytoplasma is *red*, the nucleus and the tips of the hyphae are *green* and the vacuoles are *not colored* [101]

With increasing time clumps and pellets were formed. In small pellets, the substantial part of the pellet emitted red fluorescence. Only in the center did the green fluorescence indicate the absence of mRNA (Fig. 25 a).

In large pellets only the periphery emitted red fluorescence. The largest part of the pellet emitted green and yellow fluorescence (Fig. 25 b). Also the AO-stained section of large pellets indicated, that in the periphery protein synthesis occurred (Fig. 25 c).

The propagating cells were localized with bromodeoxyuridine (BrdU), which is incorporated into the replicating DNA during the S-phase of the cell cycle. After fixation of the cells and permeabilization of the cell wall, the cells were incubated with monoclonal mouse anti-BrdU-antibody. In the second step, the samples were incubated with a fluorescein isothiocyanate (FITC)-labelled goat anti-mouse antibody, which was detected by fluorescence microscopy [39]. In Fig. 26 a – c filamentous mycelia are shown, which were cultivated in semisynthetic media. The green fluorescence indicated the presence of the replicating DNA. The yellow-orange fluorescence was caused by the autofluorescence of the fungus. These figures reveal, that the cell propagation is localized mainly in the position of the branching (Fig. 26 a), the subapical range of the hyphae (Fig. 26 b) and on the youngest hyphae (Fig. 26 c).

When the mycelium aggregates and clumps were formed, the cells propagated only in the periphery (Fig. 26 d). The center of the clumps were yellow. When large pellets were formed, in the substantial part of the pellet no propagating

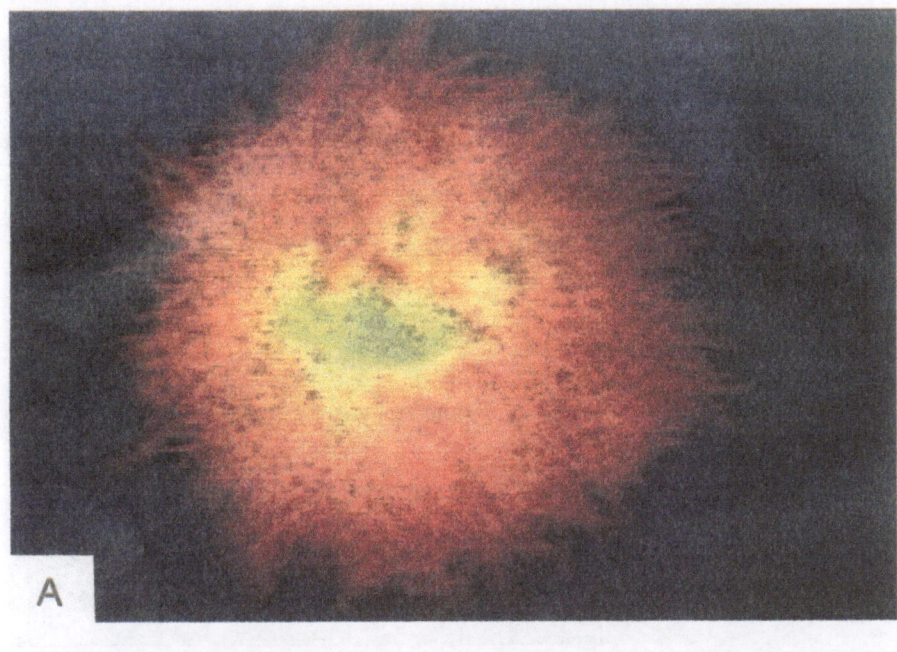

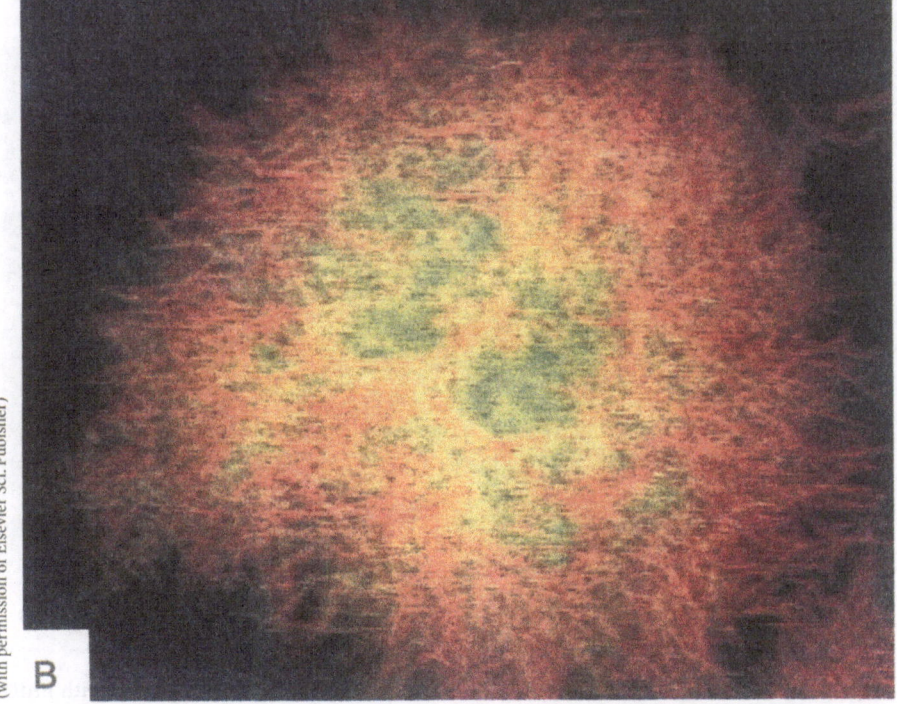

Fig. 25 A, B. AO-stained pellets [101]; A AO-stained 0.7-mm pellet; B AO-stained 1.1-mm pellet

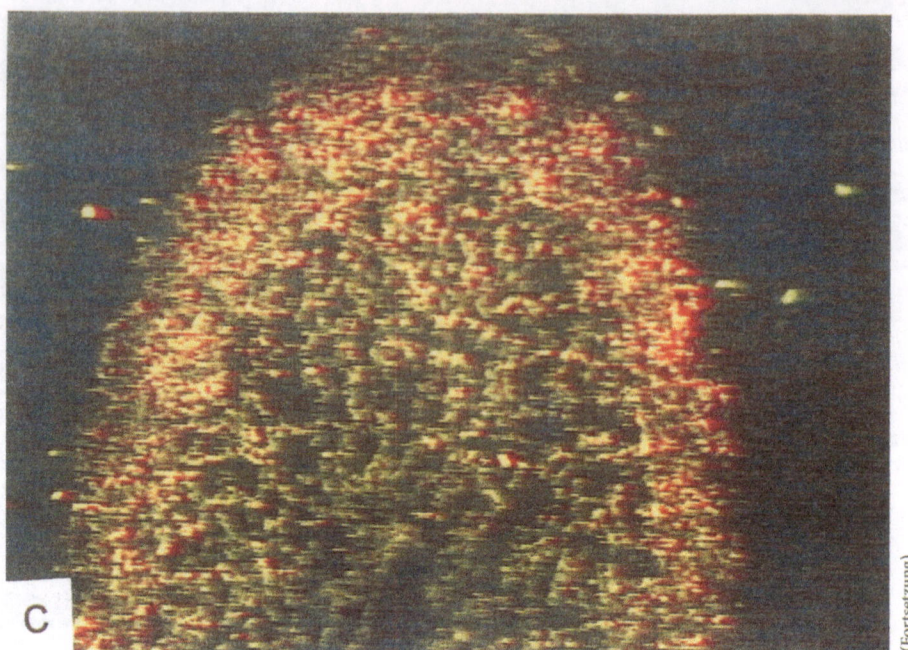

(Fortsetzung)

Fig. 25 C. C AO-stained section of a 115 h old pellet

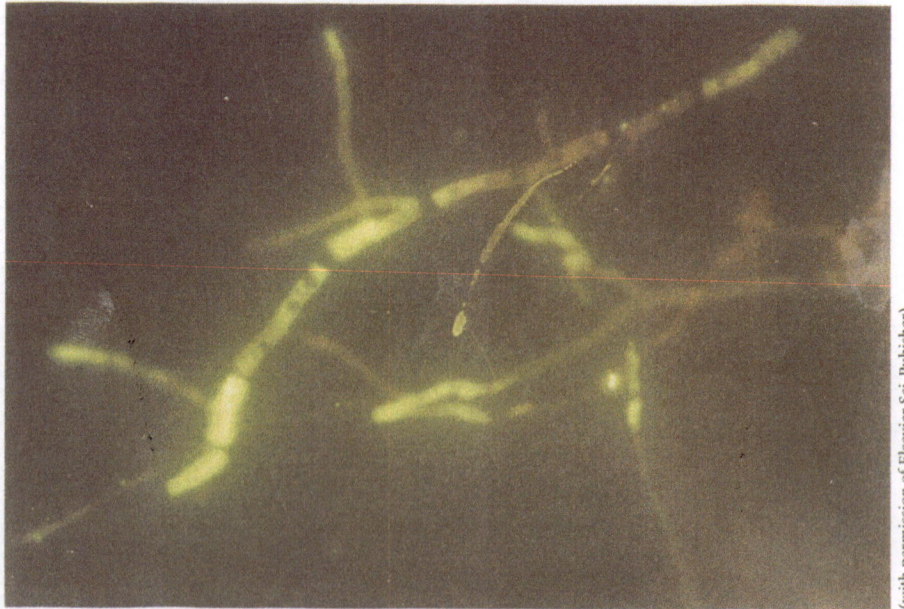

(with permission of Elsevier Sci. Pubisher)

Fig. 26 a. Filamentous mycelia and pellets cultivated in synthetic medium stained with BrdU-immuno sandwich labeling at different cultivation times [102]; **a** intensive fluorescence in subapical and branching zones (t = 14 h)

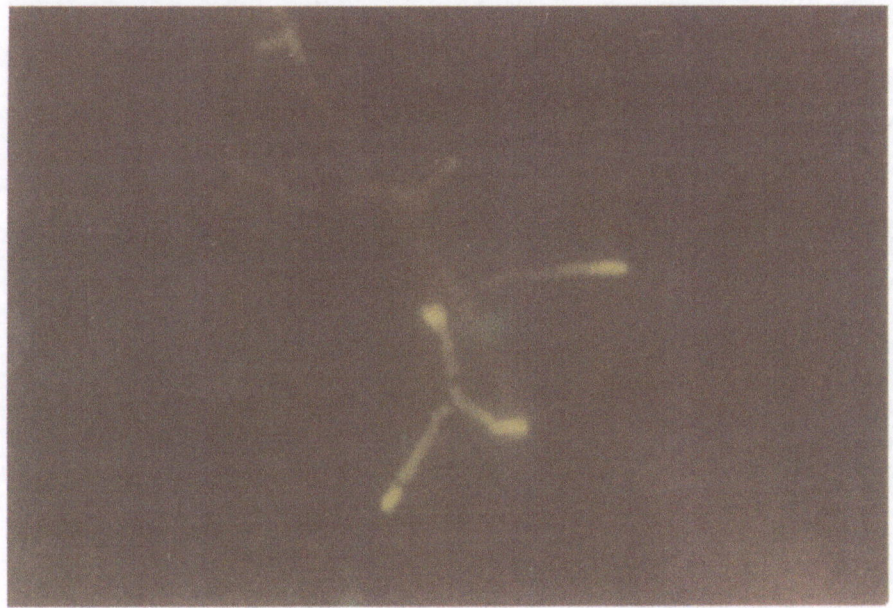

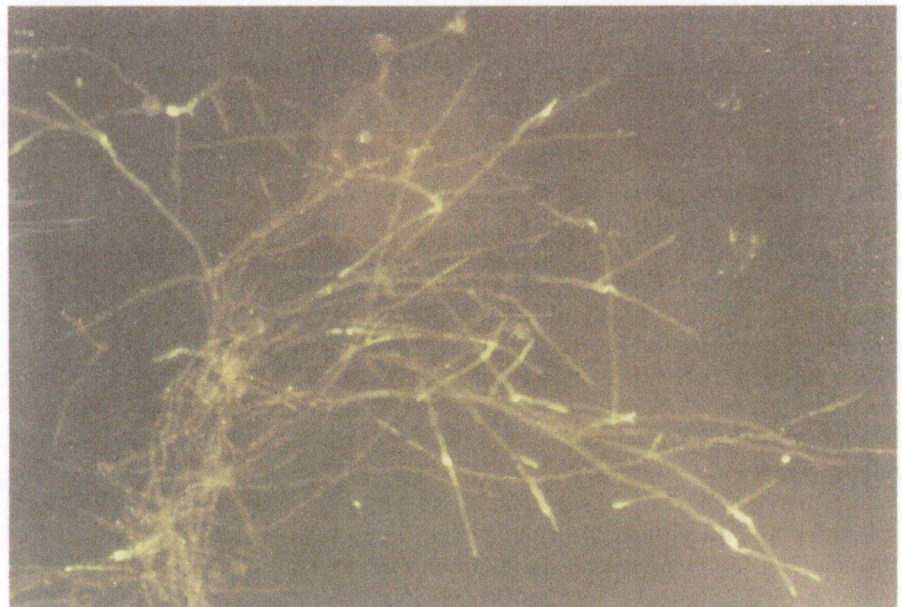

(Fortsetzung)

Fig. 26 b–c. b gradual decrease of green fluorescence intensity from the tips to the bases (t = 12 h); **c** mycelium network (t = 15 h)

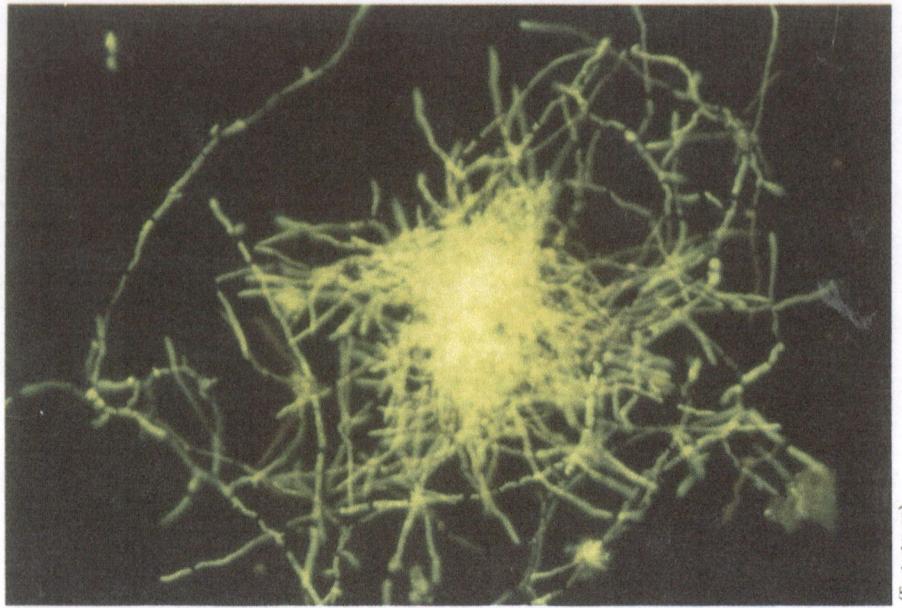

(Fortsetzung)

Fig. 26 d. **d** clump (t = 15 h)

cells were present, only in a narrow rim was green fluorescence observed, caused by replicating cells (not shown).

5.2
Effect of Spore Concentration on Morphology and Enzyme Production

The fungal morphology can be controlled by the spore concentration, as shown, e.g., by Tucker and Thomas [35], who found a transformation of pellets to clumps at $10^4 - 5 \times 10^5$ spores ml^{-1}. In the investigations reported, the spore concentrations were varied in the shake flask and stirred tank cultures. In shake flasks, at low spore concentrations large pellets and at high spore concentrations small pellets were formed in semisynthetic medium. The highest specific enzyme activity was obtained at the highest spore concentration, at which small pellets were formed (Table 4).

In a stirred tank reactor, again, at low spore concentrations large pellets and at high spore concentrations small pellets were formed [59]. However, at the intermediate spore concentrations filamentous mycelium was observed, which is unexpected (Table 4). The highest specific enzyme activity was formed at these intermediate spore concentrations. This is in agreement with the results of the enzyme activity measurement with size fractioned mycelium, namely, the filamentous mycelium exhibited the highest enzyme activity.

According to Smith and Wood [10] the spore concentration influences the enzyme production as well. At spore concentrations above 10^7 ml^{-1} the xylanase

Table 4. Influence of spore concentration on morphology and specific enzyme activity in semisynthetic medium

	spore conc.	pellet size spec.	enzyme activity (a.u)
shake flask	10^2	4	0.2
rp	10^4	2–3 mm	0.28
	10^6	small	0.55
stirred tank	10^2	large	1.8
300 rpm	10^3	filamentous	5.5
	10^5	small	3.3

productivity diminished considerably (sevenfold). The highest activity of β-xylosidase was obtained between 10^6 and 10^7 spores ml^{-1}.

5.3
Effect of Substrate Type on Morphology and Xylanase Production

Solid substrates are often applied in complex media. In solid state cultivation (SSC) the solid particles are not destroyed by the shear stress. Therefore, the fungal decomposition of the solid substrate can be followed by measuring the particle size as a function of time, after separation from the fungus. In the case of SSC of *Aspergillus niger* on wheat bran with 60% moisture content, the size of the solid particles diminished with time by enzymatic decomposition alone. Their size was determined after wet sieving by microscopical analysis. In the initial size range between 3.8 and 0.68 mm wheat bran the fungal decomposition yielded after 60 h about the same size of fragments (0.24–0.26 mm) [103], which is probably due to the size of the fragments of the outer cuticle of the seed coat [53].

In the stirred tank reactor at 300 rpm an intact aleuron layer was observed in the medium, but at 500 rpm shortly after the cultivation started, freely suspended aleuron particles and other ingredients of the wheat bran were indentified by microscopy.

At 750 rpm the fungal pellets were partly separated from the wheat bran. These shear effects increased the intraparticle mass transfer as well. With milled wheat bran, this separation had already occurred at 300 rpm. This accelerated the growth, but reduced the xylanase production, because of the diminishing intimate contact between the fungus and the substrate. According to the immunogold assay, the xylanase activity was especially high at the interface between fungus and wheat bran (Fig. 16 f, g). With the reduction of this interface, the efficiency of the xylanase induction diminished.

With milled wheat bran filamentous mycelium was formed in both of the reactors.

In semisynthetic medium pellets were formed. In the stirred tank reactor small pellets and in airlift tower loop reactors large (6 mm) pellets were present (Table 5). The highest specific enzyme activities were obtained in both of the

Table 5. Effect of substrate type on the morphology and enzyme production with spore concentration

	operation conditions	morphology	enzyme activity (a. u.)
Stirred tank wheat bran 5 10⁴ sp ml⁻¹	300 rpm 1 vvm	attached	165
Stirred tank milled wheat bran 5 10⁴ sp ml⁻¹	300 rpm 1 vvm	filamentous	135
Stirred tank Semisynthetic medium 10⁵ sp ml⁻¹	400 rpm 1 vvm	small pellets	22.5
Airlift loop wheat bran 105 sp ml⁻¹	0.5 vvm	partly attached	190
Airlift loop milled wheat bran 10⁵ sp ml⁻¹	0.5 vvm	filamentous	60
Airlift loop semisynthetic medium 10⁵sp ml⁻¹	0.3 g l⁻¹ P 0.2 vvm	6-mm pellet	2.5

reactors with wheat bran. In the airlift tower loop reactor the enzyme activity was higher than in the stirred tank reactor, probably because of the low shear stress imposed on the fungus. By milling the wheat bran the enzyme activites decreased, but was still much higher than the activities in semisynthetic medium. In the airlift tower loop reactor with semisynthetic medium the enzyme activity was especially low (Table 5).

The applied substrate influences the enzyme composition as well. With xylan-containing solid substrates (barley straw, oat straw, oat spelt xylan, larch wood xylan) five times more xylanase is formed than β-xylosidase [10]. These substrates yielded the highest xylanase activities at low protease activities. With other substrates much lower xylanase activities were obtained.

5.4
Effect of Stirrer Speed and Specific Power Input, Respectively, on Morphology and Enzyme Production

In solid state cultivations (SSC) the specific power input is zero. Therefore, the interparticle and intraparticle mass transfer influence the growth and product formation. In stirred tank reactors the interparticle mass transfer is accelerated by turbulence. The intraparticle mass transfer is rate limiting.

In the presence of solid substrate, the fungus is usually attached to the substrate surface. In solid state cultures (SSC) this phenomenon was observed at particle size well below 1 mm as well [103]. In stirred tank reactors the fungus-solid particle clumps were broken up at low specific power input already when using solid particles 1–3 mm in size, which did not protect the fungus from the shear stress any more and the fungus separated from the particle formed dense filamentous mycelium.

Several researchers investigated the pellet size in synthetic medium at various stirrer speeds. According to Taguchi et al. [104] no change of the pellet size of *A. niger* was observed at 100 rpm. The pellet size reduced to half at 300 rpm in 60 min and at 500 rpm in 20 min. However, the shear stress is usually imposed on the fungus already at the beginning of the cultivation. Therefore, the fungi adapt their morphology as a reaction to the shear stress imposed on them from the beginning. Only under growth limitation or at the end of the cultivations, when pellets become too large and their center cells die away and lyse, does pellet destruction occur.

The stirrer speed influences the specific power input, the shear rate, the apparent viscosity of non-Newtonian media and the oxygen transfer into the medium, and likely several other parameters.

At 200 rpm, the specific power input P/V was determined to 0.05 kW m^{-3} in non-aerated low viscosity medium, P/V = 0.5 kW m^{-3} in non-aerated high viscosity medium and P/V = 0.22 kW m^{-3} (see Sect. 4.3.3) in aerated high viscosity medium. The shear rate was low: 36.3 s^{-1} and the shear factor high: dVF = 130 μm. The specific enzyme activity was the lowest: 7.5 arbitrary units (a. u.). In spite of the low shear stress, the specific enzyme activitiy was low because of the oxygen limitation, caused by the low mass transfer rate.

At 400 rpm for the specific power input following data were evaluated: P/V = 0.5 kW m^{-3} non-aerated, 1.42 kW m^{-3} in non-aerated high viscosity medium and P/V = 0.85 kW m^{-3} in aerated high viscosity medium. The shear rate: 63.2 s^{-1} and the shear factor: dVF = 100 μm, respectively, intermediate and the volumetric mass transfer coefficient high: 0.007 s^{-1}. The specific enzyme activity was the highest (22.5 a. u.) under these conditions. At 400 rpm, the shear stress was moderate and the mass transfer rate sufficient for high enzyme productivity.

At 600 rpm the specific power input was too high: P/V = 2.0 kW m^{-3} in non-aerated low viscosity medium, P/V = 3.0 kW m^{-3} in non-aerated high viscosity medium and P/V = 2.40 kW m^{-3} in high viscosity aerated medium. On account of the high shear rate (110 s^{-1}) and low shear factor (dVF = 80 μm), respectively, and in spite of the high volumetric mass transfer coefficient (0.010 s^{-1}) the specific enzyme activity was rather low: 8.0 a. u. At 600 rpm the shear stress was obviously too high for enzyme production.

In the shake flask cultures under standard conditions (at initial pH 4.5, 150 rpm and 104 spores ml^{-1}) partial spore agglomeretion, pellets 2–3 mm in size and enzyme specific activities of 0.28 a. u. were obtained. At 150 rpm, the specific power input was determined to P/V = 0.25 kW m^{-3}, when 60% of the shake flask volume is filled with cultivation medium.

These examples reveal that the effect of the stirrer speed on the morphology and enzyme production in the stirred tank reactor depend on the substrate used. With non-milled wheat bran the fungus attached to the solid wheat bran particles and partly formed pellets inside of the bran (Fig. 16b). The morphology of the attached fungus was not influenced by the stirrer speed in the range 300–750 rpm. The highest enzyme activity was obtained at 300 rpm.

In semisynthetic medium at pH 4.5, the average pellet size diminished with increasing stirrer speed but, at 600 rpm, clumps were formed as well as filamentous mycelia (Table 6). The highest enzyme activity was obtained at 400 rpm, at which only filamentous mycelia and small pellets were observed.

In the airlift tower loop reactor under standard conditions (0.5 vvm, non-milled wheat bran) the specific power input was very low: 0.0162 kW m^{-3}, the shear rate intermediate: 48.7 s^{-1} and the volumetric mass transfer coefficient low: 0.0028 s^{-1}. The fungus was attached to the substrate particles. Under these conditions, high enzyme activity (190 a.u.) was gained. With milled wheat bran filamentous mycelium was formed and an enzyme activity of 60 a.u. was obtained.

Large pellets (6 mm) were formed in synthetic medium. Accordingly low enzyme activity of 2.5 a.u. was obtained.

5.5
Effect of pH Value and Temperature on Process Performance

The pH value and the temperature are important process variables. According to Carlsen et al. [105], who carried out batch cultivations of A. oryzae a stirred tank reactor with an inoculum consisting of 4×10^8 spores ml^{-1}, at very low pH (< 2.5), the mycelium was very vacuolated and the cell walls were swollen, resulting very poor growth. At pH 3.0–3.5 freely dispersed hyphal elements resulted. At pH 4–5 both pellets and some freely dispersed hyphae, and at pH values higher than 6 only pellets were observed. With increasing pH value, the pellet size increased. The pellet formation was caused by agglomeration of the spores at pH > 4. With inoculum consisting of free hyphae, namely, no pellets were observed at these pH values. Similar observations were made with A. niger as well [106].

According to Smith and Wood [10], the enzyme production of Aspergillus awamori a strong function of the pH value. Cultivation at pH 5 and above causes a dramatic reduction of the enzyme activity. The maximum xylanase activity was obtained between 3.5 and 4.0. The β-xylosidase activity had a sharp

Table 6. Effect of stirrer speed on morphology and enzyme activity in stirred tank reactor

Medium	Stirrer speed	morphology	enzyme activity (a.u.)
Complex	300 rpm	attached	160
7.10^4 sp ml^{-1}	500 rpm	attached	120
	750 rpm	attached	60
Semisynthetic	200 rpm	1-mm pellet	7.5
10^5 sp ml^{-1}	400 rpm	< 0.2 mm	22.5
	600 rpm	0.2 mm	8.0
		+ clumps	

maximum at 4.0, and below and above the activity exhibited a marked reduction. In the investigation of Siedenberg et al. [59], the cultivations in stirred tank and airlift reactors were carried out at pH 4.5, at which much more xylanase was formed than β-xylosidase. In shake flasks [88] the pH decreased during the batch cultivation from 5.5 to 2.5, and the formation of β-xylosidase was more strongly impaired than that of xylanase.

Carlsen et al. [105] investigated the influence of the pH on the α-amylase formation as well. α-Amylase production had a sharp maximum at about pH 6, in contrast to the growth rate, which had a broad maximum between pH 3 and 7.

Temperature also affected the production of xylanase, although the variation in temperature was less dramatic than the variation in pH. The productivity of xylanase was optimal at 35 °C. At lower temperature β-xylosidase activity was reduced more strongly than the xylanase activity [10]. For the xylanase production 25 °C was used, because at this temperature not only was xylanase productivity high, but the ratio of xylanase to β-xylosidase activity was too.

The specific growth rate of *A. oryzae* cultivated in a stirred tank reactor had a maximum at 35 °C. In the investigated temperature range (27–40 °C), at pH above 4 and with inoculum consisting of spores, pellets were formed and the pellet size distribution was independent of the temperature [105].

Gerlach [30] performed investigations in shake flask cultures at different temperatures, 25, 30, and 35 °C, and evaluated the effect of temperature on the morphology. According to these investigations, 25 °C was the optimal cultivation temperature. At this temperature the highest cell mass concentration was obtained [30]. During the cultivation only pellets were present, the size of which increased with time (Fig. 11 E, F). At 30 °C the pellets, which were formed at the beginning of the cultivation turned into filamentous mycelium after 50 h (Fig. 11 C), the size of which increased with time and approached a constant value (Fig. 11 F). At 35 °C at the beginning mainly filamentous mycelium (Fig. 11 C) and a only low amount of pellets (Fig. 11 B) and clumps (Fig. 11 D) were formed. However, the mycelium turned into clumps after 20 h, the size of which decreased with time (Fig. 11 E). These investigations indicate that at higher temperature the oxygen supply of the cells was inadequate. Therefore, the pellets were transformed to filamentous mycelium at 30 °C, but at 35 °C, because of the higher shear stress, clumps are formed.

5.6
Effect of Phosphate Concentration on Morphology and Enzyme Production in Synthetic Medium

The influence of phosphate concentration on fungal morphology was investigated by Siedenberg et al. [59, 88]. In shake flask cultivations pellets were formed in the orthophosphate concentration range 0.3 to 2.0 g l⁻¹. The specific enzyme activity was the highest at 0.3 g l⁻¹, and the lowest at 2.0 g l⁻¹ phosphate concentration (Table 7).

In stirred tank reactors, clumps were observed at 0.3 and 2.1 g l⁻¹ phosphate concentrations. At 1.05 g l⁻¹ pellets were formed. No enzyme activities were determined in stirred tank reactors at different phosphate concentrations.

Table 7. Effect of phosphate concentration on morphology and enzyme production in semi-synthetic medium

Reactor	phosphate conc (g l^{-1})	morphology spec.	enzyme activity (a. u.)
Shake flask	0.3	pellet	6.0
10^4	1.0	pellet	4.0
200 rpm	2.0	pellet	3.0
Stirred tank	0.3	clumps	
	1.05	pellets	
	2.1	clumps	
Airlift tower			
loop	0.3	6-mm loose pellet	2.6
10^5 sp ml^{-1}	1.05	4-mm dense pellet	1.8
	2.01	7-mm hollow pellet	1.4

In airlift tower loop reactors at 0.3 g l^{-1} phosphate concentration, 6 mm large loose pellets with low cell density in the pellet center were observed. At 1.05 g l^{-1} phosphate concentration, dense pellets of 4-mm size with uniform cell density and at 2.01 g l^{-1} phospate concentrations, 7-mm hollow pellets were formed.

This strong effect of the phosphate concentration on the morphology and enzyme production was unexpected. It is possible that the influence of the phosphate on the enzyme production is a consequence of the morphology change.

5.7
Effect of Reactor Type on Morphology and Enzyme Production

In Table 8 the process performances are compiled for the reactors applied by Siedenberg et al. [59, 88] and Gerlach [30].

In shake flask cultures of *A. awamori* with inoculum consisting of spores, at 25 °C and initial pH value of 4.5, partial agglomeration of the spores occurred. Under these conditions mainly small pellets were formed. With increasing cultivation time the pH value decreased, which did not influence the fungal morphology, only the enzyme production. At low ratio of medium volume to flask volume, the volumetric mass transfer rates were high enough to supply the cells with oxygen. At higher volume ratio the oxygen supply could become critical, but with a specific power input of 0.25 kW m^{-3} and in presence of small pellets it was sufficient for the enzyme production. The pellet morphology did not change with the rpm of the shaker. The enzyme activity was much lower than in stirred tank and airlift tower loop reactors due to the decrease of the pH in shake flasks during the cultivation.

At higher temperatures oxygen limitation occurred, which caused the formation of mycelia and clumps.

A. awamori was cultivated in the stirred tank reactor with inoculum consisting of spores, at 25 °C and pH 4.5. The specific power input was varied with the stirrer speed. At the lowest stirrer speed mainly pellets of 1 mm diameter were formed. The apparent viscosity was high. The viscosity in Table 8 was

Table 8. Comparison of performances of investigated reactors:shake flasks (SF), stirred tank (ST), and airlift tower loop reactor (ATR) in semisynthetic medium (sp. = spore concentration ml^{-1}; P = initial phosphate concentration ; vvm = medium volume in litre per gas throughput in litre/min; F = V/VSF = degree of filling of shake flask; V = liquid volume; V_{SF} = shake flask working volume

	P/V	D	η_a	morphology	d_{VF}	spec.enzyme activity	k_La
	[kW m^{-3}]	[s^{-1}]	[mPa s]		[μm]	[a.u]	[s^{-1}]
SF 10^6 sp.							
150 rpm F=0.6	0.25	n.d.	n.d.	small pellets	n.d	0.55	0.005
SF 10^6 sp F=0.3							
100 rpm	0.05	n.d	n.d	0.1-mm pellets[a]	n.d	n.d	0.008
150 rpm	0.20	n.d	n.d	0.4-mm pellets[b]	n.d.	n.d.	0.012
200 rpm	0.43	n.d	n.d	0.2 → 0.4-mm pellets[c]	n.d	n.d.	0.016
ST, 10^5 sp. 1 vvm							
200 rpm	0.22	36.3	166	1-mm pellets	130	7.5	
400 rpm	0.85	63.2	110	< 0.2-mm pellets	100	22.5	0.007
600 rpm	2.40	110	70	< 0.2-mm pellets + clumps	80	8.0	0.010
ATR 10^5 sp 0.5 vvm 0.3 g l^{-1} P	0.0162	48.7	288	6-mm loose hairy pellets	250	2.6	0.0028

[a] t=40^{-1} 20 h.
[b] t=40^{-1} 00 h.
[c] growth of pellets during t=30 and 120 h.

calculated by the Wittler [73] relationship with the parameters of Mitard and Riba [74] for smooth pellets. When using the relationship recommended by Berovic et al. [99] for hairy pellets, higher apparent viscosities were obtained: e. g., 313 instead of 166 mPa s for 200 rpm. However, the pellets in the stirred tank reactor were smooth, and therefore the Wittler relationship was used.

The volumetric mass transfer coefficient was very low. The determination of the k_La was not possible. The specific power input at 200 rpm was obviously too low for adequate supply of the cells with oxygen. Therefore, the specific enzyme production was low.

At 400 rpm, 25 °C and pH 4.5 the specific power input and the shear stress was high enough to suppress the aggregation of spores and form small pellets and filamentous mycelium without damaging the cells. The apparent viscosity was low and k_La was high enough for providing the cells with enough oxygen. According to these conditions, the specific enzyme activity was the highest.

At 600 rpm, 25 °C and pH 4.5 the aggregation of the spores was suppressed, and at first small pellets and filamentous mycelium were formed, but later they gradually aggregated to clumps. On account of the high specific power input and the low apparent viscosity, the $k_L a$ was high, but the shear stress obviously exceeded the critical value (dVF was considerably reduced) and impaired the cell physiology. Therefore, the enzyme activity and yield coefficients of the enzyme production were low.

The cultivations in ATL were performed at 25 °C, pH 4.5 and 0.2 vvm with spore concentration of 10^5 ml^{-1} semisynthetic medium. On account of the low specific power input, the aggregation of the spores was not suppressed and pellets were formed. The volumetric mass transfer coefficient ($k_L a$) and the shear stress were the lowest and the pellets were the largest in this reactor. The use of the relationship of Wittler et al. [73] yielded an apparent viscosity of 164 mPa s, which was lower than that calculated with the relationship of Berovic et al. [99] for hairy pellets: 288 mPa s. On account of the hairy pellets, which formed at 0.3 g l^{-1} phosphate concentrations, 288 mPa s was used in Table 8.

In spite of the large pellet size, the cells were obviously provided with oxygen and substrate for growth. This is due to the low density, loose and hairy pellets at 0.3 g l^{-1} phosphate concentrations, which are very flexible and therefore the mass transfer in the pellets was enhanced by the turbulence and convection caused by the bubble movement [107]. However, the enzyme activity was rather low. Obviously the oxygen supply was not sufficient for enzyme production. In semisynthetic medium the xylanase activity was always lower in airlift tower loop reactors than in stirred tank reactors.

At higher phosphate concentrations the pellet density increased, and compact and homogeneous pellets were formed, in which the mass transfer rate decreased. Therefore, the center of the pellets was not adequately supplied with oxygen, similar to the large *Penicillium chrysogenum* pellets shown in Fig. 27. In the center, where no oxygen is present, no glucose uptake occurred. In the pellet periphery, in which sufficient oxygen was present, glucose was consumed and the pH value decreased, because acids were formed. In the pellet center, the pH value increased, because no glucose was consumed due to the lack of oxygen, but ammonium was formed by consumption of proteins, which became free by cell lysis (Fig. 28) [108, 109]. The AO-staining clearly indicated that in the center of large pellets no protein synthesis occurred (Fig. 25). The BrdU-immunoassay revealed that in the pellet center no cell propagation took place (Fig. 26). In large pellets the cell density decreased in the center by lysis as shown in photograph 3 in Fig. 29 B. This figure clearly shows the smooth surface of the compact pellets formed with 1.05 g l^{-1} phosphate concentration (photograph 1 in Fig. 29 A) and the hairy surface of the low density loose pellets (photograph 2 in Fig. 29 A) with 0.3 g l^{-1} and the hollow pellet (photograph 3 in Fig. 29 B) with 2.01 g l^{-1} phosphate concentrations [110]. As a consequence of the large pellets the specific enzyme activity decreased (Table 7).

A similar comparison of the cultivations on complex medium in different reactors is impeded by the missing information on the fungal biomass concentration, which strongly influences the apparent viscosity of the cultivation media.

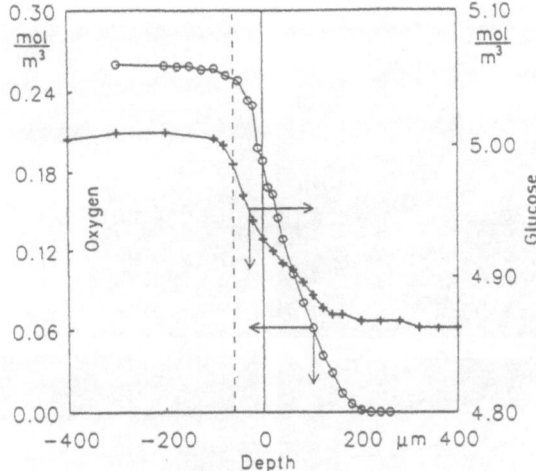

Fig. 27. Microprofiles of dissolved oxygen and glucose concentrations in a *Penicillium chryso-genum* pellet of 1.6 mm diameter at t=144 h [108]. (+) glucose, (o) oxygen. The *solid line* indicates the pellet surface in air; the *dashed line* indicates the surface of the outermost filaments with a length of 60 μm. Mixing rate 0.02 m/s

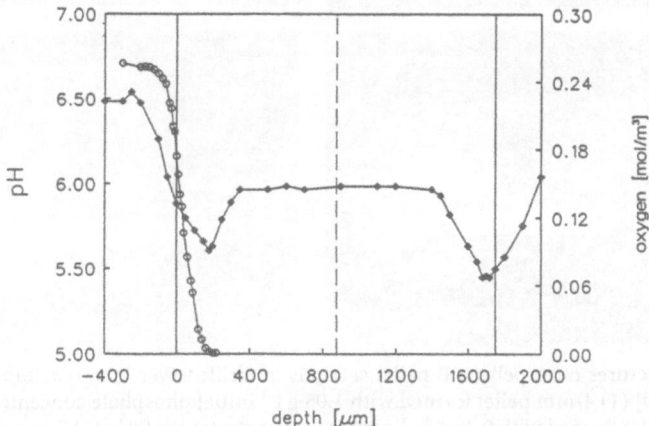

Fig. 28. Microprofiles of dissolved oxygen and pH value in a *Penicillium chrysogenum* pellet of 1.75 mm diameter at t=160 h.(♦) pH; (o) oxygen. *Dashed line* indicates the center of the pellet [109]

The comparison of the influence of the spore concentration in shake flasks at an initial pH value of 4.5 and stirred tank at pH 4.5 on synthetic medium reveals that in both of them pellets were formed at low spore concentrations. With increasing spore concentration in shake flasks the pellet size decreased. In stirred tank reactors, filamentous mycelium was formed and at still higher spore concentrations small pellets dominated. The change of filamentous mycelium to

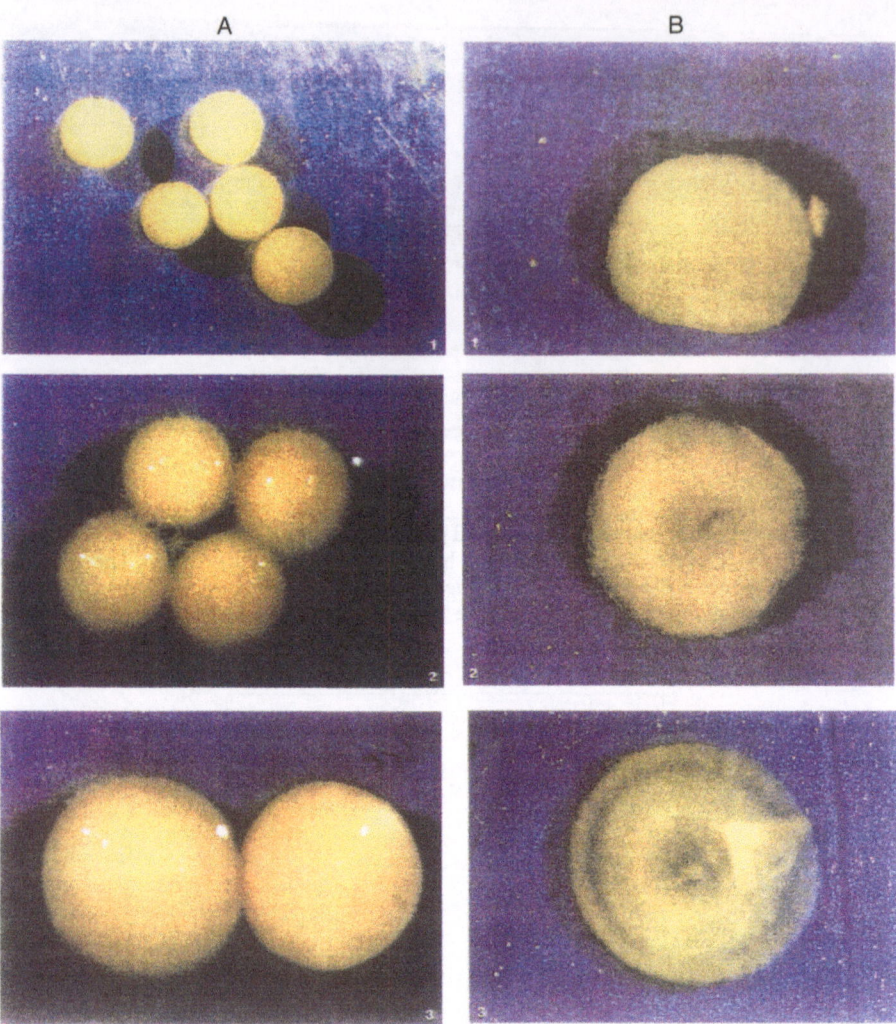

Fig. 29 A, B. Pictures of: A pellets; B pellet sections in airlift tower loop reactors on synthetic medium [110] (1) 4-mm pellet formed with 1.05 g l^{-1} initial phosphate concentration at 70 h; (2) 6-mm pellet formed with 0.3 g l^{-1} phosphate concentration at 90 h; (3) 7-mm pellet formed with 2.01 g l^{-1} phosphate concentration at 90 h

pellet with increasing spore concentration was unexpected. The highest specific enzyme activity was obtained with the smallest pellets and filamentous mycelium, as expected (Table 4).

The influence of the medium composition on the morphology was similar in stirred tank and airlift tower loop reactors with wheat bran. The fungus was attached to the wheat bran and on milled wheat bran, filamentous mycelium formed in both of the reactors. In synthetic medium in stirred tank reactors small pellets and in airlift tower loop reactors large pellets were formed, probably due to the different applied specific power inputs (Table 5). It was not

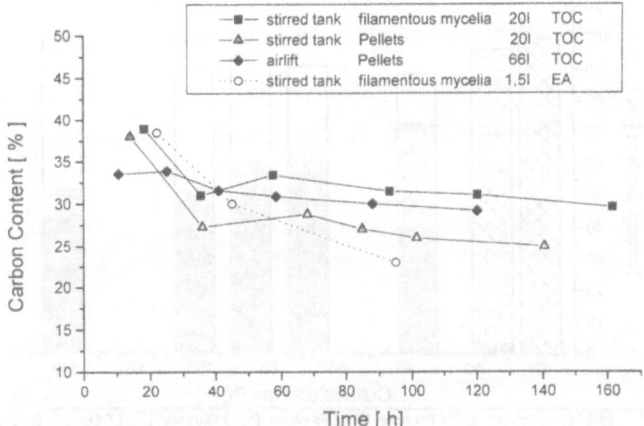

Fig.30. Carbon content of the filamentous and pellet shaped (dry) cell mass in stirred tank and airlift tower loop reactors, respectively at different cultivation times [59]. (TOC = total organic carbon, EA = elementary analysis)

possible to follow the development of the spores in complex medium. However, it is likely that in complex medium the spore aggregation at pH 4.5 is suppressed.

The fungal biomass concentration and the morphology was independent of the phosphate concentration in shake flasks, in contrast to the stirred tank reactor, in which the concentration of the fungal biomass was much lower at low phosphate concentration than at high phosphate concentration. Furthermore, in the former pellets and in the latter pellets (with 1.05 g l^{-1} phosphate) and clumps (with 0.3 and 2.1 g l^{-1} phosphate concentration) dominated. In the airlift tower loop reactor large pellets with different structures were formed. Accordingly, the specific enzyme activity was low.

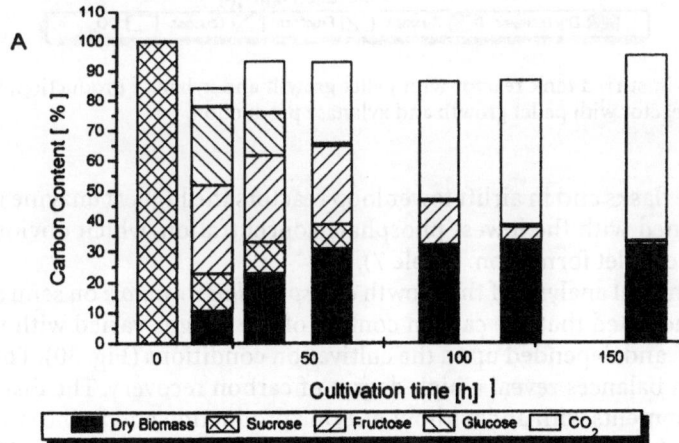

Fig. 31 A. Carbon balances of cultivations [59]; A in stirred tank reactor with filamentous mycelium and without xylanase production

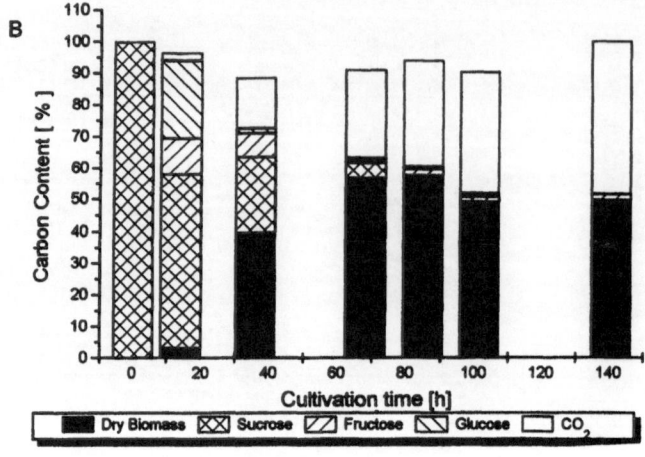

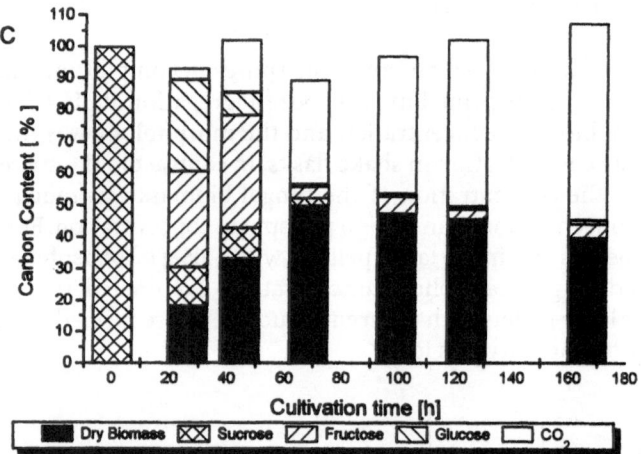

Fig. 31 B, C. B in stirred tank reactor with pellet growth and xylanase production; C in airlift tower loop reactor with pellet growth and xylanase production

In shake flasks and in airlift tower loop reactors the highest enzyme activities were obtained with the lowest phosphate concentration, which obviously promotes loose pellet formation. (Table 7).

The elemental analysis of the growth of *Aspergillus awamori* on semi synthetic medium indicated that the carbon content of the fungus varied with the cultivation time and depended upon the cultivation conditions (Fig. 30). The simplified carbon balances reveal a high degree of carbon recovery. The distinct process developments with and without enzyme production and in the stirred tank and airlift tower loop reactors are shown in Fig. 31 A, C. After 100 h, the substrates were used up, the (dry) cell mass concentration decreased and the carbon balances included only the carbon contents of the CO_2 and cell mass.

6
Conclusions

The investigations presented allow one to draw some conclusions with regard to qualitative relationships between process variables, morphology, and enzyme production of *Aspergillus awamori*.

Relationships between process variables:

- Close relationship exists between the cell growth and uptake rates of phosphate, amonium, and sugar as well as the amount of alkali, which is needed to maintain a constant pH.
- Ammonium and phosphate originating from the semisynthetic and complex medium, respectively, are consumed together with glucose. After glucose is consumed, their concentrations increase in the medium.
- α-Amylase is produced immediately after the start of the cultivation in complex medium. Maltose and glucose appear subsequently in the medium, pass a sharp maximum and are used up within 50 h. Xylanase production starts and xylose appears in the medium, after the sugars are used up.
- With increasing phosphate concentration the substrate uptake is accelerated and the polyphosphate is enriched in the vacuoles. This causes a quick reduction of the phosphate concentration in the medium, a delay of the cell growth, increase of the pellet rigidity and early cell lysis.
- The time lag is shorter, the growth rate and xylanase production rate are lower in complex medium with wheat bran containing preculture than with semisynthetic preculture.
- In spite of the use of buffer solution in shake flasks, the pH decreases with increasing cultivation time to low values (2–2.5), which reduces the enzyme production.
- At higher cultivation temperatures, oxygen transfer limitation occurs in shake flasks, which influences the fungal morphology.
- Protein synthesis is localized in the hyphae by staining of mRNA with Acridine Organge.
- The main protein synthesis occurs in young hyphae, in subapical and branching zones. In the center of large pellets no protein synthesis can be detected.
- Replicating DNA is localized with the uptake of thymidine analog bromo deoxyuridine (BrdU) during the S-phase of the growth via immunofluorescence.
- Replicating DNA is detected mainly in young hyphae, in growing and branching zones. In the center of large pellets no replication DNA is found.
- The high carbon recovery indicated the high quality of the process monitoring.

Relationships between morphology and process variables:

- At high spore concentrations filamentous mycelium and small pellets and at low spore concentration large pellets are formed in semisynthetic medium.
- Pellet size decreases with increasing spore concentration. Filamentous mycelium and small pellets have higher enzyme productivity than large pellets.

- In complex medium consisting of wheat bran, the fungus cultivated in stirred tank reactors is adhered to the wheat bran particles, penetrated into the layer between inner and outer pericarp and fills the wheat bran with dense pellet shaped mycelium.
- Wheat bran protects the fungus from the shear forces in stirred tank reactors.
- With milled wheat bran no protection exists. Filamentous mycelium is formed, which adheres to the particle surface. This causes a high growth rate, high medium viscosity but lower xylanase production than with non-milled wheat bran.
- In shake flask and stirred tank reactors pellet growth occurs with an initial pH value of 4.5 and at pH 4.5, respectively. Dense pellets with smooth surface are formed at low and moderate specific power input. With increasing power input the pellet size decreases and at high values filamentous mycelium is formed, which partly agglomerates to clumps.
- The fungal morphology is not influenced by phosphate concentration in semisynthetic medium and in shake flask cultures.
- In pneumatically mixed (airlift tower loop, ATL) reactors, large loose hairy pellets are formed with enhanced internal mass transfer due to their high flexibility. On account of this high mass transfer rate, the supply of the cells in the pellet center is satisfactory, as long as the substrate and dissolved oxygen concentrations are high in the liquid medium.
- The phosphate concentration influences the fungal morphology in airlift tower loop (ATL) reactors.
- In ATL reactors in semisynthetic medium large pellets are formed due to the low specific power input.
- In ATL reactors in complex medium with milled and non-milled wheat bran filamentous mycelium is formed.

Relationships between enzyme productivity and process variables.
- Xylan is the best inducer for xylanase.
- After induction the enzyme production with xylan, three endo-β-xylanases and β-xylosidase are formed. The three xylanases are lumped into the enzyme "xylanase". The formation of β-xylosidase is undesired. Therefore, the pH value of 4.5 was chosen, at which high xylanase to β-xylosidase ratio is obtained. However, at this pH value the agglomeration of spores and pellet formation are favored.
- Induction of xylanase is only possible after the sugars have been consumed.
- The time of the addition of the inducer to the semisynthetic medium influences the xylanase activity.
- The highest xylanase activity is obtained with filamentous mycelium and small pellets.
- The xylanase activity is high at low phosphate concentration.
- In mechanically mixed (shake flask and stirred tank) reactors the enzyme productivity is low at low stirrer speed because of oxygen transfer limitation; they are low at high stirrer speed as well, due to high stress on the fungus.
- The highest xylanase activity is obtained at intermediate specific power input (stirrer speed).

- In semisynthetic medium, the highest xylanase activity is formed in stirred tank reactors at intermediate stirrer speed. The xylanase activity is lower in ATL reactors because of the oxygen mass transfer limitation, and lowest in shake flask cultures because of the decrease of the pH during the cultivation.
- In complex medium, the highest xylanase activity is formed in ATL reactors with filamentous mycelium. Wheat bran yields higher enzyme activity than milled wheat bran and higher than stirred tank reactors, in which the fungus forms dense pellets.
- The xylanase is localized in the fungus and at the substrate with gold immunoassay. Xylanase is gradually enriched in the plasma, it migrates to the plasma membrane, where it enriches, and is secreted into the medium, where it adsorbs on the surface of the wheat bran fragments. High concentration of the xylanase is detected at the interfaces of fungus and wheat bran and around the hyphae penetrated into the wheat bran.
- The investigations with transmission electron microscopy, electron energy loss spectroscopy and gold immunoassay give invaluable information on the decomposition and utilization of the wheat bran by the fungus.

The known dependence of the fungal morphology on the spore concentration in inoculum and the pH value and the shear stress in semisynthetic medium were confirmed by the investigations presented. In addition, the influence of the oxygen transport limitation on the fungal morphology was proved. New results were elaborated with complex medium containing wheat bran particles. Wheat bran has a considerable influence on fungal morphology and enzyme production. This effect depends on the size of the wheat bran particles. By AO-staining, the mRNA and the protein synthesis and with BrdU immuno assay, the replicating DNA were localized in the hyphae and pellets. By gold immunoassay the xylanase was localized in the hyphae and at its interface with the wheat bran.

New information was gained in pneumatically mixed (ATL) reactors with synthetic and complex media. The observation that the enzyme activities were higher in the airlift tower loop reactor than in the stirred tank reactor in complex medium was unexpected. The higher enzyme activities obtained on semisynthetic medium in stirred tank reactor than those in the airlift tower loop reactor were expected. The lowest enzyme activities were obtained in shake flask cultures due to the shift of pH value during the production phase. This reveals the disadvantage of the use of shake flasks cultures for such investigations.

At present it is not possible to deduce unequivocal general relationships between process variables, enzyme production and morphology of *Aspergillus awamori*. Too many parameters influence these interrelationships.

As key parameters, the strain, spore concentration, substrate type (solid or dissolved carbon source and enzyme inductor), pH, and temperature, as well as the specific power input, which influences the shear stress imposed on the fungus, the apparent viscosity of the cultivation medium and the oxygen transfer into the medium were identified. The phosphate effect is significant. The reactor types used influenced the operation conditions, especially the specific power input and the type of shear stress.

More investigations are necessary for identifying quantitative relationships.

7
References

1. Raper KB, Fennell DI (1965) The genus *Aspergillus*. The Williams & Wilkins Co, Baltimore
2. Smith JE (1994) *Aspergillus*. Biotechnol handbooks, vol 7. Plenum Press, London and New York
3. Röhr M, Kubicek CP, Kominek J (1983) Citric acid. In: Dellweg H (ed) Biotechnology, vol 3. Verlag Chemie Weinheim, p 419
4. Jernejc K, Cimerman A (1992) J Biotechnol 25:341
5. Gomez R, Schnabel I, Garrido J (1988) Enzyme Microb Technol 10:188
6. Reddy CA, Abouzied MM (1986) Enzyme Microb Technol 8:659
7. Schmidt O, Angermann H, Frommhold-Treu I, Hoppe K (1995) Appl Microbiol Biotechnol 43:424
8. Pandey A, Selvakumar P, Ashakumary L (1996) Process Biochem. 31:43
9. Ghosh M, Das A, Mishra AK, Nanda G (1993) Enzyme Microb Technol 15:703
10. Smith DC, Wood TM (1991) Biotechnol Bioeng 38:883
11. Deschamps F, Huet MC (1985) Appl Microbiol Biotechol 22:177
12. Gokhale DV, Puntambekar US, Deobagkar DN (1986) Biotechnol Lett. 8:137
13. Manolov RJ (1992) Appl Microbiol Biotechnol 37:32
14. Metz B, de Bruijn EW, van Suijdam JC (1981) Biotechnol Bioeng 23:149
15. Adams HL, Thomas CR (1988) Biotechnol Bioeng 32:707
16. Reichl U, (1990) Einsatz eines Bildverarbeitungssystemes zur Erfassung der Morphologie und Wachstum myzelbindender Mikroorganismen in submerser Kultur. Dissertation. University Stuttgart
17. Reichl U, Buschulte TK, Gilles ED (1990) J Microscopry 158:55
18. Reichl U, Yang H, Gilles ED, Wolf H (1990) FEMS Microbiol Lett 67, 207–210
19. Yang H, Reichl U, King R, Gilles ED (1992) Biotechnol Bioeng 39, 44
20. Packer HL, Thomas CR (1990) Biotechnol Bioeng 35:870
21. Reichl U, Gilles ED (1991) Biotechn Eng Stuttgart, p 336
22. Reichl U, King R, Gilles ED (1992) Biotechnol Bioeng 39, 164
23. Tucker KG, Kelly T, Delagrazia P, Thomas CR (1992) Biotechnol Prog 8:353
24. Durant G, Cox PW, Formisyn P, Thomas, CR (1994) Biotechnol Techn 8:759
25. Olsvik E, Tucker KG, Thomas CR, Kristiansen B (1993) Biotechnol Bioeng 42:1046
26. Tucker KG, Thomas CR (1993) Trans Ichem 71:111
27. Cox PW, Thomas CR (1992) Biotechnol Bioeng 39:945
28. Treskatis SK (1995) Wachstum, Morphologie und Antibiotikaproduktion von Streptomyceten in Submerskultur, Verlag Shaker, Aachen
29. Nielsen J, Johansen CL, Jacobsen M, Krabben P, Villadsen J (1995) Biotechnol Prog 11:93
30. Gerlach SR (1996) Morphologische Untersuchungen an *Aspergillus awamori*. Dissertation. University of Hannover
31. Rumerhart DE, Hinton GE, Williams RJ (1986) Learning representations by back-propagation errors. Nature (London) 323:533
32. Gerlach D (1996) Zur Beurteilung von Meßgeräten in biotechnologischen Prozessen, Dissertation. University of Hannover
33. MacKay DJ (1992) Bayesian interpolation. Neural Comp 4:(3) 415
34. Paul GC, Kent CA, Thomas CR (1993) Biotechnol Bioeng 42:11
35a. Paul GC, Kent CA, Thomas CR (1994) Biotechnol Bioeng 44:655;
35b. Tucker KG, Thomas CR (1992) Biotechnol Lett 14:1071
36. Packer HL, Keshavarz-Moore E, Lilly MD, Thomas CR (1992) Biotechnol. Bioeng 39:384
37. Paul GC, Kent CA, Thomas CR (1992) Trans Ichem T Part C, 70:13
38. Mohseni M, Allen DG (1995) Biotechnol Bioeng 48:257
39. Freudenberg S, Fasold KI, Müller SR, Siedenberg D, Kretzmer G, Schügerl K, Giuseppin MLF (1996) J Biotechnol 46:265
40. Darzynkiewicz Z, Kapuscinski J, Traganos F, Crissman KA (1987) Cytometry 8:138
41. Monbouquette HG, Ollis DF (1988) Bio/Technology 6:1076

42. Darzynkiewicz Z, Traganos F, Sharpless T, Melamet MR (1975) Exp Cell Res 90:411
43. Darzynkiewicz Z, Traganos F, Sharpless T, Melamet MR (1975) Exp Cell Res 95:143
44. Darzynkievicz Z, Traganos F, Sharpless T, Melamet MR (1977) J Histochem Cytochem 25:46
45. Vanhoutte B, Pons MN, Thomas CR, Louvel L, Vivier H (1995) Biotechnol Bioeng 48:1
46. Matsuoka H, Yang HC, Homma T, Nemoto Y, Yamada S, Sumita O, Takatori K, Kurata H (1995) Appl Microbiol Biotechnol 43:102
47. Dien BS, Scrienc F (1991) Biotechnol Progr 7:291
48. Tanaka K, Yanagita T (1963) J Gen Appl Microbiol 9:1, 101
49. McClure WK, Park D, Robinson PM (1968) J Gen Microbiol 50:177
50. Grove SN, Bracker CE (1970) J Bacteriol 104:2, 989
51. Bradbury D, Mac Masters MM, Cull IM (1956) Cereal Chem 33:342
52. Bradbury D, Mac Masters MM, Cull IM (1956) Cereal Chem 33:361
53. Adolph S, Müller SR, Siedenberg D, Jäger K, Lehmann H, Schügerl K, Giuseppin MLF (1996) J Biotechnol 46:221
54 a. Bauer R (1988) Meth Microbiol 20:113; b Probst W, Bauer, R. (1987) Verh. Deutsch. Zool. Ges. 80, 119–128
55. Roth J (1982) The protein A-gold (pAg) technique – a qualitative and quantitative approach for antigen localization on thin sections. In: Bullock GR, Petrusz P (eds) Techniques in immunocytochemistry, vol 1. Acad Press, London, p 107
56. Jäger KM, Bergman B (1991) Planta 183:120
57. Weigel B, Hitzmann B, Kretzmer G, Schügerl K, Huwig A, Giffhorn F. (1996) J Biotechnol 50:93
58. Schmidt WJ, Kuhlmann W, Schügerl K (1985) Appl Microbiol Biotechnol 21:78–84
59. Siedenberg D, Gerlach SR, Weigel B, Schügerl K, Giuseppin MLF, Hunik J (1996) Production of xylanase by *Aspergillus awamori* on synthetic medium in stirred tank and airlift tower loop reactors. J Biotechnol 56:103
60. Burnett TJ, Chavira R, Hagemann JH (1984) Anal Biochem 136:446
61. Miller GL (1959) Anal Chem 31:426
62. Bergmeyer HU (1974) Methoden der enzymatischen Analyse. Verlag Chemie, Weinheim, vol. 1, 3rd edn
63. Zoels B (1992) Quantifizierung und Optimierung der Sceening-Bedingungen im Schüttelkolben. Master thesis, Fachhochschule Mannheim
64. Milbradt C (1966) Leistungseintrag in Schüttelkolben bei erhöhten Viskositäten. Master thesis, Technical University Aachen
65. Henzler HJ, Schedel M (1991) Bioproc Eng 7:123
66. Schultz JS (1964) Appl Microbiol 12:305
67. Schügerl K (1991) Bioreaction engineering, vol. 2:characteristic features of bioreactors. John Wiley & Sons, Chichester
68. Judat H (1977) Fortschritte der Verfahrenstechnik 15:(Abt.B) 141
69. Hoffmann J, Tralles S, Hempel DC (1992) Chem Ing Techn 64:953
70. Biedermann A (1994) Scherbeanspruchung in Bioreaktoren. Dissertation, University Köln
71. Biedermann A, Henzler HJ (1994) Chem Ing Techn 66:209
72. Patrick AJ, Kennedy MJ (1995) Biotechnol Lett 17:487
73. Wittler R, Matthes R, Schügerl K (1983) Eur J Appl Microbiol Biotechnol 18:17
74. Mitard A, Riba JP (1986) Appl Microbiol Biotechnol 25:245
75. Olsvik ES, Kristiansen B (1992) Biotechnol Bioeng 40:1293
76. Metzner AB, Feehs RH, Ramos HL, Otto RE, Tuthill JD (1961) AIChE Journal 7:3
77. Deckwer WD (1991) Bubble column reactors. John Wiley & Sons, Chicester
78. Chisti MY (1989) Airlift Bioreactors. Elsevier Science Publishing, Amsterdam
79. Shu IS, Schumpe A (1992) Chem Ing Techn 64:560
80. Deckwer WD, Nguyen-tien K, Schumpe A, Serpemen Y (1982) Biotechnol Bioeng 24:461
81. Pandey A (1992) Process Biochem 27:109
82. Lonsane BK, Ghildyal NP, Budiatman S, Ramakrishna SV (1985) Enzyme Microb Technol 7:258

83. Desgranges C, Vergoignan C, Georges M, Durand A (1991a) Appl Microbiol Biotechnol 35:200
84. Desgranges C, Georges M, Vergoignan C, Durand A (1991) Appl. Microbiol. Biotechnol. 35:206
85. Smiths JP, Rinzema A, Tramper J, Schlösser EE, Knol W (1996) Process Biochem 31:669
86. Laukevics JJ, Apsite AF, Viesturs US, Tengerdy RP (1985) Biotechnol Bioeng 27:1687
87. Aido KE, Hendry R, Wood BJ (1981) Eur J Appl Microbiol Biotechnol 12:6
88. Siedenberg D, Gerlach SR, Schügerl K, Giuseppin MLF, Hunik J (1997) Production of xylanase by *Aspergillus awamori* on synthetic medium in shake flask cultures (submitted)
89. Suijdam van JC, Metz B (1981) J Ferment Technol 59:329
90. Shamlou PA, Makagiansar HY, Ison AP, Lilly MD, Thomas CR (1994) Chem Eng Sci 49:(16) 2621
91. Priede MA, Vanags JJ, Viesturs UE, Tucker KG, Bujalski W, Thomas CR (1995) Biotechnol Bioeng 48:266
92. Nielsen J (1992) Adv in Biochem Eng./Biotechnol 46:187
93. Mitard A, Riba JP (1988) Biotechnol Bioeng 32:835
94. Herbst H, Schumpe A, Deckwer WD (1992) Chem Eng Technol 15:425
95. Schügerl K (1992) Comparison of different reactor designs and performances. In:Ladish M, Bose A (eds) Harnessing biotechnology for the 21st century. Am Chem Soc, p 232
96. Anderson NW, Clydesdale FM (1980) Food Science 45:336
97. Dr Grandel Weizenkleie (1995). Neuform Deutschland, Keimdiät GmbH Augsburg
98. Siedenberg D, Gerlach SR, Czwalinna A, Schügerl K, Giuseppin MLF, Hunik J (1996) Production of xylanase by *Aspergillus awamori* on complex medium in stirred tank and airlift tower loop reactors. J Biotechnol 56:205
99. Berovic M, Koloini T, Olsvik ES, Kristiansen B (1993) Chem Eng J 53:B 35
100. Bosch A, Maronna RA, Yantoro OM (1995) Process Biochem 30:599
101. Fasold KI (1995) Fluoreszenzmikroskopische Untersuchungen an *Aspergillus awamori*. Master Thesis, Institute for Technical Chemistry, University Hannover
102. Freudenberg S (1995) Immunofluoreszenztechniken an filamentösen Pilzen. Master Thesis, Institute for Technical Chemistry, University of Hannover
103. Nandakumar MP, Thakur MS, Raghavarao KSMS, Ghildal NP (1994) Process Biochem 29:545
104. Taguchi H, Yoshida T, Tomita Y, Teramoto S (1968) J Ferment Technol 46:814
105. Carlsen M, Spohr A B, Nielsen J and Villadsen J (1996) Biotechnol. Bioeng. 49:266
106. Galbraith J C, Smith J E (1969) Trans. Br. Mycol. Soc. 52:237
107. Wittler R, Baumgaertl H, Lübbers DW, Schügerl K (1986) Biotechnol Bioeng 28:1024
108. Cronenberg CCH, Ottengraf SPP, van den Heuvel IC, Pottel F, Sziele D, Schügerl K, Bellgardt KH (1994) Bioproc Eng 10:209
109. Cronenberg CCH (1994) Biochemical engineering on a micro-scale: biofilms investigated with needle type glucose sensors. Dissertation, University of Amsterdam
110. Siedenberg D (1996) Grundlegende Untersuchungen zur Xylanase-Produktion von *Aspergillus awamori* Rührkesselreaktoren und im Airlift-Schlaufen-reaktor. Dissertation. University Hannover

Received December 1996

Author Index Volume 1–60

Subject Index

Springer
and the
environment

At Springer we firmly believe that an international science publisher has a special obligation to the environment, and our corporate policies consistently reflect this conviction.
We also expect our business partners – paper mills, printers, packaging manufacturers, etc. – to commit themselves to using materials and production processes that do not harm the environment. The paper in this book is made from low- or no-chlorine pulp and is acid free, in conformance with international standards for paper permanency.

 Springer